Rick Tuttle

D1807088

MEDIEVAL FARMING
AND
TECHNOLOGY

TECHNOLOGY AND CHANGE
IN HISTORY

VOLUME 1

MEDIEVAL FARMING AND TECHNOLOGY

The Impact of Agricultural Change in Northwest Europe

EDITED BY

GRENVILLE ASTILL

AND

JOHN LANGDON

BRILL

LEIDEN · NEW YORK · KÖLN

1997

This book is printed on acid-free paper.

Library of Congress Cataloging-in-Publication Data

Medieval farming and technology : the impact of agricultural change in
northwest Europe / edited by Grenville Astill and John Langdon.
    p. cm. — (Technology and change in history ; v. 1)
    Includes bibliographical references and index.
    ISBN 9004105824 (cloth : alk. paper)
    1. Agriculture—Europe, Western—History. 2. Agricultural
innovations—Europe, Western—History. 3. Middle Ages—History.
I. Astill, Grenville G.   II. Langdon, John.   3. Series.
S452.M43   1997
630'.94'0902—21   97–7623
                                                                CIP

Die Deutsche Bibliothek – CIP-Einheitsaufnahme

Medieval farming and technology : the impact of agricultural
change in northwest Europe / ed. by Grenville Astill and John
Langdon. – Leiden ; New York ; Köln : Brill, 1997
    (Technology and change in history ; Vol. 1)
    ISBN 90–04–10582–4
NE: Astill, Grenville [Hrsg.]; GT

ISSN   1385-920X
ISBN   90 04 10582 4

© *Copyright 1997 by Koninklijke Brill, Leiden, The Netherlands*

All rights reserved.
No part of this publication may be reproduced,
translated, stored in a retrieval system, or transmitted in any form
or by any means, electronic, mechanical, photocopying,
recording or otherwise, without prior written permission
from the publisher.

PRINTED IN THE NETHERLANDS

CONTENTS

LIST OF FIGURES

Graphics by Steven J. Allen

LIST OF TABLES

LIST OF CONTRIBUTORS

Grenville Astill, Department of Archaeology, University of Reading, UK.

Bruce Campbell, Department of Economic and Social History, Queen's University, Belfast, UK.

Georges Comet, Centre de Lettres, University of Provence, France.

Christopher Dyer, Department of Medieval History, University of Birmingham, UK.

Peter Hoppenbrouwers, Vakgroep Geschiedenis, University of Leiden, The Netherlands.

John Langdon, Department of History and Classics, University of Alberta, Canada.

Mavis Mate, Department of History, University of Oregon, USA.

Janken Myrdal, Department of Landscape Planning, Swedish University of Agricultural Sciences, Uppsala, Sweden.

Bjørn Poulsen, Landsarkivet for Sønderjylland, Denmark.

Georges Raepsaet, Section d'histoire de l'art et d'archéologie, Free University of Brussels, Belgium.

Erik Thoen, Department of Medieval History, University of Ghent, Belgium.

Mats Widgren, Department of Human Geography, University of Stockholm, Sweden.

1. INTRODUCTION

John Langdon, Grenville Astill and Janken Myrdal

It is now over thirty years since discussion of the "agricultural revolution" of the middle ages began to take shape. In particular 1962 was a memorable year in seeing the publication of major works on the topic by Duby and White.[1] Both painted in their own way a picture of a medieval agrarian economy undergoing unprecedented growth from the ninth to the end of the thirteenth century, which to a large extent was attributed to a number of critical technological developments. White and later Gimpel[2] extended this to military and industrial activity, creating in the end an image of a vibrant medieval society and culture, heavily buttressed by new technology.

Scepticism about this aggressively technophilic interpretation was not slow in coming and was opposed on a number of fronts.[3] Marxist scholars, although certainly not denying some important technological inputs for the period, saw them very much as dependent variables, being advanced or constrained by the forces implicit in class relations.[4] In particular, Marxist historians have emphasized the low level of capital investment in medieval technology, either through the disinclination of lords or, in the case of the peasantry, the difficulties of investment in the face of excessive rents and other feudal dues.[5] The implications of a generally weak technological response in the middle ages were taken even further by the influential body of opinion forming around the neo-Malthusian theories of Michael Postan. Restricting his analysis mostly to the agricultural sphere, Postan claimed that the inexorable force of population growth leading up to the early fourteenth century eventually overwhelmed society's ability to provide adequate food supplies, a situation further aggravated by

---

[1] Duby 1962, especially Part II; White 1962.

[2] Gimpel 1977.

[3] For a recent survey of technological determinism and how it impinges upon historical understanding, see Smith and Marx 1994.

[4] Most forcefully expressed in Hilton and Sawyer 1963. Many aspects of the theme were taken up again in the famous Brenner debate: Aston and Philpin 1987.

[5] See especially Brenner 1987, 234–6.

environmental degradation from land over-use.[6] This made population decline inevitable (which Postan claimed would have happened
even without the plague) until a less stressful balance between population and land use was struck and population could rise again. According to Postan, this cyclical pattern of population rise and decline
was virtually unaltered by technological innovation. Postan's views
found acceptance on a European-wide level,[7] partly because much of
the early work on medieval crop yields seemed to bear out his conclusions. This was particularly the case with the heroic labours of
J. Z. Titow on the voluminous archives of the bishopric of Winchester in England, where the pattern of declining yields in the period
leading up to the Black Death provided Postan with his most robust
support.[8]

On the other hand, Postan's theories in particular presupposed a
remarkably passive response on the part of medieval society, in which
traditionalism, conservatism, the repressive nature of the feudal system, and the chronic lack of capital investment all played strong
roles.[9] Inevitably, such a negative (some might say condescending)
view of medieval society created its own reaction. One strand of this
opposition emphasized how people would try to innovate almost as
a natural response to the difficulties faced in their lives. Much of this
view stemmed from the work of the economist, Ester Boserup, who
linked the intensification of agricultural regimes directly to the growth
of population.[10] Although these productivity gains per unit of land
were clearly bought at the expense of a much heavier labour input,
nonetheless the compulsion to improve productivity overall when faced
with growing numbers of people was, according to Boserup, a natural and almost inevitable response. In the medieval context this has
found support in the work of several historians,[11] where a positive
correlation between the rise in population and improvements in agriculture—either by innovating from an existing pool of knowledge or
by creating entirely new agrarian innovations or formulations—has

---

[6] Postan's general thesis is most clearly articulated in Postan 1966b, 548–70; but
see also Postan 1975, ch. 4.
[7] For example, Le Roy Ladurie 1974. Wilhelm Abel, who has often been linked
with Postan in supporting Malthusian interpretations of pre-industrial economies,
was in fact rather sceptical of the strict Malthusian explanation: Abel 1980, 40–2.
[8] Titow 1972.
[9] Postan 1975, 45–9.
[10] Boserup 1965 and 1981.
[11] Hallam 1981; Campbell 1983.

been noted in the case of several areas across Europe. But the variation of response was clearly wide. Some areas showed a remarkable ability to innovate and to improve productivity, while others seemed satisfied to stick with long established regimes and to accept mediocre or even low levels of agricultural productivity.[12]

Thus, in recent years, the problem has shifted away from the question of whether rural societies in medieval Europe could innovate successfully—they clearly could in certain circumstances—to one of the motivation to do so. What would cause one area to increase productivity, while another close by would seem content to stick resolutely to traditional means? By the very nature of this question, a broader and more complex view of the subject is required. Postan, for example, tended to separate rural activities (that is, agriculture) from urban ones (that is, trade), any connection between the two for him being determined by the rise or fall in population, and he certainly did not see urban development as necessarily encouraging agricultural production (or vice versa, as Duby did).[13] As evidenced by several of the chapters of this volume,[14] such a characterization of urban-rural relations is no longer tenable. The relation between the scale and nature of agricultural production was often clearly connected to both urban and industrial development, as typified by the Low Countries in particular. If such a relation did not exist or existed weakly, it was often because specific social or political conditions were cutting across it.[15]

The result of these more recent ideas has been to see agricultural production, and thus the development of agricultural technology, as more a part of the total economy than just of agriculture itself. This has meant an inevitable shift from seeing the rural economy primarily in terms of subsistence to seeing it as part of a much wider, primarily market-driven economy. Such a "modern" interpretation of the medieval economy patently risks distortions in its own right, but the process of placing medieval agriculture and agricultural technology in a wider framework is a growing trend in the literature and is certainly a common theme in this volume.

Finally, an underdeveloped aspect of the assessment of medieval agricultural technology has been a close comparison between the

---

[12] See especially Campbell this volume.
[13] Postan 1973; Duby 1962, especially Part II, chs. III and IV.
[14] Notably Thoen, Hoppenbrouwers and Campbell.
[15] For example, Langdon this volume.

various nations of Europe. Previous surveys of agriculture (and agricultural technology) across medieval Europe have tended to treat each country as a discrete entity (for example, England, France, the Low Countries) with little direct comparison among them.[16] As long as the prevailing models of medieval agriculture stressed subsistence this made sense. But increasingly it is being recognized that the various states in medieval Europe did not exist as self-contained agricultural units but were strongly tied into wider ranging economic systems. The boundaries of these systems fluctuated over time, but by AD 1000 a number of discrete economic systems had established themselves around the world.[17] In this regard, the North Sea had been slowly developing as such an economic system from the time of the Carolingian Empire, when first the Frisians and then the Vikings established the North Sea as a major and more or less permanent trading corridor, possibly, as argued famously by Pirenne, because the spread of Islam in the Mediterranean shifted economic activity in Christian Europe further north.[18] It is important here to recognize that the North Sea and to some extent the Baltic not only functioned as a transmitting medium along which technological ideas could flow in the intellectual sense, which has long been a commonplace in the literature, but that, as in the case of urban-rural interaction mentioned above, it established another level of economic activity that again reinforced variations in technological activity from region to region.

Just as the geographical dimension in the analysis of agricultural development in the middle ages is widening, so too is the temporal dimension of analysis tending to narrow. In order to judge technological response from region to region, it is becoming more important to place such comparisons in a narrower time frame. At what period such comparisons should take place is, of course, a matter of debate. The increasing prevalence of surviving documentary records, particularly from the thirteenth century onwards, makes such a comparison easier for the later middle ages (that is, the fourteenth and fifteenth centuries), but here the advent of the plague in the middle of the fourteenth century disrupted the pattern of economic development decisively. As the various studies in this volume relate,

---

[16] For example, see Postan 1966a, 291–659.
[17] Abu-Lughod 1989.
[18] Pirenne 1957, especially 236ff.

the experience of the twelfth and thirteenth centuries was probably more consistent across Europe than that which occurred after 1350. There is also a strong sense that the entire period from the tenth to the fourteenth centuries was one of exceptional economic and technological growth, although this view carries with it the danger of underrating other periods.[19]

It was to examine these and other issues about how agricultural technology spread during the middle ages, or even how it should be studied, that the series of essays in this collection were gathered together. The genesis was a series of papers delivered at the Eleventh International Historical Congress of Economic History in Milan in September 1994 on the theme of agricultural technology in northwest Europe during the middle ages, a session organized by Janken Myrdal and John Langdon. This volume comprises the revised papers of this session plus a number of other invited chapters. The concern was to look at a particular area of Europe in a more concentrated and comparative manner than was usually the case, the geographical focus here being the lower North Sea and the western Baltic. The concern was, too, that the emphasis should not be solely upon the major agricultural areas, but also to include areas that we would consider "peripheral". Also, as a glance through the titles in this volume will indicate, the focus is not primarily on the "nuts-and-bolts" of agricultural technology—that is, a cataloguing of the various technological achievements in agriculture at the time—but rather to look at the question of medieval agricultural technology in the context of several main themes.

One of these themes concerns the problem of evidence. The various regions considered in this volume have differing access to the sources, ranging from documentary-rich areas like France and especially England, to those where archaeological and other sources form the major part of the evidentiary base (for example, Sweden and Denmark). The possible distortions to which the differing evidential makeups give rise is a concern of many of the papers, particularly as a strong documentary survival sometimes gives a spuriously "progressive" impression to a particular region or to one time period over another. On the other hand, the potential benefits from new sources of evidence or more fruitful combinations of old ones is evident in all of the contributions.

---

[19] See especially Raepsaet this volume.

Secondly, the nature of technological change is a key concern. The problem is not one of recognizing that technological change was taking place, but how and by what mechanisms was it received by the host society. Is a new piece of machinery or a change of practice a vital, reasoned response to critical societal needs, or is it simply a rather casual act to surmount a momentary inconvenience? As incremental models of early technological change have gradually become more popular recently,[20] the latter view is perhaps coming to the fore. Changes in practice or equipment come along and often slot themselves rather unobtrusively within societies without creating any great ripples.

This raises the question of technological change as revolution. The alleged transforming properties of technology have always excited people, not only of our own era. On the other hand, it seems a rather obvious statement that technological change has always had to adjust to existing social and economic conditions. This uneasy juxtaposition between the acceptance of an innovation and its transforming qualities is a feature of the middle ages as much as our own time, and the mechanism by which it works is still unclear. As Myrdal argues in this volume, it may be only when a number of changes come together to create a new technological complex or package that the technological potential of a society is suddenly raised to a new level. As a result, the vision of agricultural "revolution" has rather receded for this period, as has the profile of the inventor. Indeed, perhaps we should see the inventor as innovative combiner rather than inventive creator.

Finally, the power of social frameworks to encourage or retard technological innovation is clearly a crucial issue. Sectoral responses are clearly critical. As mentioned above, urban-rural interactions or more long-ranging economic systems were clearly important, but so too were divisions between various social groups, particularly the landholding class and peasants. Certainly the domination that lords had in economic, social and political spheres has long been recognised as being important if not crucial in the development of the medieval economy and technology.[21] According to this model, lords had great success in directing peasant activities in directions that suited the

---

[20] Notably Persson 1988 and Mokyr 1990.
[21] Some of the background to this debate can be found in Langdon 1994, especially pp. 3–4.

aristocracy, whether it was for, say, forcing peasants to transmute rents and other feudal dues into cash, thus obliging peasants to participate more strongly in market activities, or compelling them to patronize the lord's watermill or windmill, thus giving those particular technological innovations a decided boost. Over recent years, however, the ability of peasants to play a much more independent role in the promotion of medieval agriculture and technology has been increasingly recognized, whether through innovations that they themselves produced or through their consuming power, which encouraged lords and others with capital to invest.[22] It was, in fact, in the control of investment rather than from an innate interest in technology that lords, merchants, and governments had potentially the most influential impact—both negative and positive—upon the development of agriculture and agricultural technique, and many of the papers comment on this. Here, distinctions between seigneurial sectors (lay versus ecclesiastical, great magnates versus gentry) may be particularly significant. Some landlords established a particular reputation for innovation, such as the Cistercians, where the links of diffusion of innovation were more institutional than economic or social in nature (that is, travelling from one religious house to another rather than being mediated by market or other forces). Finally, the overarching role of government and hegemonic attitudes, religious or otherwise, should not be forgotten in forming a continually pervasive atmosphere in which technological change had to take place.

The approaches that the chapters in this volume have taken to these issues are varied. Some have adopted a survey mode for the various regions (notably Comet, Hoppenbrouwers, Mate, Poulsen and Thoen), while others have adopted a more thematic approach, either by examining particular aspects of medieval agricultural development (for example Widgren) or by examining issues from a more comparative, methodological, or theoretical standpoint (Astill, Campbell, Langdon, Myrdal, and Raepsaet). In this regard, this volume has aimed less at providing authoritative answers concerning the current state of knowledge about agricultural technology in the middle ages, as in providing a range of perspectives with which to highlight the rich diversity of issues and ideas underlying this complex yet critical subject.

---

[22] Langdon 1986 and 1994; Campbell 1988. See also Campbell this volume.

Bibliography

Abel, W. 1980, *Agricultural fluctuations in Europe from the thirteenth to the twentieth centuries*, trans. O. Ordish, London (originally *Agrarkrisen und agrarkonjunktur: Eine Geschicte der Land- und Ernährungswirtschaft Mitteleuropas seit dem hohen mittelalter*, Hamburg, 1966).

Abu-Lughod, J. 1989, *Before European hegemony: the world system AD 1250–1350*, New York.

Aston, T. H. and Philpin, C. H. E. (eds) 1987, *The Brenner debate: agrarian class structure and economic development in pre-industrial Europe*, Cambridge.

Boserup, E. 1965, *The conditions of agricultural growth*, London.

—— 1981, *Population and technology*, London.

Brenner, R. 1987, "The agrarian roots of European capitalism". In *The Brenner debate: agrarian class structure and economic development in pre-industrial Europe*, eds T. H. Aston and C. H. E. Philpin, Cambridge, 213–327.

Campbell, B. M. S. 1983, "Agricultural progress in medieval England: some evidence from eastern Norfolk", *Economic History Review*, 36, 26–46.

—— 1988, "The diffusion of vetches in medieval England", *Economic History Review*, 41, 193–208.

Duby, G. 1962, *L'Économie rurale et la vie des campagnes dans l'occident médiéval*, Paris.

Gimpel, J. 1977, *The medieval machine*, Harmondsworth.

Hallam, H. E. 1981, *Rural England 1066–1348*, London.

Hilton, R. H. and Sawyer, P. H. 1963, "Technical determinism: the stirrup and the plough", *Past and Present* 24, 90–100.

Langdon, J. 1986, *Horses, oxen and technological innovation: the use of draught animals in English farming from 1066 to 1500*, Cambridge.

—— 1994, "Lordship and peasant consumerism in the milling industry of early fourteenth-century England", *Past and Present*, 145, 3–46.

Le Roy Ladurie, E. 1974, *The peasants of Languedoc*, trans. J. Day, Urbana (originally *Les paysans de Languedoc*, Paris, 1966).

Mokyr, J. 1990, *The lever of riches: technological creativity and economic progress*, New York.

Persson, K. G. 1988, *Pre-industrial economic growth: social organization and technological progress in Europe*, Oxford.

Pirenne, H. 1957, *Mohamed and Charlemagne*, trans. B. Miall, Cleveland (originally *Mohamet et Charlemagne*, Paris, 1937).

Postan, M. M. (ed.) 1966a, *The Cambridge economic history of England*, vol. I, *The agrarian life of the middle ages*, 2nd edn, Cambridge.

—— 1966b, "Medieval agrarian society in its prime: England". In *The Cambridge economic history of England*, vol. I, *The agrarian life of the middle ages*, 2nd edn, ed. M. M. Postan, Cambridge, 548–632.

—— 1973, "The economic foundations of medieval society". In his *Essays on medieval agriculture and general problems of the medieval economy*, Cambridge, 3–27.

—— 1975, *The medieval economy and society*, Harmondsworth.

Smith, M. R. and Marx, L. (eds) 1994, *Does technology drive history? The dilemma of technological determinism*, Cambridge, Massachusetts.

Titow, J. 1972, *Winchester yields: a study in medieval agricultural productivity*, Cambridge.

White, L. 1962, *Medieval technology and social change*, Oxford.

2. TECHNOLOGY AND AGRICULTURAL EXPANSION IN THE MIDDLE AGES: THE EXAMPLE OF FRANCE NORTH OF THE LOIRE

Georges Comet*

Perceptions of medieval agriculture before the Black Death are associated with a growing population and a frontier mentality of clearing land.[1] Some historians claim that the simple growth of population drove land clearance;[2] some point to an overall rise in temperatures around the year 1000 which might have improved crop yields;[3] some feel that the social decay of the ancient world and the gradual advent of feudal societies can account for the agrarian development of the western world;[4] while, finally, some argue that economic development in medieval times found its origin in the development of agrarian technology.[5]

My own assumption is that there was never any real rupture in technological knowledge between antiquity and the middle ages. Ploughs, harrows, harnesses and mills spread slowly at first, then at a faster pace. People continued to apply what they already knew, but applied it in different ways, and propagated it on a larger scale. They might have worked harder, perhaps with more hope.[6] Technology accompanied them in their needs, but it is unlikely that any technological breakthrough in the middle ages ever preceded the constitution of a new society.

Sources

The range of potential sources for elucidating these issues is immense. Some, such as books, laws, deeds and accounts, were left intentionally

---

* I am grateful to Nelly Valtat and Denny Packard for translating this chapter.
[1] Duby 1962, 142.
[2] European population grew by an estimated 140 per cent (from 22 to 55 million) from around 950 to around 1300: Russell 1958.
[3] Fossier 1984, 18; Colardelle and Verdel 1993, 381.
[4] Dockès and Rosier 1977.
[5] Raepsaet 1995.
[6] Pesez 1991a; Comet 1996.

to inform future generations. Others were involuntary—human or animal bones, for example, or lost artefacts later found in excavations. Other clues still, like C^{14} and O^{18} dating, pollens or sedimentation, do not come from humans at all. These last bear testimony to the natural world in which people lived at a given moment and, at the same time, to people's effect on that world.

Texts

Normative texts by agronomists were too rare to offer a satisfactory source of agricultural knowledge available to all. Medieval copies of treatises by Latin agronomists such as Cato, Varro, Columella, Palladius and Pliny helped preserve a certain store of knowledge until the invention of printing, but only allowed a limited transmission of that knowledge. In the thirteenth century some new texts, like *Fleta*, *Senechaucy*, or others by Walter of Henley and Robert Grosseteste, appeared, reflecting practice in Anglo-Norman territories.[7] However their distribution remained limited in time and space; few of these manuscripts circulated outside England. Nevertheless, they give an account of a type of agriculture which had some similarities with that of northern France.

No new texts appeared until the early fourteenth century, notably treatises by the Italian agronomists Pietro de Crescenzi (*c.* 1304) and Corniolo della Cornia (early fifteenth-century).[8] Even though both were inspired by ancient writers, their works included a large proportion of the medieval innovations. Although they obviously concentrated on the Mediterranean world, this did not deter King Charles V from having de Crescenzi's works translated into French in 1387. The first French agronomists to deserve the name came much later, in particular Charles Estienne (1564) and Olivier de Serres (1600),[9] the latter being a southerner. More specialist works came earlier, such as *Le Bon Berger* (1379),[10] a treatise on cattle raising written by Jehan de Brie at the request of King Charles V, which is the oldest one ever written in French, while there was also a treatise on veter-

[7] Oschinsky 1971.
[8] Gaulin 1990; Bonelli Conenna 1982.
[9] Estienne 1547; de Serres 1600.
[10] Clévenot 1986.

inary practice, written in the fourteenth century but not published until very recently.[11]

Historians have wondered about this medieval silence. Does it mean people were not interested in agronomy? Was it linked to a stronger preoccupation for salvation and a broader vision of the world?[12] From the thirteenth century onwards, for example, it was mostly the task of encyclopedias to provide comprehensive surveys of the knowledge of the time, which included agrarian technology. Authors such as Brunet Latin, Gossouin de Metz, Barthélémy l'Anglais and, above all, Vincent de Beauvais[13] recapitulated part of what the ancients knew, and added what had come out of more recent local experience. Finally, let us mention some works in linguistics and literature, for example etymologies. Isidore of Seville (seventh-century), Jean de Garlande, Adam de Petit Pont, and Alexander Neckam in the twelfth and thirteenth centuries, happened to provide sometimes puzzling lists of tools, parts or goods.[14] In the field of general literature, there is also the famous thirteenth-century *De l'Oustillement au Villain*.[15]

Finally, much, of course, can also be learned through more pedestrian, day-to-day records, too numerous to itemize fully here. These might include such things as accounts, leases, trade codes, toll books, registers of witnesses, and so on. This involves not only the production of agricultural products, but also what happened afterwards. For example, a text about milling will sometimes reveal things about grains, grapes before vinification, or textiles to be dyed. Market variations are important too. In times of scarcity, the authorities tried to set the price of bread by organizing bread-making experiments called *essais du pain*,[16] the official records of which are precious sources of information.

Archaeology

Medieval archaeology in France only began to develop seriously about forty years ago, but its results are already very illuminating. Relatively few tools have been found in northern France (for example,

---

[11] Prévot and Ribemont 1994; Boulaine 1992.
[12] Comet 1992, 5–29.
[13] Picone 1994; Ribemont 1995.
[14] Scheler 1867; see also Estienne 1547.
[15] Nyström 1940; Lorcin 1985.
[16] Comet 1992, 462–71.

excavations carried out in Brebières, Pas de Calais, sixth- to seventh-centuries),[17] but the site of Dracy, in Burgundy, has proved to be much richer.[18] Thanks to excavations such as at Pen er Malo (Morbihan, twelfth-century),[19] we also know a lot about dwellings and the use of land in Brittany. Similarly, the excavations of various barns, for instance that of Vaulerent (Val d'Oise)[20] or that of Mont-Saint-Jean (Auxois),[21] both in thirteenth-century contexts, have yielded metallic tool remnants plus traces of the plants grown. But we should not limit ourselves to a small geographical area. The site of Charavines (Isère, eleventh-century),[22] that of Wharram Percy in England (from the tenth century on),[23] and certain German sites,[24] although, of course, not strictly speaking in northern France, can still supply much relevant information and insight.

Archaeology is also able to date and locate less tangible technological phenomena, such as the end of cross ploughing or the introduction of new techniques of soil improvement. It can suggest a chronology of land development, reconstruct agrarian spaces in connection with dwellings, and reveal the degree of order (or disorder) in land management. Carpology (the study of the nature, form, and even DNA of grains) and palynology (the study of pollens) are also directly derived from archaeology.

Modern reconstructions have also thrown light on the practical issues associated with medieval agriculture, such as analysing the weeds that could accompany crops and measuring out the areas in which pollens scattered. Experimental medieval gardens or farms, such as the *ferme de Melrand* in Brittany or, more recently, the excavations and experimental farming in Waldmatte (Valais, Switzerland),[25] have brought about new observations, for example on the wear and tear of tools, which seems to have been fairly considerable.[26]

---

[17] Demolon and Poulain-Jossen 1972.
[18] Beck 1987.
[19] Bertrand and Lucas 1975.
[20] Higounet 1965; Horn and Born 1968; Guadagnin 1990.
[21] Beck 1989.
[22] Colardelle and Verdel 1993.
[23] Beresford and Hurst 1971.
[24] As at Pfaffenschlag and Hohenrode: Nekuda 1975; Baumgarten 1971; Chapelot and Fossier 1980; Pesez 1991a.
[25] Lundström-Baudais 1995.
[26] Lerche 1994.

Other Sources

Iconography is clearly important. No other source can reproduce such a thing as the movement involved in guiding a plough or swinging a flail. Ethnographic investigation is also useful. It would be foolish to imagine that there still remain places in the world where time has stopped in the medieval period, but in areas such as Nepal, for example, all the traditional steps involved in processing millet, from cultivation to consumption, can be observed today and can supplement profitably the evidence from medieval sites in Europe.[27]

Altogether, it is of fundamental importance to collate all these types of information and not allow any particular one to dominate. For example, millet is almost never mentioned in medieval documents, yet is often found as grains or pollen on archaeological sites, suggesting perhaps that it was destined exclusively for peasant consumption.[28] Also, new methods of investigation, such as DNA analysis,[29] could change our views.

Cerealization

Excavations suggest that before the tenth century people's diet in France included a large proportion of meat and game, as at the site of Charavines in the south, where farm animals, especially pigs, provided a large part of daily food.[30] After that, cereal production increasingly began to dominate, such that, by the twelfth and thirteenth centuries, cereals began providing 75 to 80 per cent of the calories consumed.[31] In northern France, they were eaten essentially in the form of bread or gruel, which meant they had first to be transformed into flour. Much rarer were whole grains boiled in milk (frumenty) or used as semolina (as in pasta), which were consumed primarily in Mediterranean Europe.[32]

---

[27] Lundstrom-Baudais 1995.
[28] Ruas 1992b and 1992a.
[29] Research is being conducted in Toulouse on the molecular structure of ancient and medieval grapes. A similar project is being planned for grains.
[30] Colardelle and Verdel 1993.
[31] Uzan and Pauphilet 1955; Spooner 1961; Histoire de la consommation 1975; Braudel 1979; Hocquet 1985; Devroey 1987; Comet 1992, 305–9; Stouff, 1970.
[32] Laurioux 1995.

Grains

Apart from maize, then unknown, and rice, which was known but
not cultivated until the fourteenth century in Italy, all the principal
cereals were to be found in medieval Europe, both of the naked and
hulled variety. Naked grains, such as wheat or rye, were ready to be
ground as they fell from threshing, whereas hulled grains, such as
barley, oats and spelt, had to be husked before being taken to the
mill. Unhusked grain, however, could be used as animal fodder, and
stored better.

Of these cereals, spelt was probably the most widely cultivated
under the Carolingians, but from the tenth century it began an irreversible decline. This was probably due to the extra cost of husk-
ing,[33] because, as far as taste was concerned, Hildegarde de Bingen
still considered it as *suavior* and called it *optimum granum* in the middle
of the twelfth century.[34] Because of its superior storage characteristics,
however, it continued to function as a backup crop.

The role of spelt was taken over by wheat, which became the
grain of prestige at the dinner tables of the rich and powerful. Most
of all, wheat was the only cereal deemed worthy of representing the
body of Christ in the Eucharist, and certainly it was never used as
a fodder crop. Wheat in northern France was of a soft variety,[35] and
was invariably sown in winter, unless bad weather delayed it until
the spring. Its density remains uncertain. Today's soft wheat is around
0.75 grams per cubic centimetre and has been so since the nineteenth century. Precise (and converging) computations based on *essais
de pain* and market prices make it appear close to 0.64 in the fifteenth century, around 15 per cent lower than now. On the other
hand, computations based on Pliny's data reveal that wheat density
in ancient times was about the same as in modern times. As yet, no
valid explanation has been found for the fact that the density of
wheat declined during the medieval period and only began increasing again in the eighteenth century.[36]

Rye, arriving in Gaul in the last centuries before Christ, only began
to increase markedly in popularity from the fourth and fifth cen-

---

[33] The husks were removed either by a mortar and pestle or by a preliminary
pass through the millstones of a mill (after which the freed grains would have to
milled again to produce flour).

[34] Migne 1855; Comet 1989b.

[35] Comet 1989a.

[36] Comet 1996.

turies AD.[37] In northern France, its expansion was faster in the west than in the east. It quickly became the main cereal crop in the Armorican massif, and the lords of Brittany used to speculate on it in the fourteenth century. Along with oats, it did not require rich soil and it could resist temperatures as low as −22° C.

Barley was also a winter grain, widespread in Carolingian times, because it could grow in poor soil. It was used as fodder and often mixed with wheat when intended for human consumption. Regional and social differences are significant. Like rye, barley was consumed by the poorest people, more in Normandy and Brittany than anywhere else. It was, of course, one of the grains primarily used for the production of ale and, when hops became available, beer. The *Livre des metiers* of Etienne Boileau in 1268, for example, carefully itemized the composition of the ale made in Paris.[38]

Oats also contributed to the new crop pattern of the middle ages. In some places they were used to make bread, but they were mainly intended for cattle and, increasingly, horses.

Millet existed in Carolingian times, but its climatic demands made it more difficult to produce in northern France, so increasingly it was hardly grown at all north of Limousin, Burgundy or Normandy. Some other grains were also known. Emmer or einkorn, which was grown in antiquity, had probably not entirely disappeared in the middle ages.[39] Finally, saracen, which is not properly speaking a cereal but rather a polygonaceous plant, was used as a cereal, and bore the popular name of buckwheat. Some written documents show that it was grown in Brittany around 1450, coming from Mecklenburg, but its medieval history is still very uncertain.[40]

Much has been written on the yields of these grains. The very low ratios computed from Carolingian *Brevium Exempla* cannot be taken seriously, and only later evidence from documents or treatises are of value. These indicate a wide range of yields. Some were truly high. Thierry d'Hireçon in Artois obtained from 9:1 to 13:1 for wheat in the fourteenth century, and from 3.7:1 to 6.7:1 for oats. As early as the thirteenth century around Lille, some yields reached 10:1 to 12:1.

---

[37] Devalette, Barriere, Comet and Conte 1994, 61–74; Van Mol and Devroey 1994.

[38] "... de l'eau et du grain: c'est a savoir orge, metail et dragee": Lespinasse and Bonnardot 1859, 26.

[39] For a systematic study of medieval cereals, see Comet 1992.

[40] Visset 1985.

Near Brussels in the fifteenth century a yield ratio of 14:1 was obtained for almost 50 years in a row. But in general yields may have been more like the 4:1 on the bishop of Winchester's lands across the Channel in the thirteenth and fourteenth centuries.[41] Altogether, it would seem yields evolved from about 4:1 in ancient times to 5:1 c. 1600. This increase by a quarter was enough to alleviate malnutrition. As several studies have shown, a precise chronology of this evolution varied considerably according to local conditions.[42]

Market prices demonstrated the social value of the different grains. For example, wheat, the most sought after grain, climbed much higher than the others in times of famine. Nevertheless, price ratios between the various grains, when computed on the basis of weight,[43] remained very stable until the seventeenth century. If wheat is rated at 10, rye varied around 6–6.5, barley around 4.5–5.5 and oats around 5–5.8 during the whole of our period.[44]

Other Crops

In the Christian world, wine's symbolic importance comes second only to bread. First produced in Europe by the ancient Greeks, vineyards eventually spread to the very limits of what was climatically possible, being found even in England and Scandinavia, as well as northern France (Laon, Metz, Argenteuil, la Somme). However, after the peak geographical expansion up to the thirteenth century, certain vineyards became less profitable as transportation improved and they finally disappeared altogether from England (largely because of England's possessions in the Bordeaux area) and Brittany.[45] Others dwindled (Laon for example). In the north of France, the wines from Burgundy benefited from all this, and established a near monopoly in Paris and Flanders. Studies on wines have tended to ignore the less famous vintages, so that medieval vinification[46] in Bordeaux and

---

[41] Richard 1892; Derville 1975–6; Titow 1972; *Husbandry* (thirteenth-century) in Oschinsky 1971; Lacour 1856; Braudel 1979; Serres 1600; Slicher van Bath 1963.

[42] Sivéry 1990, 29–32.

[43] In the middle ages grains were not measured in terms of weight, but in volume. To facilitate comparisons, the equivalence in weight has been given.

[44] Curschman 1900; Fossier 1968; Neveux 1974; Comet 1992, 305–9.

[45] Lachiver 1988.

[46] Mane 1983; Pesez 1991b; Chédeville 1973.

Burgundy is better known now than that in northern France, where they probably produced a variety close to today's rosé wine.[47] As to the vine itself, written documents began mentioning varieties[48] such as Merlot or Pinot in the fourteenth century, but we cannot know if they were the same as those of today until, say, studies on the DNA of grapes provide further evidence.

Other food crops accompanied grain production, particularly legumes (peas, beans, vetches and lentils), which not only provided another food source for both humans and animals, but also supplied valuable fertilizing properties. Other vegetables (cabbages, onions, carrots and parsnips), particularly from gardens, also yielded both variety and supplementary nutrition, as did orchard fruits (plums, pears, apples, peaches and cherries).[49]

Oleaginous plants such as rape, false flax, hempseed, flax, colza and poppy were grown too.[50] Certain plants had several uses. For example in Flanders from the thirteenth century on, colza was used for sheep fodder and for making oil. It probably disappeared from the area in the fifteenth century, but reappeared at the end of the sixteenth century. Rape could be sown as a catch-crop on fallow land for green fodder. If sown before winter, it produced fodder or food, or it could be left to grow oil-giving seeds.[51]

"Industrial" crops were particularly important for the cloth industry. Textile fibres principally came from flax, although hemp was more likely to be found near the sea (in Brittany for example).[52] Carding thistles were also grown. Finally, some plants were used for dyeing, such as woad for blue, madder for red, or weld for yellow. Woad was a great speciality in the Somme area.[53]

Finally, as suggested above, most of these crops were probably not grown in fields but in rural or urban garden patches. The gardens of medieval northern France are still not well understood, but clearly grew in number from the thirteenth century onwards, especially in urban settings.[54] They were seemingly places where intensive

---

[47] Le Mené, 1991.
[48] de Serres 1600, 151.
[49] Fossier 1968, 424; Delmaire 1995; Higounet-Nadal 1989; Derville 1987.
[50] Fossier 1968, 429; Bois 1976, 179.
[51] Thoen 1992; Delmaire 1995.
[52] Delmaire 1995; Bois 1976, 178; Plaisse 1961, 185.
[53] Fossier 1968, 422.
[54] Higounet-Nadal 1989; Fossier 1968, 392, 424.

agriculture was performed, often with irrigation. They were also generally given first priority for manure and other fertilizers.[55] Gardens were also likely places for experimentation. For example, we know that crop rotation was abandoned early in gardens, and that people turned progressively to the growing of new produce, such as spinach, carrots and asparagus. According to the area, flax, hemp, sometimes vines, and, of course, fruit trees, were grown in gardens.

Livestock

Domestic animals provided power, food (meat and dairy products), manure and manufactured goods (wool, hides, parchment and horn). Oxen, horses and donkeys performed most of the heavy work in the fields or on the roads. Donkeys and mules were valued for their easy maintenance, and oxen for their strength and durability. Horses were fast but costly. Altogether, since documentary evidence deals mainly with the results rather than the techniques of animal production, little is known of medieval stock-breeding or veterinary science.[56] It is only archaeology which can give precise information on the types, ages, or sizes of these animals.[57]

Animals in the middle ages were smaller than those of today. In Charavines, for example, around the year 1000, an adult pig weighed around 70 to 80 kg, a sheep 20 to 30 kg, and a cow or ox 200 to 250 kg.[58] Archaeozoologists have recently discovered that animals' wither-heights varied through the ages. They increased between the Iron Age and the Roman period, decreased in the middle ages, and increased again in the sixteenth century. This variation in size of about 100 to 150 mm may have been related to population density and the proportion of cultivated land,[59] and it is interesting that these variations in animal size are roughly parallel to the variations in grain density noted above.

Up to the tenth century, horses were mainly used for war or riding. After that, in the plains of northern France, they gradually replaced

[55] Bois 1976, 177.
[56] As mentioned above, only one copy of a single medieval treatise has survived: Prévot and Ribemont 1994.
[57] Audoin-Rouzeau 1993; Beck 1989; Delort 1984.
[58] Colardelle and Verdel 1988, 44–5. In comparison, at the beginning of the twentieth century, an ox weighed in the region of 650 kg, a sheep from 50–150 kg, and a pig from 100–200 kg: *Larousse Agricole* 1921; Chavard and Gau 1915.
[59] Audoin-Rouzeau 1991a; 1991b.

oxen as draught animals.[60] In neighbouring Flanders their use as draught animals became widespread as early as the twelfth century.[61] Around Chartres at the end of the thirteenth century, some areas began to gain a reputation for breeding horses, and farmers often brought in good breeders from far away in order to improve the stock.[62] However, the progression of the use of horses was slow and uncertain, and sometimes interrupted. In the fourteenth century, oxen were still used in Champagne, Brie and Nivernais, which suggests that the eastern part of the Paris basin used a similar mode of traction observed in most of England at this time.[63]

Breaking in, feeding and harnessing also changed. The relationship between the cultivation of oats and the prevalence of horses is well known. As for harnessing, its evolution was seemingly gradual from antiquity to the middle ages, and such things as the padded horse-collar did not appear suddenly.[64] For oxen, neck yokes tended to replace horn yokes from the eleventh or twelfth century onwards.[65]

New orientations in stock-breeding appeared in the fourteenth and fifteenth centuries.[66] In Normandy, the income earned from acorn grazing decreased by 90 per cent, as pigs were increasingly confined to pigsties for fattening.[67] At the very period when iconographic representations of acorn-gathering under oak trees became particularly numerous, this activity had virtually disappeared, particularly as woodland was increasingly taken over for the hunting and other pursuits of the well-to-do.

Tools and Technology

Ploughing Tools

Used in its broadest sense, "ploughing", or turning over the earth for cultivation, could be done by hand or by using animals. In the

---

[60] Bautier and Bautier 1978; Fossier 1968, 379; Fourquin 1964, 77–80; Contamine 1995.
[61] Thoen 1994.
[62] Chédeville 1973, 209–10.
[63] Contamine 1995; Langdon, 1986.
[64] Amouretti 1991; Raepsaet and Rommelaire 1995.
[65] Haudricourt and Delamarre 1955, 145–68; de Serres 1600, II, 84–5.
[66] Sivéry 1976b.
[67] Plaisse 1961; Fourquin 1969, 338.

former case, a spade was used. Generally the spade was made entirely of wood, save for an iron sleeve that covered the cutting edge
of the blade. The use of iron helped to prolong the life of the implement, but it was expensive and had to be economized on as much
as possible. Hoes could also be used for turning over soil, but they
were primarily weeding tools.

Animal-drawn implements were more frequent in larger expanses
of land, particularly those owned by noblemen or wealthier peasants.
A preliminary distinction can be made here between ards and ploughs
(see figure 2.1). An ard is a symmetrical instrument that tends to
tear up the soil more than it turns it over. There were two main
types of ard commonly in use in the medieval period—the ard with
a sloping sole, which could only be used for straightforward tearing
up of the soil, and that with a rather flat horizontal sole, which
could to some extent cut underneath the earth and throw it to one
side, depending on which way the ard was tilted. A plough is an
asymmetrical instrument, being fitted with a mouldboard and having
a much greater ability to cut under the soil and toss it to one side.
Here again, two main types of plough can be discerned for the
medieval period. The first was the plough with a mouldboard fixed
to one side (mostly the right), such that the soil could be thrown in
one direction only. This was best for soils which had to be thrown
up in ridges (to aid drainage), where the ploughman would plough
the land from one side to the other in a sort of circular fashion,
throwing up the earth towards the centre. The second type was the
"swivel-plough", where the mouldboard could be shifted to the other
side of the plough when the end of the field was reached. In this
way, the furrows could all be turned in one direction as the plough
went up and down the field, a method that was preferred on well-
drained land that did not require ridging.

In northern France, asymmetrical ploughing with ards had been
performed commonly in antiquity, but ploughs were also evident in
the very first centuries AD. The gradual evolution from one to the
other can be related as follows:[68] first the ancient ard was fitted with
a coulter and a wheeled fore-carriage, which made it heavier and
required more draught animals. The farmer could lean on the forecarriage, so that the ard became easier to steer and could be tilted
to one side. With the addition of a mouldboard and the development

---

[68] Amouretti and Comet 1993, 133.

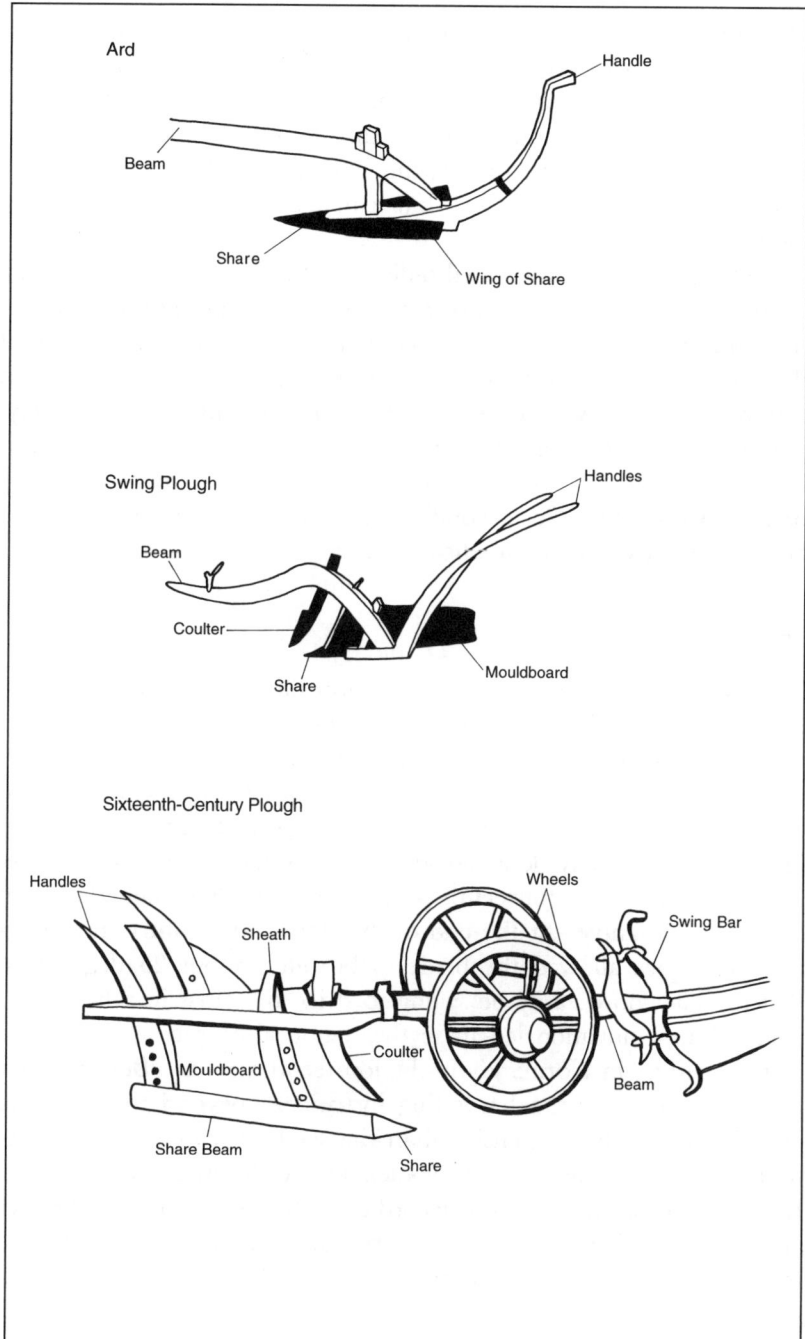

Figure 2.1. An ard (top), swing plough (middle) and sixteenth-century wheeled plough (bottom) (Comet 1992).

of asymmetric shares, the transition to the plough was made. Eventually, ploughs existed in both wheeled and unwheeled varieties, the latter being known as "swing ploughs". Along with swivel-ploughs, swing ploughs were known from at least the early thirteenth century. Without a wheeled forecarriage, this type of plough was smaller and easier to handle and transport, but it demanded more skill from the ploughman.

If ploughs clearly functioned differently from ards, it would still be a mistake to contrast the two instruments too radically, and to deduce that the use of the one rather than the other was a criterion of development. It is true that ploughs were more efficient than ards in turning soil, but it was probably only of critical importance in heavy, poorly drained clay soils. Otherwise, ploughs replaced ards in specific pedo-climatic and social contexts, and farmers opted for ploughs or ards according to local conditions, working hard to improve and adapt these instruments to their needs.

Other Tools

Specific tools were needed for grain and hay harvesting, as well as for grape picking. For harvesting, both toothed and non-toothed sickles were used, although it is not clear why one type or the other predominated in any particular area. In contrast to sickles, which used a slicing or short chopping motion, scythes worked with a long swinging stroke. Originally designed solely for hay harvesting, the blade of the scythe, up to the thirteenth century, was simply extended straight out from its handle. At the end of the thirteenth century, new forging techniques allowed the blade to be tilted about 25 degrees, so that it could more easily be swung to strike the grass in a horizontal plane. This innovation led to scythes being also used in barley and oats harvesting in Normandy in the fourteenth century. But the sickle remained the favourite harvesting instrument beyond the medieval period.[69] It was less expensive than the scythe because less iron was needed. It could be used by women and children as well as men. And it could cut high or low according to the need for straw. Above all, scything meant that many more ears fell to the ground, thus resulting in a grain loss of around 10 per cent.[70] Given these speci-

---

[69] In Diderot's *Encyclopaedia*, the illustration of the plate entitled "Agriculture" shows a harvesting scene featuring sickles: Diderot 1762, figures 2 and 3.
[70] As calculated from Steensberg 1943.

ficities, people opted for scythes or sickles according to the relative costs of labour and grain.

One harvesting tool was really new—the Flemish hook or short-handled scythe.[71] At the end of the thirteenth century, it spread in Flanders, Artois and Hainault, but hardly anywhere else. It allowed faster harvesting, saved labour, made straw-cutting easier, was well adapted to narrow ridged fields (which were easier to drain), and it did not lose as much grain as the scythe. Finally, the famous Roman *vallus*, or harvesting cart, which was present in northeastern Gaul at the end of antiquity, was never mentioned later and had most likely disappeared.[72]

Harrows were used for weeding in ancient times, but they did not spread throughout Europe until the tenth century. Harrows could even the ground before sowing, but above all they were used to cover the seeds after sowing.

The tools used for crops other than cereals tended to be of a more general-purpose type—spades and hoes for soil preparation, and billhooks and pruning knives for harvesting. Most of these tools were helved in the same way as sickles (the head of the iron blade being wedged in the wooden handle), and they were all more or less curved. The wine grower's pruning knife, which pruned the vine and cut off the grapes, was the only specialized tool. It came from the Romans, was used until the invention of the pruning shears and varied somewhat according to local requirements.

Techniques

Probably even more important than the development of new tools was the ongoing development of new practices to use these tools more effectively. Ploughs and ards, for example, generally made furrows about 300 mm apart,[73] but within that rather consistent characteristic there were a multitude of possibilities. Fields could be cross-ploughed, ploughed flat, or thrown up in ridge and furrow to aid drainage. The ridges themselves were generally between three to six metres in width, but could have extremes as wide as 25 m or as narrow as a metre. Very narrow ridged fields, known in German as

---

[71] Thoen 1992; see also Thoen this volume.
[72] Amouretti and Comet 1993, 94; Raepsaet this volume.
[73] Precisely 27 cm, according to the computations of Walter of Henley (Oschinsky 1971).

bifange, were preferred on wetter lands, where the narrow strips of ground formed much better seed beds and allowed a more effective concentration of seeds and fertilizers.

Similarly, soil improvement generally consisted of adaptations of the age-old practices of burning, liming and marling.[74] Thus ashes began to be scattered on thirteenth-century fields, first for rye, then for other cereals. The application of manure similarly evolved. Initially, in the eleventh century, animal dung was simply spread directly onto gardens or fields.[75] In the thirteenth century, however, Walter of Henley recommended spreading clayey earth in animal sheds for two weeks, and then spreading the earth and dung mixture onto the fields. In the fourteenth century, a 5:1 ratio of straw and dung was fermented to produce a more potent fertilizer. This new technique brought about a modification in the cutting of cereals, so that straw was no longer left in the fields but brought instead to the barns for the creation of the new "manure". This also favoured different tools for harvesting, mainly Flemish hooks or scythes.

There were variations even in activities that appeared very straightforward. For example, broadcast sowing (as opposed to dibbling or drilling) was the only common propagation technique in Europe, despite its high cost in seeds. There were, however, two different procedures. Sowing could be done *after* ploughing, with the seeds being covered afterwards with a harrow. Or it could be done *before* ploughing, where the ploughing itself would cover the seeds. In this case, the sower followed the ploughman and dropped the seeds into the open furrow that the plough-share would cover in its next passage; but this particular technique required much more care and time. Similarly, harrowing was very effective on flat fields, but much less so (or even non-existent) on narrow ridged fields. Thus, harrowing became associated with certain types of crops, particularly oats which tended to be sown on flat fields (in contrast to, say, wheat, which was mostly sown on ridges).

The separation of grains from their ears also brought several techniques over the centuries. Threshing by articulated flails, for instance, had been known since the end of antiquity.[76] Treading by animals

---

[74] Plaisse 1961, 155–6.
[75] Chédeville 1973, 212.
[76] Sigaut 1988, 18–19; Bentzien 1980, 45, 78; Nekuda 1975; Parain 1979, 17–28; Mane 1983; Comet 1992, 344–51; Horn and Born 1968.

(often brood mares) was practised in Brittany and even in the Île de France during the early modern period.[77] *Chaubage*, the beating of small sheaves against a wall or plank, was preferred for small crops, perhaps as a tax evasion device, since it could be done more secretly.

Altogether, taxation and social relationships in general had a great influence on the adoption of tools and techniques. It is most likely, for instance, that labour *corvées* favoured the spread of drawn implements. It was in the interest of the nobleman to encourage or even demand tenant ploughs, since one peasant with a plough could put in the same amount of work as several with spades. Changes in tools in turn had a fundamental impact on the evolution of the tax system. However, the most important shift was not from the ard to the plough, or from a given type of plough to an improved one, but rather the replacement of spades by drawn implements. The substantial number of days of *corvée* labour required on Carolingian demesnes reflects a time when the spade probably played a much larger role in cultivation.[78] When these same tasks came more and more to be performed by ards or ploughs, the required number of days per person could be considerably reduced. However, such a shift often took centuries to complete.

Tilling the land with ploughs and ards also only required a half to a third of the iron needed for spades. Altogether, ploughs (or ards) provided a sensible solution in the face of manpower and/or iron shortages that were possibly encountered in Carolingian times. They would then have spread faster in large estates, and there is no doubt they were more frequently represented in the illustrations of works destined for the aristocracy.[79]

Soil and Landuse

Fields

The fundamental composition of soils in northern France has probably not changed much since the eleventh century, but it would be wrong to take continuity for granted. The centuries-old process of

---

[77] Comet 1992, 335–44; Parain 1979, 92–3.
[78] Comet 1992, 135–6.
[79] Amouretti and Comet 1993, 135.

clearing, tilling and improving has more or less leached the soil and led it to react differently to climatic conditions.[80] What stands out is the frequent irreversible degradation of soil.[81] The clearing of land facilitated erosion, and ploughing displaced earth from the slopes. All these resulted in a topography that gradually softened, particularly as the bottoms of valleys began to fill up.

Ploughing created landmarks still visible today, such as ridges and furrows, headlands and lynchets.[82] The particular characteristics of these landmarks was often linked to the tools used and in turn affected yields. For example, long and narrow fields often owed their shape to the use of mouldboard ploughs. It is also well known that what grows at the edge of a field generally develops better. So small fields, where boundaries were proportionally greater, were likely to have better yields than larger fields generally found on great estates.

Landscapes

In northern France, geographers distinguish two main types of landscape, which correspond more or less to two types of soil and human organization.[83] Hamlets predominated in the west, whether in *bocage* (that is, surrounded by a network of enclosed fields, often a feature of hilly or wooded terrain) or open-field. The northeast was definitely open-field and remained so for a long time, with collective three-field crop rotation and bay houses predominating.[84]

Bocage itself dates to two main periods. A first type occurred when land was first cleared, either in the twelfth century, as in Saint Martin of Fouilloux (Maine et Loire),[85] or even earlier (ninth or tenth centuries in some places). A second type appeared much later with a complete reorganization which began in the thirteenth century (in the Thiérache area) and continued into the fourteenth and fifteenth centuries.[86] This second phase was linked to the development of stock-breeding and the desire to evade the constraints of open-fields.

---

[80] Audouze and Fiches 1993.
[81] Revel 1991.
[82] Zadora-Rio 1991; Bertrand and Lucas 1975; Lerche 1986; Roupnel 1974, 144.
[83] Trochet 1993; Pitte 1983, 107–26; Bloch 1952, 49–65; Sivéry 1990, 40–9; Zadora-Rio 1991; Meynier 1970, 143ff.
[84] de Planhol 1959, 418.
[85] Zadora-Rio 1991, 180.
[86] Sivéry 1976a; Flatrès 1976.

In contrast, the term "open-field" not only designated a type of landscape, but also a system in which compulsory crop rotation brought about the multiplication and scattering of plots and the adoption of commons. Archaeology and other evidence neither gives a clear answer as to the dating of open-fields—they existed at least as early as the thirteenth century—nor as to whether open-fields appeared predominantly upon the initial clearing of land or were a result of later reorganization. In Beauce and Picardy, however, we do know that some areas of scattered settlement had open-fields as a consequence of radical reorganization.[87]

Crop Rotation and "Assolement"

Crop rotation is the order in which different crops are grown every year on a given plot of land so as not to exhaust the soil in the absence of regular improvements.[88] A two-course crop rotation meant one year of planting cereals and one year of laying fallow; a three-course rotation meant winter grain, spring grain and fallow. Stock animals grazed on the fallow land and fertilized it. A large part of the European continent had adopted two- and three-course rotations. In northern France, three-course rotation seems to have been more frequent,[89] possibly because of the need for oats in a more horse-dominated agriculture. Three-course rotations, however, were by no means the "revolution" they are often claimed to be. On any given cultivated surface, it did not increase the value of production, for the somewhat greater proportion of spring grains on three-course rotations carried a lesser overall market value. However, it did allow the division of risk over several seasons and, through fodder crops such as oats, made easier the feeding of animals for which the "cerealization" of agriculture had deprived pasture.

In any case, the French word *assolement*, which is commonly used in such matters, means more than just "crop rotation". *Assolement* refers to an overall organization of the whole of the cultivated land in a given *finage* or parish.[90] The village lands were divided into "breaks", and each break was devoted to one compulsory crop every

---

[87] Ferdière and Fourteau 1979.
[88] Sivéry 1990, 33.
[89] Bois 1976, 176.
[90] Sivéry 1990, 38; Higounet 1965.

year. Each peasant had to till at least one plot in each break. This
collective organization of agriculture does not seem to have existed
before the thirteenth century, and its constraints appeared at the same
time as the intensification of cereal cultivation.[91] As a result of crop
rotation, *assolement*, and the obligation to devote certain lands to livestock (grazing, hay, fodder crops), the notion of yields becomes problematic. It is arguably much more correct to consider production
from the perspective of an entire community rather than on the basis
of few fields, particularly as the production of grains was increasingly
directed towards supporting livestock.[92]

Cereals and Mechanization

Production cannot be dissociated from the techniques used to transform grains into food, and it is fundamental to realize that the earliest mechanization sprang from agriculture. It was for grinding grains
and pressing grapes and olives that people invented machines such
as mills and presses, and endeavoured to capture water, wind and
animal power.

Mills

The most important of these mechanisms was the mill, which arguably provided the medieval foundations for modern engineering.[93]
As large-scale investments (both in money and skill), mills were beyond the means of most individuals, although some types of horizontal mills might have been constructed by peasant families or small
communities.[94] Generally, however, it was lords and squires who took
advantage of their economic and political power to establish a monopoly in the building and, above all, the operation of those implements. They initiated the right of *banalité*, as a kind of tax people
had to pay when they used the mill.

---

[91] Chédeville 1973, 213–15.
[92] Sigaut 1976; for England, these issues concerning the nature of medieval agrarian
productivity and how to measure and standardize it have very much been the focus
of the writings of B. M. S. Campbell (especially 1983).
[93] Gille 1964; Amouretti and Comet 1993.
[94] Moog 1994.

Mills soon became places where, and the means through which, power was wielded.[95] They were turned into objects of conflict in which technological, economic, political and social interests were tightly interwoven. On the other hand, they also provided places and occasions for people to meet and socialize. Folklore has represented mills as open places, sometimes favourable to frivolous behaviour. Saint Bernard, in the twelfth century, raged about Cistercian mills where prostitutes came to tout for customers.[96]

The chief function of mills, which remained the most important, was to grind cereals. Yet it was not the only one.[97] Malt and hops were ground in brewing regions. Poppy, hemp and other seeds were crushed for oil. Dye plants such as madder and, from the fourteenth century on, woad, were brought to the mill, as well as tanner's bark. Mills were used to sharpen blades (thirteenth century) and operate turning lathes (fourteenth century). In the beginning of the twelfth century, the convent of Saint-Bertin (near Saint-Omer) was supplied with water by a water-raising wheel powered by a hydraulic mill. With the fixing of cams to the mill-wheel axle, first identified around 1040 in Dauphiné,[98] it was possible to operate tilt-hammers and other mechanisms requiring linear rather than rotary motion. This was extended to a host of activities, such as fulling (for cloth), forging (for iron), sawing and paper-making.

Altogether, mills domesticated natural movement, such as that of rivers, or even the sea, since tide-mills were in use in the Nantes area around 1120.[99] Wind was harnessed later by the first windmills of the western world in the late twelfth century.[100] They first appeared along shores, perhaps because people there were already accustomed to using the wind for navigation purposes, and then spread everywhere. They quickly complemented, but did not replace water power, because neither wind nor water flow was guaranteed. Watermills and windmills together helped ensure reasonably continuous milling capacity through periods of floods or low waters, as well as storms or dead calms.

---

[95] Fossier 1981.
[96] Leclercq and Rochais 1974, VII, 212.
[97] Bautier 1960.
[98] Sclafert 1926; Lohrmann 1990b.
[99] Minchinton 1979.
[100] Philippe 1982.

Assessing the Repercussions

Mills and the consequences of mechanization have been at the centre of many discussions and controversies. I, for my part, would like to emphasize a few general points. First of all, mills were exceptionally pervasive. Grain mills, whether powered by water or wind, were found everywhere. From at least the eleventh century, mills also began to play an important role in other activities, as industrial activity began to appear in all rural areas. As in the case of metalworking, it was organized in small units near rivers, and sometimes a mere brook was enough. Rural craftsmanship then bloomed, and many peasants became part-time artisans, features that remained constant until the nineteenth century.[101]

As much as any other innovation during our period, mills have also highlighted the relationships between technology and social organization. The links that Marx established between the two are of course well known: "social relations are closely bound up with productive forces. In acquiring new productive forces men change their mode of production, in changing the way of earning their living, they change all their social relations. The hand-mill gives you society with the feudal lord; the steam-mill, society with the industrial capitalist".[102] Marc Bloch thought that slavery had impeded the spreading of mills in antiquity.[103] According to him, mills had appeared with the first slave shortage in the fourth century AD, and the sudden rise in the number of mills in the eleventh and twelfth centuries was due to the desire of lords to derive more profit from their domains through *banalité*.[104]

For Charles Parain,[105] the ancient social system was no incentive for invention and, as most of the cultivated cereals then had hulled grains, mills were not as advantageous as in the middle ages, when naked grains became much more prominent. Feudal masters could impose the right of *banalité* because peasants accepted it as a means of saving working-time, which was so precious in periods of expansion. But it can be objected that that particular type of work was usually performed by women, whose working-time was highly under-

[101] Guillerme 1990.
[102] Marx 1847, 100 (translated into English in 1984, VI, 166).
[103] Bloch 1935.
[104] Lohrmann 1990a.
[105] Parain 1979, 305–23.

valued. More recently, historians such as Dockès and Rosier have asserted that "technological progress is a by-product of the class struggle".[106]

Technological evolution, however, is not created by social struggles alone. As with other innovations considered in this study, technology was developed to deal with a whole host of social, economic and environmental situations. It is vital that historians of agrarian technology do not oversimplify their subject and try to squeeze it into narrow theoretical boxes. Given the relative youth of the study of medieval technology,[107] and the difficulties in drawing together diverse and often contradictory sources of evidence, both caution and imagination must be used to draw out patterns that often only become clear over centuries. It is not an easy task. As Lynn White, Jr, wrote over half a century ago: "the medievalist who ventures into this study [history of technology] must arm himself with the palm and cockle-shells of pilgrimage and be prepared to journey through strange lands in peril of his scholarly reputation; the perfectionists had best stay in the shadow of Notre-Dame, where their vanity will be less vulnerable."[108]

Bibliography

Amouretti, M.-C. 1991, "L'attelage dans l'antiquité, le prestige d'une erreur scientifique", *Annales E. S. C.* 46, 219–32.

Amouretti, M.-C. and Comet, G. 1993, *Hommes et techniques de l'antiquité à la Renaissance*, Paris.

Audoin-Rouzeau, F. 1991a, "La taille du bœuf domestique en Europe de l'antiquité aux temps modernes". In *Fiches d'ostéologie animale pour l'archéologie* 2, eds J. Desse and N. Desse-Berset, Juan les Pins, 1–40.

—— 1991b, "La taille du mouton en Europe de l'antiquité aux temps modernes". In *Fiches d'ostéologie animale pour l'archéologie* 3, eds J. Desse and N. Desse-Berset, Juan les Pins, 1–36.

—— 1993, *Hommes et animaux en Europe de l'époque antique aux temps modernes. Corpus de données archéozoologiques et historiques*, Paris.

Audouze, F. and Fiches, J.-L. 1993, "L'archéologie Française et les paleoenvironnements", *Annales E. S. C.* 48, 219–32.

Baumgarten, K. 1971, "Ethnographische Bemerkungen zum Grabungsbefund Hohenrode", *Ausgrabungen und Funde* 16, 49–53.

---

[106] Dockès and Rosier 1977, 78.
[107] Thoen 1995.
[108] White 1940, 157.

Bautier, A.-M. 1960, "Les plus anciennes mentions de moulins hydrauliques industriels et de moulins à vent", *Bulletin Philologique et Historique*, 567–626.

Bautier, R.-H. and Bautier, A. M. 1978, "Contribution à l'histoire du cheval au moyen âge—l'élevage du cheval", *Bulletin Philologique et Historique*, 9–75.

Beck, P. (dir.) 1987, *Bourgogne médiévale, la mémoire du sol: vingt ans de recherches archéologiques*, catalogue d'exposition, Dijon.

Beck, P. 1989, *Une ferme seigneuriale au XIVe siècle. La grange du Mont (Charny, Côte d'Or)*, Paris.

Bentzien, U. 1980, *Bauernarbeit im Feudalismus*, Berlin.

Beresford, M. and Hurst, J. G. 1971, *Deserted medieval villages*, London.

Bertrand, R. and Lucas, M. 1975, "Un village côtier du 12 siècle en Bretagne: Pen-er-Malo en Guidel (Morbihan)", *Archéologie Médiévale* 5, 73–101.

Bloch, M. 1935, "Avènement et conquête du moulin à eau", *Annales E. S. C.* 7, 538–63.

—— 1952, *Les caractères originaux de l'histoire rurale Française*, Paris (first edn, Oslo, 1931).

Bois, G. 1976, *Crise du féodalisme*, Paris.

Bonelli Conenna, L. 1982, *La Divina Villa di Corniolo della Coznia Lezione di agricoltura tra XIV e XV secolo*, Siena.

Boulaine, J. 1992, *Histoire de l'agronomie en France*, Paris.

Braudel, F. 1979, *Civilisation matérielle, économie et capitalisme, vol. I: les structures du quotidien*, 2nd edn, Paris.

Campbell, B. M. S. 1983, "Agricultural productivity in medieval England: some evidence from Norfolk", *Journal of Economic History* 48, 379–404.

Chapelot, J. and Fossier, R. 1980, *Le village et la maison au moyen âge*, Paris.

Chavard, A. and Gau, L. 1915, *Quarante leçons d'agriculture*, Paris.

Chédeville, A. 1973, *Chartres et ses campagnes (XIe–XIIIe siècles)*, Paris.

Clévenot, M. 1986, *Le vrai règlement et gouvernement des bergers: le Bon Berger de Jean de Brie*, Paris.

Colardelle, M. and Verdel, E. (dir.) 1988, "Un village de l'an Mil retrouvé sous les eaux (Charavines)", *Dossiers Histoire et Archéologie* 129, 4–85.

Colardelle, M. and Verdel, E. (eds) 1993, *Les habitats du lac de Paladru (Isère) dans leur environnement. La formation d'un terroir au XIe siècle*, Paris.

Comet, G. 1989a, "Dur ou tendre? Propos sur le blé médiéval", *Médiévales* 16–17, 103–12.

—— 1989b, "La perception des vertus de l'épeautre aux époques médiévale et moderne". In *L'épeautre (triticum spelta), histoire et technologie*, eds J.-P. Devroey and J.-J. Van Mol, Treignes, 149–64.

—— 1992, *Le paysan et son outil, essai d'histoire technique des céréales (France VIIIe–XVe siècle)*, Rome.

—— 1996, "Productivité et rendements céréaliers: de l'histoire à l'archéologie". In *L'homme et la nature au moyen âge, paléoenvironnement, des sociétés occidentales* (Actes du 5e Congrès International d'Archéologie Médiévale), ed. M. Colardelle, Paris, 87–91.

Contamine, P. 1995, "Le cheval dans l'économie rurale d'après les archives de l'Ordre de l'Hôpital (France du nord, XIVe siècle)". In *Campagnes médiévales: l'homme et son espace, études offertes à Robert Fossier*, ed. E. Mornet, Paris, 163–75.

Curschman, F. 1900, *Ungersnöte im mittelalter (Leipziger Studien*, VI, 1), Leipzig.

Delmaire, B. 1955, "Note sur la dîme des jardins, 'mes' et courtils dans la France du nord au moyen âge". In *Campagnes médiévales: l'homme et son espace, études offertes à Robert Fossier*, ed. E. Mornet, Paris, 231–46.

Delort, R. 1984, *Les animaux ont une histoire*, Paris.

Demolon, P. and Poulain-Jossen, T. 1972, *Le village Mérovingien de Brebières (VIᵉ–VIIᵉ siècles)*, Arras.

Derville, A. 1975–6, "Le rendement du blé dans la région Lilloise, 1285–1541", *Bulletin de la Commission Historique du Département du Nord* 40, 23–39.

—— 1987, "Dîmes, rendements du blé et "révolution agricole" dans le nord de la France au moyen âge", *Annales E. S. C.* 42, 1411–32.

Devalette, J., Barriere, B., Comet, G. and Conte, P. 1994, *La peste de feu. Le miracle des Ardents et l'ergotisme en Limousin au moyen âge*, Limoges.

Devroey, J.-P. 1987, "Units of measurement in the early medieval economy: the example of Carolingian food rations", *French History* 1, 68–92.

Diderot, D. 1762, *Encyclopédie ou dictionnaire raisonné des sciences, des arts et des métiers, Planches, 1, Agriculture*, Paris.

Dockès, P. and Rosier, B. 1977, "Questions aux historiens", *Cahier de l'Institut d'Études Économiques de l'Université de Lyon II* 11, 1–172.

Duby, G. 1962, *L'économie rurale et la vie des campagnes dans l'occident médiéval*, Paris.

Estienne, C. 1547, *De Latinis et Graecis nominibus arborum, fruiticum, herbarum*, Paris.

Ferdière, A. and Fourteau, A.-M. 1979, "Gestion des archives du sol en milieu rural. Expérience de prospection systématique à Lion-en-Beauce (Loiret)", *Revue d'Archéométrie* 3, 67–96.

Flatrès, P. 1976, "Rapport de synthèse". In *Les bocages, histoire, écologie, économie* (round-table held at Rennes), Rennes, 21–30.

Fossier, R. 1968, *La terre et les hommes en Picardie jusqu'à la fin du XIIIᵉ siècle*, 2 vols., Paris.

—— 1981, "L'équipement en moulins et l'encadrement des hommes". In *L'histoire des sciences et des techniques: doit-elle intéresser les historiens?* (Colloque de la Société Française d'Histoire des Sciences et des Techniques), ed. J. Bouvier, Paris, 230–48.

—— 1984, *Paysans d'Occident*, Paris.

Fourquin, G. 1964, *Les campagnes de la région Parisienne à la fin du moyen âge, du milieu du XIIIᵉ au début du XVIᵉ siècle*, Paris.

—— 1969, *Histoire économique de l'occident médiéval*, Paris.

Gaulin, J.-L. 1990, *Pietro de Crescenzi et l'agronomie en Italie (XIIᵉ–XIVᵉ siècles)*, unpublished thesis from the University of Paris I.

Gille, B. 1964, *Les Ingénieurs de la Renaissance*, Paris.

Guadagnin, R. 1990, "Une exploitation Cistercienne en plaine de France, la grange de Vaulerent", *Vivre en Val d'Oise* 3, 16–26.

Guillerme, A. 1990, *Le temps de l'eau: la cité, l'eau et les techniques, nord de la France, fin IIIᵉ–début XIXᵉ siècle*, Seyssel.

Haudricourt, A. and Delamarre, M. J.-B. 1955, *L'homme et la charrue à travers le monde*, Paris; reprinted in 1986.

Higounet, C. 1965, *La grange de Vaulerent*, Paris.

Higounet-Nadal, A. 1989, "Les jardins urbains dans la France médiévale", *Jardins et vergers en Europe occidentale (VIIIᵉ–XVIIIᵉ siècles), Flaran 9, 1987*, Flaran, 115–44.

Histoire de la consommation, 1995, *Annales E. S. C.*, 30.

Hocquet, J.-Cl. 1985, "Le pain, le vin et la juste mesure à la table des moines Carolingien", *Annales E. S. C.* 40, 661–86.

Horn, W. and Born, E. 1968, "The barn of the Cistercian grange of Vaulerent (Seine-et-Oise), France", *Festchrift Ulrich Middeldorf*, eds A. Kosegarten and P. Tigler, Berlin, 24–31.

Lachiver, M. 1988, *Vins, vignes et vignerons. Histoire du vignoble Français*, Paris.

Lacour, L. 1856, "Traité d'économie rurale composé en Angleterre au XIIIᵉ siècle", *Bibliothèque de l'École des Chartes* 17, 123–41.

Langdon, J. 1986, *Horses, oxen and technological innovation*, Cambridge.

Larousse agricole 1921.

Laurioux, B. 1995, "Des lasagnes Romaines aux vermicelles Arabes: quelques réflexions sur les pâtes alimentaires au moyen âge". In *Campagnes médiévales. L'homme et son espace, études offertes à Robert Fossier*, ed. E. Mornet, Paris, 199–215.

Leclercq, J. and Rochais, H. (eds) 1974, *Sancti Bernardi opera*, Rome.

Le Mené, R. 1991, "Le vignoble Français à la fin du moyen âge". In *Le vigneron, la viticulture et la vinification en Europe occidentale au moyen âge et à l'époque moderne (Flaran 11, 1989)*, Flaran, 189–205.

Lerche, G. 1986, "Ridged fields and profiles of plough-furrows", *Tools and Tillage* 3, 132–56.

—— 1994, *Ploughing implements and tillage practices in Denmark, from the Viking period to about 1800, experimentally substantiated*, Herning.

Lespinasse, R. and Bonnardot, F. (eds) 1859, *Le livre des metiers d'Etienne Boileau*, Paris.

Lohrmann, D. 1990a, "L'histoire du moulin à eau avant et après Marc Bloch". In *Marc Bloch aujourd'hui: histoire comparée et sciences sociales*, eds H. Atsma and A. Burguière, Paris, 339–347.

—— 1990b, "Travail manuel et machines hydrauliques avant l'an Mil". In *Le travail au moyen âge*, eds J. Hamesse and C. Muraille-Samaran, Louvain, 35–47.

Lorcin, M. T. 1985, "De 'l'oustillement au villain' ou l'inventaire sans raton laveur", *Revue Historique* 556, 321–39.

Lundström-Baudais, K., Schneider-Rachoud, A.-M., Baudais, D., Nightengale, A. and Jacquot, K. 1995, "Les millets: recherche ethno-botanique et culture expérimentale. Brig-Glis 'Waldmatte', Valais, CH, 1992–94", unpublished research report.

Mane, P. 1983, *Calendriers et techniques agricoles (France-Italie, XIIᵉ–XIIIᵉ siècles)*, Paris.

Marx, K. 1847, *Misère de la philosophie*, Paris; translated in *idem*, *Collected Works* (Moscow, 1984).

Meynier, A. 1970, *Les paysages agraires*, Paris.

Migne, J.-P. 1855, *Patrologie cursus completus*, CXCVII (Hildegarde de Bingen, *Physica*, I, V), Paris; reprinted, Turnout, 1958.

Minchinton, W. 1979, "Moulins à marée: une étude préliminaire", *Les moulins de France* 6, 1–12.

Moog, B. 1994, *The horizontal watermill, history and technique of the first prime mover*, The International Molinological Society.

Nekuda, V. 1975, *Pfaffenschlag, zanikla stredveka ves u Slavonic*, Brno.

Neveux, H. 1974, *Les grains du Cambrésis, fin du XIV*ᵉ*–début du XVII*ᵉ *siècle*, Lille.

Nyström, U. 1940, *Poèmes français sur les biens d'un ménage depuis "l'Oustillement au Villain" du XIII*ᵉ *siècle jusqu'aux "Controverses" de Gratien du Pont*, Helsinki.

Oschinsky, D. 1971, *Walter of Henley and other treatises on estate management*, Oxford.

Parain, C. 1979, *Outils, ethnies et développement historique*, Paris.

Pesez, J.-M. 1991a, "Outils et techniques agricoles du monde médiéval". In *Pour une archéologie agraire*, ed. J. Guilaine, Paris, 131–64.

—— 1991b, "Témoins archéologiques de la viticulture médiévale". In *Le vigneron, la viticulture et la vinification en Europe occidentale au moyen âge et à l'époque moderne (Flaran 11, 1989)*, Flaran, 241–46.

Philippe, R. 1982, "Les premiers moulins à vent", *Annales de Normandie* 32, 99–120.

Picone, M. 1994, *L'enciclopedismo medievale: atti del Convegno Internazionale di Studi su "l'enciclopedismo medievale", San Giminiano, 1992*, Ravenna.

Pitte, J.-R. 1983, *Histoire du paysage français*, Paris.

Plaisse, A. 1961, *La Baronnie du Neubourg: essai d'histoire agraire, économique et sociale*, Paris.

de Planhol, X. 1959, "Essai sur la genèse du paysage rural de champs ouverts". In *Géographie et histoire agraires*, (actes du colloque international organisè par la Facultè des Lettres de l'Universitè de Nancy (2–7 September, 1957)), Nancy, 414–24.

Prévot, B. and Ribemont, B. 1994, *Le cheval en France au moyen âge. Sa place dans le monde médiéval; sa médecine: l'exemple d'un traité vétérinaire du XIV*ᵉ *siècle, "La Cirurgie des Chevaux"*, Orleans.

Raepsaet, G. 1995, "Les prémices de la mécanisation agricole entre Seine et Rhin de l'antiquité au 13ᵉ siècle", *Annales H. S. S.* 50, 911–42.

Raepsaet, G. and Rommelaire, C. (eds) 1995, *Brancards et transport attelé entre Seine et Rhin de l'antiquité au moyen âge*, Treignes.

Revel, J.-C. 1991, "Pédologie et archéologie". In *Pour une archéologie agraire*, ed. J. Guilaine, Paris, 323–43.

Ribemont, B. 1995, *D'Isidore de Séville (VII*ᵉ *siècle) à Jean Corbechon (XIV*ᵉ *siècle): études des encyclopédies sur la nature dans l'occident médiéval latin*, unpublished thesis, Université d'Orléans, Orleans.

Richard, J.-M. 1892, "Thierry d'Hireçon, agriculteur artésien", *Bibliothèque de l'École des Chartes* 53, 383–416, 517–614.

Roupnel, G. 1974, *Histoire de la campagne française*, new edition, Paris.

Ruas, M.-P. 1992a, "Les plantes exploitées en France au moyen âge d'après les semences archéologiques", *Flaran 12*, 9–35.

—— 1992b, "The archaeobotanical record of cultivated and collected plants of economic importance from medieval sites of France", *Festschrift for Professor W. Van Zeist*, ed. J. P. Pals, *Review of Palaeobotany and Palynology* 73, 301–14.

Russell, J. C. 1958, *Late ancient and medieval population* (Transactions of the American Philosophical Society, 48, pt. 3), Philadelphia.

Scheler, M.-A. (ed.) 1867, *Lexicographie Latine du XIIe et du XIIIe siècle. Trois traités de Jean de Garlande, Alexandre Neckam et Adam de Petit Pont*, Leipzig.

Sclafert, T. 1926, *Le Haut-Dauphiné au moyen âge*, Paris.

de Serres, O. 1600, *Théâtre d'agriculture et mesnage des champs*, Paris; reprinted in Geneva, 1991.

Sigaut, F. 1976, "Pour une cartographie des assolements en France", *Annales E. S. C.* 31, 631–43.

—— 1988, "L'évolution technique des agricultures européenes avant l'époque industrielle", *Revue archéologique du Centre de la France* 27, 8–41.

Sivéry, G. 1976a, "Les noyaux de bocage dans le nord de la Thiérache à la fin du moyen âge". In *Les bocages, histoire, écologie, économie*, (round-table held at Rennes), Rennes, 69–74.

—— 1976b, "Les profits de l'éleveur et du cultivateur dans le Hainaut à la fin du moyen âge", *Annales E. S. C.* 31, 604–30.

—— 1990, *Terroirs et communautés rurales dans l'Europe occidentale au moyen âge*, Lille.

Slicher van Bath, B. H. 1963, *The agrarian history of western Europe (AD 500–1850)*, London.

Spooner, F. 1961, "Régimes alimentaires d'autrefois: proportions et calculs de calories", *Annales E. S. C.* 16, 568–74.

Steensberg, A. 1943, *Ancient harvesting implements: a study in archaeology and human geography*, Copenhagen.

Stouff, L. 1970, *Ravitaillement et alimentation en Provence aux XIVe et XVe siècles*, Paris.

Thoen, E. 1992, "Technique agricole, cultures nouvelles et économie rurale en Flandre au bas moyen âge", *Flaran 12 (1990)*, 51–67.

—— 1994, "Le démarrage économique de la Flandre au moyen âge". In *Économie rurale et économie urbaine au moyen âge*, eds A. Verhulst and Y. Morimoto, Ghent/Kyushu, 165–84.

—— 1995, "L'influence de l'histoire rurale française à l'étranger", *Histoire et Sociétés Rurales* 3, 31–48.

Titow, J. 1972, *Winchester yields: a study in medieval agricultural productivity*, Cambridge.

Trochet, J.-R. 1993, *Aux origines de la France rurale: outils, pays et paysages*, Paris.

Uzan, M. and Pauphilet, D. 1955, "Aperçu d'ensemble sur le problème alimentaire en Tunisie", *Cahiers de Tunisie* 12, 627–42.

Van Mol, J.-J. and Devroey, J.-P. 1994, *Le seigle. Histoire et ethnologie*, Brussels.

Visset, L. 1985, "Palynologie des sites préhistoriques et protohistoriques dans le district phytogéographique de Basse-Loire (Massif armoricain-France)", in *Palynologie archéologique: actes des journées de Valbonne*, eds J. Renault-Miskovsky, M. Bui-Thi and M. Girard, Paris, 443–61.

White, L. 1940, "Technology and invention in the middle ages", *Speculum* 15, 141–59.

Zadora-Rio, E. 1991, "Les terroirs médiévaux dans le nord et le nord-Ouest de l'Europe". In *Pour une archéologie agraire*, ed. J. Guilaine, Paris, 165–92.

3. THE DEVELOPMENT OF FARMING IMPLEMENTS BETWEEN THE SEINE AND THE RHINE FROM THE SECOND TO THE TWELFTH CENTURIES

Georges Raepsaet

For Jean-Jacques Van Mol

The tenth century has traditionally been regarded as *the* pivotal period in the economic and agricultural evolution of western Europe. From Lefèbvre des Noëttes to Marc Bloch, from Lynn White to Finley, the orthodox view established a dichotomy: on the one hand, antiquity is characterized by technological stagnation (shown in particular by the lack of a productive technology) and a rudimentary economy; on the other hand, the tenth century was a period of development and growth, due mainly to the appearance of the horse collar. In the last few years, however, such views have been repeatedly qualified or questioned: both a technological deficit of antiquity and the concept of a revolutionary leap associated with medieval agricultural expansion have been challenged as a result of more rigorous methodologies and new sources of evidence, especially archaeology.

The main issue can be outlined as follows: did farming technology in the Gallo-Roman world make an effective contribution to the economy? If it can also be established that technological innovations developed in the area between the Seine and the Rhine during the second and third centuries, a period of economic growth, then it is necessary to discard the breaks in the traditional chronology and consider the reality and role of technology at all times and in all areas. The aim is not to claim continuity where breaks had previously existed, or to replace revolutions by periods of gradual transition, but to establish a number of technological facts in both a synchronic and diachronic perspective. As the essays in this collection demonstrate, there are innumerable technological improvements and variants according to period, geographical location, and context.

Thus, it is easy to deconstruct the orthodox model and expose its defects, but it is impossible to replace it by one or even several other paradigms. Even if this were the right approach, it would be premature in the present state of research. However practical and wide in its scope, our approach is faced with numerous methodological difficulties. Documents are scattered, incomplete and often second-hand, and the traditional method of comparing various periods of time is inadequate for our purposes, if only because of discrepancies in the evidence. When studying the technology of the plough, for example, it is not easy to compare a seventeenth-century drawing of a second-century funerary carving in low relief, the corroded share of an ard, a reference to a *carruca* in an eleventh-century demesne book, and a thirteenth-century church painting.

Janken Myrdal's and Kevin Greene's use of such concepts as context and complex in their assessments of rural technology are important because they help to establish a technology's value, even if they tend to make research more intricate. Particular implements are significant not in themselves but when related to regional typo-chronologies and to the surrounding modes of agricultural exploitation and socio-economic structures. The incompleteness of our sources, especially for antiquity, means we cannot hope for anything better than approximations and must seek to cross-check our hypotheses. However, the situation is not hopeless: as Grenville Astill and Janken Myrdal show in this volume, archaeology is a "very helpful source" and recent research also demonstrates its rich potential for the study of antiquity. Iconography must be incorporated into our work, and the usual distrust of illustrations must be replaced by an appropriate methodology, for medieval church paintings or the Roman bas-reliefs in Trier possess as much historical value as a demesne book.

If antiquity is to break out of its imposed confinement, an assessment of technological innovations must be attempted, and this is what this essay is about. We have deliberately chosen to focus on the sets of implements that represent an actual—and usually mechanical—improvement on traditional, generally hand-held, tools. Any improvement in a tool requires a certain amount of time, money and knowledge, and, above all, an innovative mindset, which may be as significant as the innovation itself.[1]

Let us begin by reviewing the agricultural mechanization in northern Roman Gaul in terms of the three most studied topics: cultivation, harvesting and transport.

---

[1] For the previous two paragraphs, see Raepsaet 1995; Greene 1994, and Myrdal in this volume.

Cultivation

A huge body of literature exists about the invention of the plough, and most of it relies on classical texts.[2] After having described the various ard shares and defined the coulter, Pliny notes:

> Not long ago, people from the Raetia of Gaul hit upon the idea of fitting two small wheels on such an implement, which people call a *plaumoratum*; the pointed end resembles a spade. Only cultivated land and plots that are practically fallow are sown in this fashion. The ploughshare is of a such a size as to turn over the clods. People immediately sow seeds and drag the harrow over the land. Land that is thus seeded does not need weeding, but it is with ox teams of two or three pairs that they plough in this manner. For each pair of oxen, on normal ground, forty or so *jugers* can be counted on each year; on uneven ground, thirty or so.[3]

This tilling implement—the *plaumoratum*—described by Pliny is usually described as a wheel-ard because it had produced asymmetrical furrows, which are regarded as the principal characteristic of the plough.[4] However, the matter seems far from settled, even at the terminological level, since the lack of a beam may also be considered a characteristic of the ard.[5] It is more important to establish if this implement scratched and broke up the soil, or if it had a mouldboard and was thus able to turn the soil over. Pliny's text is ambiguous, a mouldboard is not mentioned—to our knowledge it has no Latin term—but according to Pliny it was the ploughshare (of unknown type) that turned over the clods. One could not interpret *versat* any other way. Columella corroborates this possibility about tilling on fallow land: *mox conversa vomeribus*.[6] But ethnography shows (and indeed some ancient writers) that asymmetrical work can be performed by an ard, even without the addition of an "ear" or a "wing", by pressing on the handle and by tilting it.[7] In the northern and western

---

[2] See, for example, Haudricourt and Jean-Brunhes Delamarre 1955, 107–15, and its bibliography.

[3] Pliny, *HN*, XVIII, 48 [171–3].

[4] Ferdière 1988, II, 23–40; Comet 1992, 113–17.

[5] "L'araire est la plus simple de toutes les charrues, caractérisée par l'absence de toute pièce de soutien de l'âge à son extrémité antérieure" in *Larousse agricole* 1921, 87 and 291–8, fig. 291.

[6] *De re rustica*, II, 1.

[7] The ard plough with ears (*aures*) is quoted by Virgil, *Georgics*, I, 172. Compare Haudricourt and Jean-Brunhes Delamarre 1955, 97; see Comet 1992, 113–17 for archaeological traces of asymmetrical ploughing between the first and fifth centuries.

provinces of the Empire, numbers of coulters and asymmetrical shares continue to be found.[8] Thus, by putting together a front axle, a coulter, a large and asymmetrical share and an ear or wing, we have the components which define the plough in the modern sense of the word. In spite of the lack of comparative iconography of antique ploughing,[9] one suspects that some illustrations of ploughs of the thirteenth and the fourteenth centuries,[10] even of the eighteenth century,[11] must not be very different from the implement described by Pliny (figure 2.1).

Historians and ethnologists agree on the importance of the *rotulae*. The front axle (with wheel(s) or a runner), in its modern version, brings stability, control of ploughing depth and work comfort for both animals and ploughman. On the basis of Pliny's single text it is difficult to understand how the implement was used. The implication is that it was efficient, with a fairly sophisticated draught-pole-axle-beam combination which avoided the traditional rigidity of antique harnesses.[12] If this were the case, I should consider the implement as important an innovation as asymmetrical ploughing.

A further hypothesis, developed by François Sigaut,[13] connects the adoption of the plough in northern Europe with the cultivation of oats. This springtime cereal, used mainly as fodder, was almost always sown after one ploughing and buried by harrowing. Oats thus became a key element in the introduction of a new farming system which was based on an alternation of arable and fallow, and on the use of a new tool to clear the ground. One difficulty remains, however: where did these new techniques originate? This Raetia "of Gaul" which, in Pliny's time, stretched over a vast territory, on the south-eastern edges of *Gallia Belgica*,[14] had soils which were more suited to spelt than oats. This would displace the origin of the plough further north, towards the coastal region of the North Sea. But such exclusive interpretations are perhaps unnecessary. The ard appears, from

---

[8] Balassa 1971; Pohanka 1986, pls. 1–8; Rees 1979, I, 154–76, pls. 49–70; Ferdière 1988, II, 27–9. The asymmetrical plough share was probably used by the Gauls: Audouze and Buchsenschutz 1989, 199–200.

[9] Espérandieu, V, 4092, 4243 and XIV, 8387; see also Renard 1959, 53–7.

[10] Verhulst and Bublot 1980, 6–7; *Entre les foins*, 91.

[11] *Entre les foins*, 187.

[12] This mobile joint is clearly shown in a fourteenth-century French manuscript at the Bibliothèque Royale Albert Ier, Brussels: Verhulst and Bublot 1980, 6.

[13] Sigaut 1985, 23–33.

[14] Kolendo 1979, 61–73.

our limited evidence, to have been the preferred implement in Gaul and the Rhine and Danubian areas.[15] Following this hypothesis, ards were more suitable for winter corn cultivation and cross-ploughing on fallow land. But is there a real contradiction? The (limited) palaeo-botanical analyses do not show a monoculture but regimes of mixed farming where particular cereals dominated according to region.[16] It is also important to remember that the distributions of the plough and the ard were not mutually exclusive. For various reasons—economic, geographic, botanic—the improved implement could exist alongside the traditional ard, even allowing for the wealth of improvements—asymmetrical share, coulter, heavier yoke—around the third and fourth centuries.[17] Similarly Janken Myrdal (this volume) has found that several types of ards and ploughs were used at the same time in neighbouring regions in medieval Scandinavia.

Harvesting

Harvesting techniques are no less problematic. Apart from picking, almost every way of cutting involves the sickle,[18] whether it be at ground level, mid-way up the stalk, or under the ear. A few representations are known for the curved or toothed sickle (*falx rostrata* or *denticulata*) often mentioned by the *Rei Rusticae Scriptores* (figure 3.1).[19]

Much attention has been given to the animal-driven, mechanical harvester called a *vallus* mentioned by Pliny (XVIII, 72 [296]) and Palladius (VII, 21) and represented in the reliefs of Buzenol, Arlon, Trier, Reims and Koblenz.[20] The *vallus* only harvested the ears, and

---

[15] Renard 1959, 53–7.

[16] Compare in particular Körber-Grohne 1989, 51–9.

[17] See Pohanka 1986, 304–15; Rees 1979, I, 6–69, especially 57–9; compare Roesener 1986, 74–6.

[18] Ferdière 1988, 49–54; Kolendo 1960; Sigaut 1985, 33; Pohanka 1986, 128–75; Rees 1979, II, 438–509; White 1969, 804–9. Harvesting using sickles has been tried by Reynolds 1979, 64–5.

[19] The scythe with a long blade and tall handle is the haymaking implement—*falx faenaria*. The Italian scythe is shorter. The Gaulish implement, according to Pliny, is longer and cuts grass at mid-height, allowing farmers to work faster in the areas where the climate imposes a short period for haymaking: White 1970, 448.

[20] Fouss 1958; Renard 1959; Kolendo 1960; Müller 1985, 191–6. Interpretations vary from "one of the Roman Empire's rare and technological inventions and innovations" (Drinkwater 1983, 167) to an implement hardly more efficient than the *mergae* and perhaps less than the sickle (Sigaut 1985, 34–6), via the straw- and grain-wasteful implement (Martin 1971, 73–80; Gille 1978, 391).

may be comparable to the *mergae* or *mergites* of Pliny and the Latin agronomists,[21] and with the *mesorias* still used in Catalan country, both of which had a small cutting fork or "comb".[22] Its cutting edges were well adapted to harvesting spelt, whose ear separates easily from the stalk, and neither did it involve a substantial loss of grain.[23] The *vallus* was an important innovation not only because it was mounted on wheels, but also because it was pushed, rather than pulled, on a shaft. While this in itself was not unusual, the arrangement required an important modification of the usual Gallo-Roman traces. The efficiency of the traditional double-shafted carriage pulled by a single animal was generally reduced by the use of a rigid harness mounted on a small yoke. In this case, however, as can be seen from the relief

Figure 3.1. The *vallus*, a Gallo-Roman harvesting machine (Raepsaet 1982, 263).

---

[21] Columella, II, 20, 3.
[22] Sigaut 1982a.
[23] Kolendo 1960.

of Koblenz, the traction device of the *vallus* included supple traces, which marks a break with former practices.[24]

According to Palladius, oxen were used as draught animals, but the iconographic sources invariably showed the horse and the mule.[25] The *vallus* existed from the first century and, to judge from the iconographic sources, its use was limited to the Seine-Rhine region. Pliny (XVIII, 30), however, mentions only Gaul, and Palladius the *pars Galliarum planior* (VII, 2, 2–4), adding that the implement *campestribus locis vel aequalibus utile est, et iis quibus necessaria palea non habetur*. The original idea that the *vallus* was primarily used in spelt growing regions seems to be confirmed by the palaeo-botanical analyses from the excavations of military sites in Upper Germany, while wheat dominated in as many sites in Lower Germany where the soils, in for example the Hesbaye region, traditionally grew this cereal.[26]

Stubble got trampled underfoot with the *vallus*, but this was not necessarily a handicap because flattened straw could still be mowed and it can also be burnt and used as fertilizer; or used as feed. The suggestion that the *vallus* is only suitable for extensive agriculture misunderstands the situation in these provinces. Evidence for intensification, such as the use of fertilizer, exists in Gaul, but this does not preclude attempts to increase the arable area.

Neither the "first harvester in the world", nor a simple *mesoria* on wheels, the *vallus* is an original cutting implement that is a constituent part of a technological complex, well adapted to spelt, and developed in the context of an increased need for cereals, and in the wake of innovations in rural transport.

One may wonder if the relationship between the scythe, the sickle and the *vallus* should be viewed in terms of competition, rather than seeing each as adaptations to different cereals. Archaeological

---

[24] The harvester rectifies the height of the cut by pressing or by raising the shafts, according to the text of Palladius. Compare Raepsaet 1982, 263. In the modern carriage the notion of (supple) traces is associated with the swingletree and of shoulder collar or breast harness ("bricole" or "tablier", in French).

[25] Iconography on funerary monuments has a tendency to highlight the horse, a "noble" animal which emphasizes the status of the deceased, more than the ox would do, but which, however, does not negate, in the Gallo-Roman context, the value of historical representation on illustrations. Compare Raepsaet 1982. Experiments in Virton (Belgium) seem to confirm the rapidity of the work mentioned by Palladius, but it would be worth doing more extensive trials: Fouss 1958.

[26] Ferdière 1988, II, 61–72. On the density of the villas in the Hesbaye country, see: Wightman 1985, 119–28.

discoveries suggest increasingly diverse typologies of cutting implements, some of which even show a definite analogy with the Flemish
hook, and tend to prove, at least for central Europe and England[27]
from the first century AD, that scythes, sickles or reaping hooks were
all used for harvesting at the same time. And it is likely that the
same is true for the area between these two regions, but we lack
systematic data.

Transport

Land transport, dependent more on the ox and the donkey than on
the horse, was again characterized by some special features or technical innovations.[28] There was a wide variety of carts—some with
two or four wheels and a draught pole, some drawn by oxen with a
double yoke on the head or nape, or by mules and donkeys with
either a neck or back yoke. But there was also a double-shafted
carriage for one horse or mule with a small neck yoke. This is a
surprising innovation and without immediate precedent. Horses (or
mules) were also depicted pulling heavy transport in teams and in
line, as shown on the relief of Langres, and, while this document is
unique, the practice was perhaps not exceptional.[29] Between the Seine
and the Rhine, and in the Danubian provinces, representations of
vehicles with shafts are, however, frequent and show a similar range
of light carts (figure 3.2). The arrangement of two animals at a single
pole with a back yoke and "shortened" breast harness, which was
common in the Mediterranean, is of great technical simplicity and
was used for the transport of people as well as objects. It better
suited animals with short and low necks, such as oxen, rather than
horses or mules.[30]

   The double-shafted carriage in Gaul shows a special harness, a
kind of small yoke, laid on a padded prong saddle. When the point

---

   [27] Pohanka 1986, 147–75 and pl. VIII (Sense); 128–46 and pl. VII (Sichel); Rees
1979, II, 438–86.
   [28] Raepsaet 1987; 1979. For the historiography of the subject, Amouretti 1991;
Molin 1987–8; Burford 1960; Spruytte 1977.
   [29] Raepsaet 1982, 271; Molin 1984.
   [30] Besides, when working the horse has a tendency to bend the neck so as to
bring down the holding point at chest level and help itself by the engagement of the
back, as shown in the realistic iconography of northern Gaul.

Figure 3.2. A typology of Roman carts and wagons from northern Gaul
(Raepsaet 1982, 253).

of power hold is located at mid-height on the neck, as suggested by the iconography, the traction which is exerted on the small yoke from front to back mobilizes the strap of support and imposes some constraint. But we must point out that, in most cases, it concerns light carts—judged to carry a maximum load of half a tonne—which does not normally generate a traction handicap.[31] Besides, it is not excluded that some training and adequate rein structure allow adjustment of the animal's involvement in function of the required effort.[32] The technical configuration of this original harness and the way it functioned certainly deserve re-examination.[33] This might be done on a new experimental basis and from the discoveries of Wange and Hondelange of metallic pieces of supposed small yokes in Belgium.[34] In fact, this small yoke appears to be the driving force behind the light double-shafted carts in northern Gaul, offering the users, in addition to "heavy" cartage with four wheels and a pole, an alternative form of transport for various purposes, service and freight, in an economic system which did not lack commercial ambition. In addition, the Gallo-Roman carriage structure which, in its specifity and diversity, was developed in Gaul and the Rhineland provinces is certainly not unrelated to the emergence, improvement, and adoption of the agricultural mechanized tools considered above.

In fact, everything seems to indicate that the mechanical and pre-mechanical agricultural improvements in the northern provinces, in particular *Gallia Belgica*, were linked to the increasing economic development over at least three centuries. This process started at the time of Augustus, under a remarkable convergence of factors[35] which created an increased demand for food, and the conditions and means to meet that demand: good natural conditions (soils and climate); the expertise and knowledge of the local Celtic population;[36] the stabilization of the *Limes* on the Rhine, causing a considerable civilian and military population to settle there; the setting up, by Agrippa and Augustus, of an efficient road infrastructure; the rapid, whether

---

[31] It is not impossible that when the strap maintaining the yoke is represented at mid-height of a neck turned up as a swan-neck, seeming to heighten the pressure backwards and forwards, it is a question of aesthetic bias.

[32] *Entre les foins*, 258–63.

[33] Raepsaet and Rommelaere 1995.

[34] Fairon 1992, 29–35; Raepsaet and Rommelaere 1995, 57–66.

[35] For a comprehensive study of these questions, see for example Wightman 1985.

[36] Audouze and Buchsenschutz 1989, 24–5, 160–70, 198–200.

planned or organic, urbanization; the local population retained a large degree of freedom in terms of economic and commercial initiatives; a significant demographic rise; and also, under Claudius, the conquest of Britain, which was largely provisioned from the continent. During the whole Early Empire, and also during the Late Empire in regions such as the *Civitas Treverorum*, archaeology shows that urban and rural areas developed through the growing influence of the market. The problem is complex and, of course, impossible to quantify. However, one must note that the commercial network spread far beyond the hinterlands of towns and that rural surpluses, via road or river markets and *negotiatores* and *mercatores*, were absorbed into middle- and long-distance distribution circuits.[37] It is of course tempting to link the technical innovations we have discussed, which all tend to make labour and transport more efficient, with the intense development of this area, in particular in the basin of the Sarre and the Mosel.[38]

Is there, then, any link between these technological innovations and attempts to meet a shortage of labour or to make work more efficient? In the case of the classical and early medieval periods this question is bound to receive only a conjectural answer. Because of the documents available, the issue of the labour force cannot be dealt with precisely, whether qualitatively or quantitatively. But it is now clear that there was no causal relationship between the existence of an enslaved labour force and the technological level of a given tool. While it is hard to defend the notion of technical stagnation with regard to the Graeco-Roman world, it is even more perilous to do so in the case of farming tools in northern Gaul. However, it is still difficult to assess the technological level reached: are the *vallus*, the *plaumoratum*, or the double shafts new mechanisms? The question remains open, the answer depending on the criteria adopted. The assessment of an innovation will depend on whether or not one considers that a technological complex marks a genuine mechanical advance.

---

[37] Schlippschuh 1974; Wierschowski 1995; Jacobsen 1995; Greene 1986, 164–7; compare Raepsaet-Charlier 1988, 45.

[38] Bonn and Paris exhibition catalogue: *La civilisation romaine de la Moselle à la Sarre* 1983; Burnand 1990.

The Fourth to Ninth Centuries

The study of medieval agrarian techniques has in the past been selective. There has, for example, been a tendency to concentrate exclusively on the problems, however fundamental, of crop rotation, cultivated plants and yields. Similarly, any consideration of farming tools is usually slight and limited to the invention of the horse collar and the mouldboard plough. In the complex and muddled political and social situation that existed between the fourth and seventh centuries, the study of techniques might appear inappropriate or superfluous, but this is actually not the case. The data only allow us to make very general remarks or suggest working hypotheses, and we cannot give as precise an assessment as for earlier or later periods.

The period from the Late Roman Empire to Carolingian times was long considered to have been a time of crisis and decline, but recently it has undergone a reevaluation. The accepted interpretation of a Europe with mediocre tools and poor crop yields has been challenged by a more optimistic assessment of production techniques, a view, however, that is still not shared by everyone.[39] The new ideas have been helped by several recent works which show that the transition from Roman to Merovingian times, although of great complexity, should no longer be considered in terms of a break.[40] The Roman rural economy, upset by the invasions of the third century, recovered. A certain number of villas resumed their activities, although on a reduced scale because the marketing of surplus on the Rhenish market had become less certain in the Late Empire, especially in *Germania Secunda*.[41]

Yet it is still difficult both to comprehend the reality of the landuse from the fourth to seventh centuries, and to establish connections

---

[39] Duby 1962, I, 77–8; Duby 1966, 267–84; Duby 1973, 22–40 and 218–20 (more moderate). Fossier 1982, 614–799, especially 621–625; compare Fourquin 1989, 22: "l'agriculture durant tous les siècles du haut moyen âge est demeurée une agriculture très extensive, défavorisée par la médiocrité des outils et l'insuffisance du gros bétail".

[40] The thesis of a cultural break was formerly defended by De Laet *et al.* 1952. Verhulst wavers between a structural non-continuity and elements of continuity particularly at the archaeological level: Verhulst 1985a and b; a moderate position is also taken by Devroey and Zoller 1991, 225–8. Very moderate positions are developed in Drinkwater and Elton 1992.

[41] Van Ossel 1992, 171–84. However, a relative desertion of rural settlements is highlighted in the Rhineland hinterland by Gechter and Kunow 1988, 109–28.

between the Roman villa and the first medieval villages. Tamara Lewit and Paul Van Ossel's analyses[42] show, however, that it is not a hopeless problem. There was a connection between the *fundus* and the Merovingian estate, and between Gallo-Roman settlement and the proto-medieval nucleated settlement, but with many regional, or indeed micro-regional, variants. Even in a land as badly disrupted by successive waves of invasions as *Gallia Belgica*, the progressive disappearance of the traditional forms of settlement is accompanied by a shift towards new forms of settlement which emerge in the Merovingian period. According to Paul Van Ossel, this is another form of continuity, it signals "the gradual transition from one type of landuse to another". John Percival advances a related hypothesis when he suggests "at least a possibility that the people who lived on a villa estate, the people who buried their dead in the villa ruins, and the people who ultimately came together in a village, were the same people".[43] With the fourth-century introduction of groups of farmer-soldiers of Germanic origin to northern Gaul, the official settlement in the Empire of certain trans-Rhenish populations during the first half of the fifth century, and the rise of the Frankish aristocracy in the military and administrative hierarchy, these areas made a transition toward a Frankish monarchy which achieved a kind of "symbiose effective gallo-romano-franque".[44] The persistence of craft technologies, noticeable in for example the applied arts, suggests that farming tools did not change fundamentally.[45] Moreover, in the field of transport, the Merovingian period can no longer be regarded as a *wagenlose Periode*; the tradition of cartage existed and establishes a connection between the Roman and the Carolingian worlds, even if it seems restricted to the transport of people rather than things.[46] All in all, the concept of functional continuity,[47] is relevant to production techniques, although we cannot be certain that there was "aucun

---

[42] Lewit 1991; Van Ossel 1992, 171–84.

[43] Percival 1992, 164.

[44] Werner 1984, 255–92. The global continuity is also defended by Durliat 1989, 137–54 and Lebecq 1990, 15–41. A remarkable example of continuity in farming is shown at Mondeville: Lorren 1989, 439–66.

[45] Dasnoy 1975, 55–6; Claude 1981, 230–3. For the use of the sickle in the Merovingian period, see Henning 1991.

[46] Janssen 1989, 174–228. A recent book demonstrates the extraordinary refinement reached during the Merovingian period in the technology and morphology of the horse bit: Oexle 1992.

[47] Verhulst 1990, 18.

hiatus entre le Bas-Empire et le début du haut moyen âge" as boldly put forward by M. Banniard.[48]

The watermill represents one of the few innovations where continuity has been reasonably established. The contexts in which mills were used, however, were different in antiquity and the middle ages, as shown by Kevin Greene.[49] Since many Germans had settled between the Seine and the Rhine, the development of technology must also be approached in terms of the relative density of the various populations and of their interaction, as well as in terms of their possible acculturation. Future research will have to analyse tools with much the same methodology that has been applied to arms, jewels or ceramics.

The surviving Carolingian sources are ambiguous and thus sometimes misinterpreted as far as tools are concerned.[50] For example, the mention of six oxen at the yoke, far from being a proof of weakness,[51] probably indicated a real capacity for traction, especially as ploughing or the carrying of heavy goods with harnesses and head or nape yokes have for the past two millennia normally required four to six oxen. The evidence of regular long-distance transport of agricultural surplus,[52] is an indication of the techniques and rural production that then existed. The regular use of heavy transport implies the continued existence not only of craftsmen, in particular cartwrights and harness-makers, but also of good draught animals, in this case oxen,[53] and a road network that is sufficiently safe and in good repair. The recent reevaluation of the use of the watermill in the Merovingian and Carolingian periods makes a similar point.[54] All these technical elements relate to a general tendency toward growth that most historians recognize, although there are some dissenters.[55]

---

[48] Banniard 1991, 37.

[49] Greene 1994.

[50] *Un village au temps de Charlemagne* 1988, 214–18.

[51] Fourquin 1989, 18–22; Fourquin 1975, 331–3.

[52] Devroey 1984; Devroey 1979; Lebecq 1989; compare Verhulst 1990, 24 and 36–41.

[53] The oxen of Saint Denis Abbey seem to be of a small size compared to the Roman oxen: compare *Un village au temps de Charlemagne* 1988, 226–41.

[54] Lohrmann 1989, 367–404.

[55] So R. Fossier in *La croissance agricole du Haut Moyen Age. Chronologie, modalités, géographie* 1990, 182–4.

The Tenth to Thirteenth Centuries

The economic growth which is visible from the tenth century onwards is seen as a reflection of structural change within Carolingian society, and not the result of autarkic "gardening",[56] and confirmed by the increase in commercial exchange between cities and the countryside in the ninth and tenth centuries.[57] Most historians agree about the importance of agricultural growth between the tenth and the thirteenth centuries, "the most significant agricultural expansion since the Neolithic".[58] But we are far from reaching a consensus about the role of tools and technical innovations in the process. There are those who, following Lynn White and the first works of G. Duby, were influenced by Lefèbvre des Noëttes and accordingly place special importance on the "revolution" of the horse collar in the tenth century.[59] Others rejected this "extremely tempting hypothesis" and tried to show that growth resulted from many modifications in the agrarian system,[60] including connections with the urban markets; and that the impact of the technological innovations on productivity was negligible before the twelfth or the thirteenth century.[61] The few sources available—written, archaeological and iconographic—suggest, however, that from the tenth century onwards some new improvements in tools and techniques occurred[62] but, as the data are not very explicit, these are difficult to interpret and date. This information also causes us to consider whether the tenth century is of technological importance.

As Grenville Astill has shown in the case of England (in this volume), signs of expansion emerge as early as the seventh and eighth centuries, while indications of a joint technical and economic development were present in the tenth and eleventh centuries; but the later twelfth century was a "time of immense innovation" linked to widespread, accelerated, growth.

---

[56] Schweiter 1984, 187–8.
[57] Despy 1968; Verhulst 1990, 40–1; Watson 1981, 65–82.
[58] Verhulst in Verhulst and Bublot 1980, 7.
[59] For example: Duby 1962, 190–4; Fourquin 1989, 81–91; Gimpel 1975, 49–78. It is also the argument of White 1962. A moderate stance on the "agrartechnische Revolution" is defended by Roesener 1986, 118–33.
[60] Sigaut 1985, 65–7 also mentions the three-yearly rotation of crops, the development of a full-time and more professional agriculture.
[61] Verhulst 1988.
[62] Pesez 1991, 131–64.

While there is evidence for the continued exploitation of wind and water as sources of energy,[63] the horse assumed, for the first time since the Roman period, a greater role for transport from about the tenth century, but also for ploughing.[64] The animal's importance continued throughout the eleventh and twelfth centuries, but there was great regional diversity,[65] and was partially achieved through the notable modifications in the harness and the adjustments of the horse collar which produced innovations such as the saddlepad and various girths.[66]

This new horse equipment needs some comment because it is twenty years since Lefèbvre des Noëttes' theory of a horse collar revolution was questioned and we can reexamine the issue in a less polemical spirit. On its own the padded horse collar was, from a technological point of view, of peripheral importance. It is the combination of all the elements of the equipment—collar, shafts, girths, followed by the swingletree—which constitute a remarkable innovation whose significance should not be underestimated. The small ancient yoke, associated with the double shaft in front of the withers or on the lower part of the neck, suits the shape of oxen and donkeys and permits a transfer to the shafts of power generated by the animal through its shoulders and back. The neck of the horse is, however, longer and helps the animal to retain its balance when in motion.

For the antique small yoke to reach its full efficiency, the shafts had to be located at the lowest possible point, as shown on Gallo-Roman reliefs, and this can be achieved through training and a balancing of the shafts.[67] As long as the small yoke system is retained with its rigid connection to a shaft with turned up crook, it is impossible to get the shafts low enough to exploit the maximum potential energy from the horse's chest. It is only possible to do this with a harness which covers the shoulder, and also with traces attached to the "collier", in the middle of the chest. From a typological point of

---

[63] Sigaut 1985, 55–6; Comet 1992, 388. See also: Oleson 1984; Wikander 1984; Polge 1967b, 85–120. On the windmill: Rivals 1987.

[64] Hoebanx 1975, 164.

[65] Hoebanx 1975, 180–1; Fourquin 1969, 155. The general need to harness a horse to the plough did not occur before the fourteenth century in the northern countries according to Slicher van Bath 1960, 196, and Thoen, this volume.

[66] Clutton-Brock 1992. On the typology of road transportation in medieval France, see from now on: Girault 1992.

[67] Compare Raepsaet 1982.

view, the Gallo-Roman small yoke, with its thick pad, seems strikingly similar to the collar, but the functional difference lies in the lowered power hold, the balance of shafts, the traces and the swingletree. The "modern" horse collar combines in one traction apparatus two antique harnesses, the "shortened" breast harness attached to the chest, and the small rigid yoke which rests on the shoulder.[68]

The invention of the horseshoe, which has long been associated with the "revolution" of the harness need not detain us. The shoeing of horses already existed among the Celts and was widespread, with the hipposandal, in the Roman period.[69] If the practice was abandoned during the early middle ages—which has yet to be proved—it is because it was not absolutely indispensable for the use of the horse. The shoe is a protection and a reinforcement of the hoof, especially useful on hard or rocky ground, on roads rather than on tracks. It is hoof care, not shoeing, which is indispensable in all circumstances. The removable Roman hipposandal was perhaps designed to allow the horse to be used on hard and soft ground. If it is proved that the shoeing of the horse reappears and becomes widespread during the tenth and eleventh centuries, it could be due to a more frequent use of the horse for transport on roads with a hard surface.

The new harnesses also permitted heavier types of horses ("Kaltblut") to replace the ox, even for heavy traction.[70] However, it is difficult to assess not only when, but where, how, to what extent and in what circumstances, this replacement took place. It seems that it is not until the late twelfth century that the more rapid draught horse gradually replaced the ox, and then mainly in the Flemish area—where the intensification and commercialization of agriculture are well attested (Thoen, this volume).[71] Such a development will neither be systematic nor continuous.[72] In some industrialized countries

---

[68] For the mechanics of traction, see Abeels 1995. The metallic frame recently unearthed in La Grande Paroisse (Seine-et-Marne) may well be the splint of a horse collar. As the artefact has been dated to the tenth century, it might be the oldest archaeological evidence of the new harnessing system, Petit 1993.

[69] See, for example, Ferdière 1988, 143.

[70] It is possible that the Franks introduced heavier horses: Nobis 1973.

[71] Compare Verhulst 1985. One could mention the possibility of a relation between the adoption of the horse in Flanders and the predominance of small farms on light soils with high working productivity where only one horse may be enough and profitable. Compare Verhulst 1990, 64–7.

[72] Horse and ox have their own particular qualities which, as far as the management of an estate is concerned, may be complementary and therefore replacement of one by the other is not necessarily required. One finds some traces of this old debate on

oxen tended to be replaced by draught horses only a few decades before the horse itself was superseded by motorized tractors.[73] The marked increase in the use of the horse has been studied for medieval England by John Langdon.[74] Horses were used for cartage more than ploughing, but there are great variations from region to region, depending on the soil, relief, and the organization of production and of farming. The harnessing of horses was linked to the growing sophistication of market transactions and to the development of urban communities. The impact of the horse was greatest for transportation at the level of exchange between town and country, but it is impossible to quantify the global contribution to economy, and it is even difficult to prove the phenomenon. Is the improved horse carriage the— or one—cause of the development of agricultural commercialization, or is the growth of the market economy responsible for the increase in the number of carriages? And, Bruce Campbell asks, can we really consider that the horse was "technologically the most active member of medieval society"?[75]

Harvest tools were not, as in Roman times, apparently influenced by the new harnessing techniques. The scythe and sickle both seem to be used for cutting hay and cereals, with perhaps an improvement in the quality of the metal used. The Flemish hook or pick, twice as efficient as the sickle, may have appeared in Flanders before the end of the thirteenth century[76] and progressively replaced the sickle for the cereal harvest by the fifteenth century. Cereals were cut high in the twelfth century and by the fifteenth century were cut low. Despite much work by Georges Comet there are no clear patterns[77] and even in the fifteenth century the sickle was favoured for

---

the respective qualities of the horse, of the mule and of the ox in all the agronomical literature, from Roman times up to the last issues of *La Maison Rustique* and, even today, in the agricultural manuals and guides intended for tropical countries. For example: *Techniques rurales en Afrique* 1971, 25–8.

[73] Liebowitz forthcoming. On the general use of the horse in the nineteenth century, see Lizet 1982.

[74] Langdon 1986, and 1984. Compare Sigaut 1985, 54–5.

[75] Campbell 1988, 87–98, especially 93.

[76] Verhulst 1990, 128–9; Hoebanx 1975, 164–5; Lindemans 1952, II, 50. See also Sigaut 1985, 37–42. For Slicher van Bath (1960, 207–8), one should wait until the fourteenth century before the *zicht* or *pik* replaces the sickle. Then, in only one movement, the stems are cut by armfuls and laid down on the ground in swathes. Hence, most of the straw is used for litter, giving a manure of quality and thus increasing the productivity of the fertilized soil.

[77] Comet 1992, 192–4.

the harvest but some original, intermediate, forms between the hook and the sickle continued in use.[78]

Ploughing techniques were also modified. The "modern" plough together with the horse collar are for some the driving forces behind the "agricultural revolution" (figure 3.3). But when attempts are made to discover the circumstances and rate of replacement of the ard by the plough, the effort is almost doomed because the terminological confusion between *carruca* and *aratrum* remains. The complete plough, with its three fundamental parts—coulter, symmetrical or asymmetrical ploughshare, and mouldboard—is well attested in the thirteenth century, so it was most probably perfected well before. Moreover, as we have noticed, a ploughing instrument with coulter, wide ploughshare and "wings" able to turn over the clods, was known in Roman Gaul. The front-axle wheel was also in use in the Early Roman Empire but has been a neglected element in the study of the plough—it may be because from the end of the middle ages, it was sometimes replaced by the runner and henceforth considered as secondary.

Two particular points must be emphasized. The typological variety of the "plough" is unquestionable in the thirteenth century and is even more obvious, thanks to iconography, in the fourteenth and fifteenth centuries. Heavy, light (Flemish *eenstaartploeg*), mobile, with runner or *tourne-oreille* ploughs, which all have their own variants, especially in the adjustment and assembly systems, deserve a careful

Figure 3.3. A thirteenth-century depiction of a wheeled plough drawn by a horse with a collar (Vieil Rentier des Seigneurs de Pamele-Audenarde, Bibliothèque Royale, MS 1175, fol. 156, Brussels).

[78] Collins and Davis 1991.

and detailed inventory on the chronological and regional level,[79] as Janken Myrdal has done for Scandinavia (this volume). There is also the problem of the front axle: it was an instrument of work comfort, of control and of setting the depth of ploughing, but it is also a required articulation in a carriage with traces. Finally, it is, especially in the case of the light plough, a counterbalance which prevents the beam rising as a result of slanting traction. The representation of the plough in the *Vieil Rentier d'Audenarde* is particularly realistic in this respect (figure 3.3).

The diffusion of the harrow[80] from the (?) eleventh century on, would also be linked to the new harnessing, the rapidity of traction being, according to some, an essential advantage in the use of this instrument. We are not convinced that the harrow is a medieval "invention",[81] nor that its development was the consequence of the new utilization of the horse. The framed harrow is still used today in numerous countries, pulled by oxen, whose speed is hardly less than a quarter of a horse's.[82] There are thus some missing links between the Roman *irpex* and the harrow of the Bayeux Tapestry.

Georges Comet, like François Sigaut, rightly emphasizes the importance of the interrelationship between plough, sowing and harrow, and the order in which they were used for sowing.[83] It is interesting to note that this effort of improving, of innovating or extending the use of original techniques between the tenth and the thirteenth centuries is consonant within a context of dynamic rural expansion and intensification of the town and country exchanges as happened in the same regions in the Early Roman Empire.[84] Evidence is lacking but, as cultural gaps shrink and the great "rupture" of civilization eases, it is no longer forbidden to think that a connection exists between the small padded yoke and the horse collar, between the *plaumoratum* and the plough.

---

[79] See Lindemans 1952, I, 164–203. For plough traction in France, see Comet 1992: he argues that the front axle precedes the monolateral mouldboard and that the "moving" plough appears in the thirteenth century.

[80] Sigaut 1985, 54–5; Fourquin 1969, 157.

[81] In the sixteenth century, Olivier de Serres will still have to plead in favour of the interest of harrowing: Gorrichon 1976, 174–6.

[82] *Techniques rurales en Afrique* 1971, 39–48: Hopfen 1970, 10–13.

[83] Comet 1992, 157–64; Sigaut 1982b, 40–3.

[84] The development of exchange between the city and the country in the eleventh to thirteenth centuries is a recurrent theme, for example, Billen 1991.

Conclusion: The Imaginary Revolution

The change that took place between two periods of great economic vitality—the second and third centuries, and the twelfth and thirteenth centuries—appears to have had a more halting than gradual development. It was an age when there were not a great number of technological innovations, but those which did occur were fully exploited. In the tenth to the thirteenth centuries, relatively simple farming tools evolved in numerous variations in terms of materials, shapes and uses, and mechanical improvements were combined in ways that were indicative of better farming performances, such as the heavy plough with forewheels drawn by a horse equipped with a collar, or the double-shafted cart (figure 3.4). The part these innovations played in increasing productivity and in enhancing effective transport and commercialization of goods should not be overestimated. Erik Thoen

Figure 3.4. A thirteenth-century depiction of a cart drawn by a horse with a collar (Vieil Rentier des Seigneurs de Pamele-Audenarde, Bibliothèque Royale, MS 1175, fol. 156, Brussels).

shows in this volume that, as early as the twelfth century, agriculture in some regions of Flanders took advantage of the superiority of the horse over the ox, but also that this superiority fitted into a complex technical framework where mechanical developments were neither more nor less important than farming practices (three-year rotation of crops), manuring, industrial plants, livestock management. John Langdon comes to a similar conclusion about thirteenth-century England, where a "horse-based" farming economy also existed, but with less uniformity. In Scandinavia, especially Sweden, Janken Myrdal has also noticed technological innovations such as heavier shares or the exploitation of horses, in technological complexes where the two-year rotation of crops was prevalent—sometimes with significant regional variation. Whether technological innovation is a cause or an effect is a question which has to be treated with great caution. As most of the essays in this volume indicate, the current tendency in the study of the middle ages seeks to relativize technological impact, and to examine the rate at which innovations were adopted and adapted, to study other prevalent systems as well as the interdependency of regional and general contexts. Relativizing the role of innovation does not, however, imply neutralising or obscuring it. For example, Myrdal has cogently shown that a two- or four-fold increase of the weight of shares corresponds to substantial changes in ploughing techniques. Likewise, the development of the horse collar is far from insignificant; it was a complex and original mechanical system which made horses extremely efficient within the framework of an economy where trade is more widely spread and exchanges are accelerated.

But it is now time to return to our starting point: the area between the Seine and the Rhine was not a technological wasteland. Productivity had reached a respectable level, growth was tangible and continuous in the second and third centuries, and there were a number of original technological innovations. This alone is enough to invalidate the notion of a chasm between the classical and medieval periods, and thus we are led to reconsider the nature of the connections between the two eras.

As we have already pointed out, the denial of an arbitrary historical break does not necessarily imply the prevalence of continuity. Antiquity is not necessarily the background or point of departure for later technical innovations. Even if a practice has disappeared, the knowledge, techniques and traditions that surround it can be handed down over several generations, regardless of political and social changes.

From the Early Empire to the "long thirteenth century" a number of scenarios are possible: from the continuity of a technique to its complete abandonment, to various forms of adaptations. In that respect, the watermill is a very enlightening example, whose interpretation is currently subject to thorough revision. It is almost certain now that watermills never ceased to exist between the classical and medieval periods. It appears that the mills' uses developed extensively from the tenth century within different socio-economic contexts, but we must be aware of the importance of the Roman background. Yet, in most other fields of investigation, substantial research still needs to be carried out—especially in archaeology—before one can define a classification of farming tools and contextualize them as precisely and relevantly as Myrdal has done for Sweden.

We initially chose to focus on long-term historical developments. This slightly risky option has nevertheless yielded a few simple observations. Technical innovations, which archaeology can substantiate, are reliable indications of wider economic phenomena; indeed the value of innovations can be measured by the way in which they fit into a technological complex or assembly and by the socio-economic conditions surrounding it, and this is true for both the classical and medieval periods. And lastly, it will be necessary, whenever one studies the medieval farming implements of northern Roman Gaul, to take account of the technological stock of knowledge without any prejudice as to the prevalence of historical continuity or discontinuity.

Bibliography

Abeels, H. 1995, "Les configurations de la traction aux brancards: forces et effets". In *Le transport attelé entre Seine et Rhin de l'Antiquité au Moyen Age. La révolution du brancard*, eds G. Raepsaet and C. Rommelaere, Brussels, 13–29.

Amouretti, M-C. 1991, "L'attelage dans l'antiquité. Le prestige d'une erreur scientifique", *Annales E. S. C.*, 219–32.

Audouze, F. and Buchsenschutz, O. 1989, *Villes, villages et campagnes de l'Europe celtique*, Paris.

Balassa, I. 1971, "The appearance of the one sided plough in the Carpathian Basin", *Acta Ethnographica Academiae Scientiarum Hungaricae* 20, 411–37.

Banniard, M. 1991, *Le Haut Moyen Age occidental*, Paris.

Billen, C. 1991, "Binche et sa campagne: des relations économiques exemplaires (XIIe–XIIIe siècles)". In *Villes et campagnes au Moyen Age. Mélanges Georges Despy*, eds J-M. Duvosquel and A. Dierkens, Liège, 87–109.

La civilisation romaine de la Moselle à la Sarre 1983. Bonn and Paris exhibition catalogues, Mainz.

Burford, A. 1960, "Heavy transport in ancient Greece", *Economic History Review* 13, 1–18.

Burnand, Y. 1990, *Les Temps Anciens. 2. De César à Clovis*, [Histoire de la Lorraine. I], Nancy-Metz.

Campbell, B. 1988, "Towards an agricultural geography of medieval England", *Agricultural History Review* 36, 87–98.

Claude, D. 1981, "Die Handwerker der Merowingerzeit nach den erzählenden und urkundlichen Quellen". In *Das Handwerk in vor- und frühgeschichtlichen Zeit*, ed. H. Jankuhn, Göttingen, 230–3.

Clutton-Brock, J. 1992, *Horse power. A history of the horse and the donkey in human societies*, Cambridge, Mass.

Collins, M. and Davis, V. 1991, *A medieval book of seasons*, London.

Comet, G. 1992, *Le paysan et son outil. Essai d'histoire technique des céréales (France, VIIIe–XVe siècle)*, Rome.

La croissance agricole du Haut Moyen Age. Chronologie, modalités, géographie, 1990, Xe Journée internationale d'Histoire, Abbaye de Flaran, Auch.

Dasnoy, A. 1975, "Les Germains dans la romanité". In *La Wallonie. Le pays et les hommes*, I, ed. H. Hasquin, Brussels, 37–60.

De Laet, S. J., Dhondt, J. and Nenquin, J. 1952, "Les laeti du Namurois et l'origine de la civilisation mérovingienne". In *Mélanges F. Courtois*, Namur, 149–72.

Despy, G. 1968, "Villes et campagnes aux IXe et Xe siècles: l'exemple du pays mosan", *Revue du Nord* 50, 145–68.

Devroey, J. P. 1979, "Un service de transport à l'abbaye de Prüm au IXe siècle", *Revue du Nord* 61, 543–69.

—— 1984, "Un monastère dans l'économie d'échanges", *Annales E. S. C.*, 570–89.

Devroey, J. P. and Zoller, C. 1991, "Villes, campagnes, croissance agraire dans le pays mosan avant l'an mil. Vingt ans après. . . ." In *Villes et campagnes au Moyen Age. Mélanges Georges Despy*, eds J-M. Duvosquel and A. Dierkens, Liège, 223–60.

Drinkwater, J. F. 1983, *Roman Gaul. The three provinces, 58 BC–AD 260*, London.

Drinkwater, J. F. and Elton, H. (eds) 1992, *Fifth century Gaul: a crisis of identity?*, Cambridge.

Duby, G. 1962, *L'économie rurale et la vie des campagnes dans l'Occident médiéval*, Paris.

—— 1966, "Le problème des techniques agricoles", in *Agricoltura e Mondo Rurale in Occidente nel alto Medioevo*, Spoleto, 267–84.

—— 1973, *Guerriers et paysans, VIIe–XIIe siècle*, Paris.

Durliat, J. 1989, "Qu'est-ce que le Bas-Empire?", *Francia* 16 (1), 137–54.

Entre les foins et la moisson, 1984, Marloie.

Espérandieu, E. 1907–, *Recueil général des bas-reliefs, statues et bustes de la Gaule romaine*, Paris.

Fairon, G. 1992, "Un petit bâtiment rural d'époque romaine près de Hondelange (Messancy)", *Cahiers du groupe de recherches aériennes du Sud Belge* 8, 29–35.

Ferdière, A. 1988, *Les campagnes en Gaule romaine*, Paris.

Fossier, R. 1982, *L'enfance de l'Europe. X^e–XIIe siècles. Aspects économiques et sociaux. II. Structures et problèmes*, Paris.

—— 1990, intervention in *La croissance agricole du Haut Moyen Age. Chronologie, modalités, géographie*, 182–4.

Fourquin, G. 1969, *Histoire économique de l'Occident médiéval*, Paris.

—— 1975, "Le premier Moyen Age. Le temps de la croissance. Au seuil du XIVe siècle". In *Histoire de la France rurale*, I, eds G. Duby and A. Wallon, Paris, 291–611.

—— 1989, *Le paysan d'Occident au Moyen Age*, Paris.

Fouss, E. P. 1958, "Le vallus ou la moissonneuse des Trévires", *Le Pays Gaumais* 19, 125–36.

Gechter, M. and Kunow, J. 1988, "Zur ländlichen Besiedlung des Rheinlandes vom I. Jahrhundert v. bis ins V. Jahrhundert n. Chr." In *First millennium papers. Western Europe in the first millennium AD*, ed. R. F. Jones, Oxford, 109–28.

Gille, B. 1978, *Histoire des techniques*, Paris.

Gimpel, J. 1975, *La révolution industrielle du Moyen Age*, Paris.

Girault, M. 1992, *Attelages et charrois au Moyen Age*, Nîmes.

Gorrichon, M. 1976, *Les travaux et les jours à Rome et dans l'ancienne France. Les agronomes latins inspirateurs d'Olivier de Serres*, Tours.

Greene, K. 1986, *The archaeology of the Roman economy*, London.

—— 1994, "Technology and innovation in context: the Roman background to medieval and later developments", *Journal of Roman Archaeology* 7, 22–33.

Haudricourt, A. and Jean-Brunhes Delamarre, M. 1955, *L'homme et la charrue à travers le monde*, Paris.

Henning, J. 1991, "Fortleben und Weiterentwicklung spätrömischer Agrargerätetradition in Nordgallien. Eine Mähsense der Merovingerzeit aus Kerkhove", *Acta Archaeologica Lovaniensia* 30, 49–59.

Hoebanx, J.-J. 1975, "Seigneurs et paysans". In *La Wallonie. Le pays et les hommes*, I, ed. H. Hasquin, Brussels, 161–211.

Hopfen, H. J. (ed.) 1970, *L'outillage agricole pour les régions arides et tropicales*, Rome.

Jacobsen, G. 1995, *Primitiver Austauch oder Freier Markt? Untersuchungen zum Handel in den gallisch-germanischen Provinzen während der römischen Kaiserzeit*, St. Katharinen.

Janssen, W. 1989, "Reiten und Fahren in der Merowingerzeit". In *Untersuchungen zu Handel und Verkehr der vor-und frühgeschichtlichen Zeit in Mittel- und Nordeuropa. V. Der Verkehr*, ed. H. Jankuhn, Göttingen, 174–228.

Kolendo, J. 1960, "La moissonneuse antique en Gaule romaine: son emploi", *Annales E. S. C.* 15, 1099–114.

—— 1979, "Origine et diffusion de l'araire à avant-train en Gaule et en Bretagne", *Cahiers d'histoire* 24, 61–73.

Körber-Grohne, U. 1989, "The history of spelt". In *L'Epeautre. Histoire et ethnologie*, eds J. P. Devroey and J. J. Van Mol, Treignes, 51–9.

Langdon, J. 1984, *Horses, oxen and technological innovation*, Cambridge.

—— 1986, "Horse hauling: a revolution in vehicle transport in twelfth- to thirteenth-century England", *Past and Present* 103, 37–66.

Larousse agricole 1921, eds E. Chancrin and R. Dumont, Paris.

Lebecq, S. 1989, "La Neustrie et la mer". In *La Neustrie. Les pays au Nord de la Loire de 650 à 850*, ed. H. Atsma, Sigmaringen, 405–40.

—— 1990, "Les origines franques (V^e–IXe siècles)". In *Nouvelle histoire de la France médiévale*, I, Paris, 15–41.

Lewit, T. 1991, *Agricultural production in the Roman economy, AD 200–400*, Oxford.

Liebowitz, J. J. forthcoming, "The persistence of draft oxen in the west". Paper presented to the Tenth International Economic History Congress, 1990, Louvain.

Lindemans, P. 1952, *Geschiedenis van de landbouw in België*, Antwerp.

Lizet, B. 1982, *Le cheval dans la vie quotidienne*, Paris.

Lohrmann, D. 1989, "Le moulin à eau dans le cadre de l'économie rurale de la Neustrie (VIIe–IXe siècles)". In *La Neustrie. Les pays au Nord de la Loire de 650 à 850*, ed. H. Atsma, Sigmaringen, 367–404.

Lorren, C. 1989, "Le village de Saint-Martin de Trainecourt à Mondeville (Calvados) de l'Antiquité au Moyen Age", in *La Neustrie. Les Pays au Nord de la Loire de 650 à 850*, ed. H. Atsma, Sigmaringen, 439–66.

Martin, R. 1971, *Recherches sur les agronomes latins*, Paris.

Molin, M. 1984, "Quelques considérations sur le chariot des vendanges de Langres", *Gallia* 32, 97–114.

—— 1987–8, "La faiblesse de l'attelage antique ou la force des idées reçues en histoire ancienne", *Bulletin Archéologique du Comité des Travaux Historiques* 23–4, 39–84.

Müller, H. H. 1985, "Zur Rekonstruktion des gallo-römischen Ernten-maschine", *Zeitschrift für Agrargeschichte und Agrarsoziologie* 19, 191–6.

Nobis, G. 1973, "Die Pferde aus dem Frankische Gräberfeld von Rübenach". In *Die Frankische Gräberfeld von Rübenach*, eds C. Neuffer-Müller and H. Ament, Krefeld, 275–82.

Oexle, J. 1992, *Studien zur merowingerzeitlichem Pferdegeschirr am Beispiel der Trensen*, Mainz.

Oleson, J. P. 1984, *Greek and Roman mechanical waterlifting devices: the history of a technology*, Buffalo.

Percival, J. 1992, "The fifth-century villa: new life or death postponed?". In *Fifth century Gaul: a crisis of identity?*, eds J. F. Drinkwater and H. Elton, Cambridge, 156–64.

Pesez, J.-M. 1991, "Outils et techniques agricoles du monde médiéval". In *Pour une archéologie agraire*, ed. J. Guilaine, Paris, 131–64.

Petit, M. 1993, "Production et techniques agricoles". In *L'Ile-de-France de Clovis à Hugues Capet du Ve au Xe siècle*, Guiry-en-Vaast, 266–71 and 310.

Pohanka, R. 1986, *Die eiserne Agrargeräte der römische Kaiserzeit in Österreich*, Oxford.

Polge, H. 1967a, "Les modalités, les étapes et les limites de la substitution du travail mécanique au travail humain". In *Etudes et documents. Etudes de technologie rétrospective*, Auch, 103–20.

—— 1967b, "Notes de cinématique rétrospective: la production artificielle de mouvements alternatifs, son invention, sa signification et sa portée". In *Etudes et documents. Etudes de technologie rétrospective*, Auch, 85–102.

Raepsaet, G. 1979, "La faiblesse de l'attelage antique: la fin d'un mythe?" *L'Antiquité Classique* 48, 171–6.

—— 1982, "Attelages antiques dans le Nord de la Gaule. Les systèmes de traction par équidés", *Trierer Zeitschrift* 45, 215–73.

—— 1987, "Archéologie et iconographie des attelages dans le monde gréco-romain: la problématique économique". In *Histoire économique de l'Antiquité*, eds T. Hackens and P. Marchetti, Louvain-la-Neuve, 29–48.

—— 1995, "Les prémices de la mécanisation agricole entre Seine et Rhin de l'Antiquité au XIIIᵉ siècle", *Annales. Histoire, Sciences Sociales*, 911–42.

Raepsaet, G. and Rommelaere, C. (eds) 1995, *Le transport attelé entre Seine et Rhin de l'Antiquité au Moyen Age. La révolution du brancard*, Treignes.

Raepsaet-Charlier, M.-T., and G. 1988, "Aspects de l'organisation du commerce dans le nord de la Gaule", *Münstersche Beiträge zur Antiken Handelsgeschichte* 7 (2), 45–69.

Rees, S. 1979, *Agricultural implements in prehistoric and Roman Britain*, Oxford.

Renard, M. 1959, *Technique et agriculture en pays trévire et rémois*, Brussels.

Reynolds, P. J. 1979, *Iron-Age farm. The Butser experiment*, London.

Rivals, C. 1987, *Le moulin à vent et le meunier dans la société française traditionnelle*, Paris.

Roesener, W. 1986, *Bauern im Mittelalter*, Munich.

Schlippschuh, O. 1974, *Die Händler im römischen Kaiserreich in Gallien, Germanien und den Donauprovinzen Rätien, Noricum und Pannonien*, Amsterdam.

Schweiter, J. 1984, *L'habitat rural en Alsace au Haut Moyen Age*, Riedisheim.

de Serres, O., 1600, *Le théâtre d'agriculture et le mesnage des champs*, Paris.

Sigaut, F. 1982a, *La "moissonneuse" gauloise et les techniques apparentées de récolte des grains*, Paris.

—— 1982b, "Les débuts du cheval de labour en Europe", *Ethnozootechnie* 30, 33–46.

—— 1985, *L'évolution technique des agricultures européennes avant l'époque industrielle*, Paris.

Slicher van Bath, B. H. 1960, *De Agrarische Geschiedenis van Westeuropa (500–1850)*, Utrecht.

Spruytte, J. 1977, *Etudes expérimentales sur l'attelage. Contribution à l'histoire du cheval*, Paris.

Techniques rurales en Afrique, 1971, Paris.

Van Ossel, P. 1992, *Etablissements ruraux de l'Antiquité tardive dans le Nord de la Gaule*, Paris.

Verhulst, A. 1985a, "L'intensification et la commercialisation de l'agriculture dans les Pays-Bas méridionaux au XIIIᵉ siècle", in *La Belgique rurale. Mélanges J.-J. Hoebanx*, Brussels, 89–100.

—— 1985b, *Le grand domaine aux époques mérovingienne et carolingienne*, Ghent.

—— 1988, *Agrarische revoluties: mythe of werkelijkheid*, Ghent.

—— 1990, *Précis d'histoire rurale de la Belgique*, Brussels.

Verhulst, A. and Bublot, G. 1980, *L'agriculture en Belgique*, Brussels.

Un village au temps de Charlemagne. Moines et paysans de l'abbaye de Saint-Denis du VIIe siècle à l'an mil, 1988, Paris.

Watson, A. M. 1981, "Towards denser and more continuous settlement: new crops and farming techniques in the early middle ages". In *Pathways to medieval peasants*, ed. J. A. Raftis, Toronto, 65–82.

Werner, K. F. 1984, "Les origines". In *Histoire de la France*, I, ed. J. Favier, Paris, 255–92.

White, K. D. 1969, "The economics of the Gallo-Roman harvesting machines". In *Hommages à M. Renard*, Brussels, 804–9.

—— 1970, *Roman farming*, London.

White, L. 1962, *Medieval technology and social change*, Oxford.

Wightman, E. 1985, *Gallia Belgica*, London.

Wierschowski, L. 1995, *Die regionale Mobilität in Gallien nach den Inschriften des 1. bis 3. Jahrhunderts n. Chr.*, Stuttgart.

Wikander, O. 1984, "Exploitation of water power or technological stagnation?", *Scripta Minora* 111, Lund.

4. THE BIRTH OF "THE FLEMISH HUSBANDRY": AGRICULTURAL TECHNOLOGY IN MEDIEVAL FLANDERS

Erik Thoen

Since the work of Paul Lindemans in 1956[1] and the internationalization and partly new interpretation of his results by Bernard Slicher van Bath in 1960,[2] the progressive character of agriculture in the former county of Flanders from at least the fourteenth century has been well understood. Indeed, only in the last ten to fifteen years has significant new work in the topic been carried out, where the views of Lindemans and Slicher van Bath have been nuanced, changed and supplemented with new material. Also, there has been much more exploration of the period before the late middle ages. A new *status questionis* is thus emerging, although it also shows that much new research still needs to be done.

Sources

The sources concerning Flemish medieval agricultural technology tend to be patchy and varied. One of the richest sources are demesne accounts, containing data on the direct exploitation of large rural demesnes. These are rather scarce for medieval Flanders, partly because the tradition of making exhaustive demesne accounts started later here. The earliest surviving accounts of this type date to the thirteenth century and, for most institutions, only to the late fourteenth century or even later. On the other hand, the late medieval European trend of leasing out the large demesnes started earlier in Flanders, becoming fairly general by the last decades of the thirteenth century. So the great majority of demesne accounts give only the name of the leasehold farmer, the leasing term, and the rent. Most institutions kept only a few plots of arable land in the immediate

---

[1] Lindemans 1952.
[2] Slicher van Bath 1960 and 1963.

neighbourhood of their headquarters under direct exploitation, but nothing concerning these was recorded in the accounts, because they were used solely for direct consumption.

Some accounts, however, are very rich. These include not only the rare accounts of private persons, such as the notebook of the Bruges bourgeois, Simon de Rikelike (1323–36),[3] but also the accounts of several hospitals, which apparently preferred to produce their own foodstuffs for much longer than other institutions, thus providing useful accounts from as early as the beginning of the fourteenth century.[4]

In compensation for the relative lack of detailed demesne accounts, certainly in relation to countries like England, there are a number of other sources peculiar to Flanders. Thus, Flemish archives contain an enormous amount of leases, especially from the fourteenth century onwards. A lot of these contain agricultural information on the way the farmer was to operate the demesne, especially concerning crop rotations, and how he was to leave it at the end of the lease. These sources have the advantage that they give information on the demesnes of a much broader sector of society, including the minor nobility and the bourgeoisie. On the other hand, these contracts often had an antiquated character that was probably a far cry from reality.[5] This was especially true for large farms, which were often leased out for centuries with the same conditions. Small farms and plots of land on short-term contracts, especially those temporarily leased out by institutions taking care of the inheritances of orphans—a common feature of town life during the later middle ages—are more reliable. Such contracts sometimes give very useful details about the sown acreages and crop rotations on these small farms and plots of land. Other information, such as lists of sales of crops on the fields, is sometimes given in the records of these institutions.[6] Altogether, these data often reveal important differences between the techniques of small farms and demesnes. Finally, information on both small and large farms can be found in documents dealing with the forfeiture of estates during the many civil wars in late medieval Flanders.

For the period before the late middle ages, we must fall back on inventories and lists of rents from the estates of large abbeys. Also of

---

[3] De Smet 1933.
[4] Mertens 1970.
[5] Thoen 1988, II, 659–69.
[6] As from the town archives of Audenarde: Thoen 1988, 678–9.

great importance is the oldest general account of the count of Flanders, known as the *Gros Brief*, which gives an (unfortunately partly incomplete) overview of the income of the count in 1189, but which has until recently been relatively ignored by rural historians.[7] Finally, there are the charters for this period, the most interesting often being the arrangements made between clerical institutions and the stewards or bailiffs (*maires, maiores, villici, advocati*) who took care of (or usurped) the various manors.

From Specialization to Intensification: Flemish Agriculture in the Eleventh and Twelfth Centuries

It is likely that the most important transformations of Flemish agriculture took place in the period of least documentation, that is before 1300. Urban development accelerated from the second half of the eleventh century and reached its peak two centuries later. As I have argued elsewhere, the evolution of the Flemish urban economy was firmly grounded on the rural economy.[8] Major imports were unnecessary before the thirteenth century, and from at least the eleventh century Flanders was seemingly made up of a few areas, each specializing in various food products and raw materials necessary both for the region's subsistence and for its developing manufacturing base (upon which its urban population was increasingly dependent). Altogether, it is likely that the rural economy of Flanders in the eleventh and twelfth centuries was based on a macro-economic agrarian "ecosystem" with regional specialization. This structure was, generally speaking, based on the two pillars of soil and seigneurial management.[9]

First, concerning soils during the period, Flanders was divided, *grosso modo*, into three different areas (figure 4.1) as follows.

(a) There was first of all the very important and very commercialized coastal area, specializing in products coming from stock-breeding: wool, cheese, butter, meat and hides. These all derived from cattle and sheep pastured on saline clay soils, dunes and moorlands. Increasingly, however, the economic function of the moorlands changed from stock-breeding to peat digging. From at least the twelfth century, this

---

[7] Verhulst and Gysseling 1962.
[8] Thoen 1993a and 1994a.
[9] For what follows in this paragraph, see Thoen 1993a.

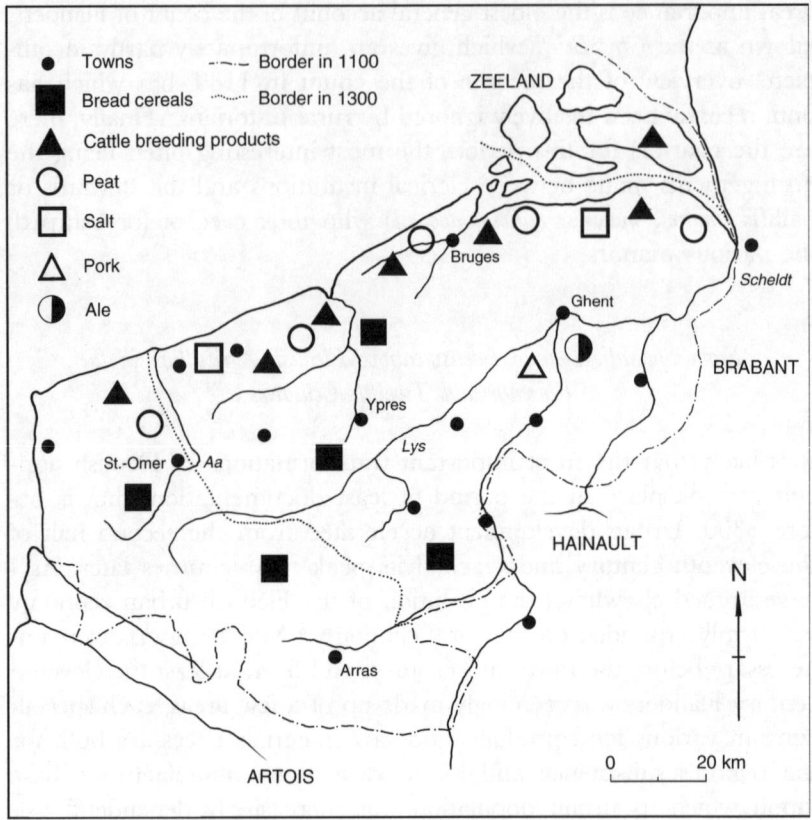

Figure 4.1. Principal rural products exported to towns in Flanders between the eleventh and thirteenth centuries.

became of very great importance for the fuel and salt (made from burning peat) consumption of the emerging towns.

(b) Southern Flanders, with its loamy soils, was important for the export of cereals (especially wheat), via the Scheldt and the Lys rivers to the towns in the northern part of the county.

(c) The sandy soils of central Flanders from at least the twelfth century partly supported an extensive commercial agriculture based on the production of ale. Here ale was made out of oats, which until the late thirteenth century was probably more important than beer or ale made from barley. Oats played an important role in the relatively slow reclamation process of these poor and light soils and in the development of "up and down" husbandry (see below). The importance of beer and ale consumption in the Flemish towns is illustrated by the oldest accounts of Bruges in which the income from beer/ale taxes was among the most important revenues of the town.[10] In the same area, until the eleventh century, the as yet unreclaimed forests and heathlands were probably also important for pig-breeding and the delivery of pork to the towns, as the historical and archaeological sources seem to imply.

Second, this rather extensive commercial agriculture, producing surpluses based on regional specialization, was the consequence not only of soil conditions, but also of the activities of great landowners and lords. Initially for the provisioning of their institutions and/or courts, great landowners, and in the first place the count of Flanders himself, had from the early middle ages established their estates over the various regions of the county. There they collected surpluses by direct management of their demesnes or by receiving rents according to their needs, and in that way probably also stimulated production and specialization. This is most clear for the domain of the count, who had specialized collecting centres spread over the whole county (except the extreme south), certain centres sometimes having names which referred to the kinds of rents collected (for example, *lardaria, spiceria*).

In brief, it is likely that the shift of the balance of power from the Rhine-Meuse area to the borders of the North Sea in the post-Carolingian period, coincidental with the early development of the county of Flanders in the tenth century and the creation of a rich nobility already before AD 1000,[11] stimulated the increasing production of agricultural surpluses and the creation of a regional specialization even before the real take-off of urban growth. The subsequent growth of the towns, beginning somewhere in the second half of the eleventh century, was also stimulated by lords and especially the count.[12] Of course subsequently the increasing demand from the towns accelerated that process.

---

[10] It is often difficult to make a clear linguistic distinction between ale and beer in the sources. Because (probably a soft variety of) oats was the basic crop for early inland drink production it might be supposed that most of this was for ale. From the fourteenth century much beer was imported from Holland and the Baltic Sea area. Hops, however, were introduced to Flemish agriculture rather late. In contrast to Holland and Brabant, where hops production is already evident in the fourteenth century, the earliest recorded instances of Flemish hops production date only to the fifteenth century: Van Uytven 1973.

[11] Warlop 1975; Thoen 1993a, 261.

[12] Thoen 1993a; Verhulst 1994.

It would appear that from the second half of the twelfth century, and certainly at an accelerated pace during the thirteenth, the "eco-system" described above was already fragmenting. The extensive agriculture became increasingly intensive. Contrary to Slicher van Bath's views of the 1960s,[13] this process was well under way before the fourteenth century.[14] Although much more research needs to be done, it is likely that the famous "Flemish husbandry" was born during "the long thirteenth century" (which I label as such because some innovations found their origin earlier or later). This, of course, does not mean that this agriculture was static in the following centuries.

Features of Intensification since the Thirteenth Century

Cropping Systems and Crop Rotations

The nature of the cultivated crops and crop rotation are essential elements in the evolution of agriculture towards intensification. Although the antiquity of a three-course rotation is currently under debate,[15] it is clear that it was a common practice on large demesnes from at least the twelfth century and that it continued to gain importance.[16] From late medieval sources, however, it appears that a strictly topographical three-field system—in which the complete arable acreage of a village or a manor was divided into three, more or less equal parts—did not exist, and according to some, had never existed.[17] Instead, in areas such as south Flanders, there often occurred a more fragmented system known as *Flurzwang*, with smaller sections of open fields being divided from each other by roads, brooks or other natural boundaries. This created a kind of patchwork over the village, which gave peasants much more flexibility in crop cultivation.[18] In central Flanders, *Flurzwang* existed as micro-openfields amidst a landscape of enclosures. From the fourteenth century, these islands of micro-openfield, created for the most part from some large parcels of land, were called *kouters* (*culturae* in Latin).[19] They had a

---

[13] Slicher van Bath 1960 and 1963, 122–5, 178–80.
[14] Thoen 1988, II, 863; Verhulst 1985 and 1990a.
[15] Derville 1988; Morimoto 1994, especially 121.
[16] Verhulst 1966; Thoen 1993b.
[17] Derville 1988.
[18] For the area south of Ghent, see Thoen 1988, II, 761–75; for northern France, where it was much more common, see Derville 1988.
[19] Verhulst 1966; De Vos 1991, 118–55.

surface area of about 15 to 50 hectares (40–120 acres) and formed the infield of demesnes or villages.[20] From the twelfth century most of these were organized into three-field systems. Similarly, particularly in sandy areas of central Flanders, until the thirteenth century new infields—*kouters*—were sometimes constructed from new reclamations.[21] Outside of such areas, a regimented system of *Flurzwang* was seldom introduced, and it probably never existed in the coastal area. In the later middle ages, furthermore, the restrictive character of *Flurzwang* relaxed on most of these "infields".

In short, classic three-field systems, although known from the twelfth century, were not at all common in Flanders, even in the modified *Flurzwang* version. On the other hand, probably again from the twelfth century but not necessarily applied to a specific field structure, three-course crop rotations became common practice on many large demesnes all over Flanders, except perhaps in the coastal area.[22] This may also have applied to smaller holdings.[23] In later leasing contracts, stretching into the early modern period, three-course rotations dominated as the prescribed way in which farmers had to leave demesnes at the end of their contracts. But this practice was not necessarily uniform. Deviations from three-course rotations are known for the thirteenth and fourteenth centuries,[24] and the number of cases increases in the better documented later middle ages. Generally demesnes steadily reduced the amount of fallow, especially by sowing the fallow with legumes (see below). Most of the early examples concern southern and coastal Flanders, but there are none for "sandy" central Flanders (the large area around Ghent).

This does not mean, however, that three-course crop rotations were solidly entrenched in central Flanders. As we have already indicated, this area was characterized by land reclamation, which occurred at varying times and was of varying intensity. While a strict, topographically organized three-field system was likely on the *kouters*, which were used as infield and received a lot of stable manure,[25] the more difficult soil conditions made other crop rotations necessary for the reclamation of areas outside these open fields. Indeed, although the data are still limited, there are indications that "up and down husbandry"

---

[20] On the *infield* character of the *kouters*, see Thoen 1993b.
[21] Verhulst 1995, 157.
[22] Lindemans 1952, 2, 95–101; Tits-Dieuaide 1984.
[23] Thoen 1993a, 265.
[24] Thoen 1992b.
[25] Thoen 1993b, 81–91.

(where the cropping of land alternated between long spells of fallow) was used on these soils as an intermediary for more intensive cultivation, maintaining in that way the fertility of the soil. These plots, which were probably enclosed, alternated between being ploughed for a number of years and then being used as pasture land. During the latter pasture phase, it was called *dries*-land. While the land was under the plough, different crop rotation systems were used, in which the cultivation of oats was apparently very prominent.[26] The prevalence of the *dries* form of cultivation seemingly increased in periods of growing agricultural activity and decreased when that activity diminished in the later middle ages.[27] The system may have been particularly common on the reclaimed lands of the twelfth and thirteenth centuries, as some recently discovered sources seem to prove.[28] The *dries* system also gave farmers much greater possibilities for experimentation, so that, already from the end of the thirteenth century, such areas had changed from rather "conservative" regimes into much more "progressive" ones as regards crop rotations and the intensification of husbandry. When evidence becomes more abundant in the later middle ages, it is clear that four-course rotations were practised on many large farms, and, when seen in the leases, were even then considered as being of long standing and accepted by customary farming law.[29] A typical rotation involved winter grains being sown for two consecutive years (mostly rye, which was more suitable than wheat in light soils).[30] Rye sown after rye was called *stoppelkoren* ("stubble grain": figure 4.2).

The information for crop rotations on small holdings is scarce and dates only from the later middle ages. This information, however, tended to reflect actual practice much more than the leases of large demesnes, as in the case of the short-term leases for the lands of orphans around Audenarde in the fifteenth century. Altogether, these smaller farms tended to adopt more flexible rotations, as on the Audenarde orphans' plots, where winter fallow was reduced to a minimum.[31] Despite the lack of earlier sources, there is little reason to suppose that the situation was substantially different for those small

---

[26] Thoen 1994b.
[27] Thoen 1994b, 151.
[28] Thoen 1994b, 136–42.
[29] Thoen 1995, especially 110–11.
[30] Lindemans 1952, II, 110–11.
[31] Thoen 1988, II, 742ff.

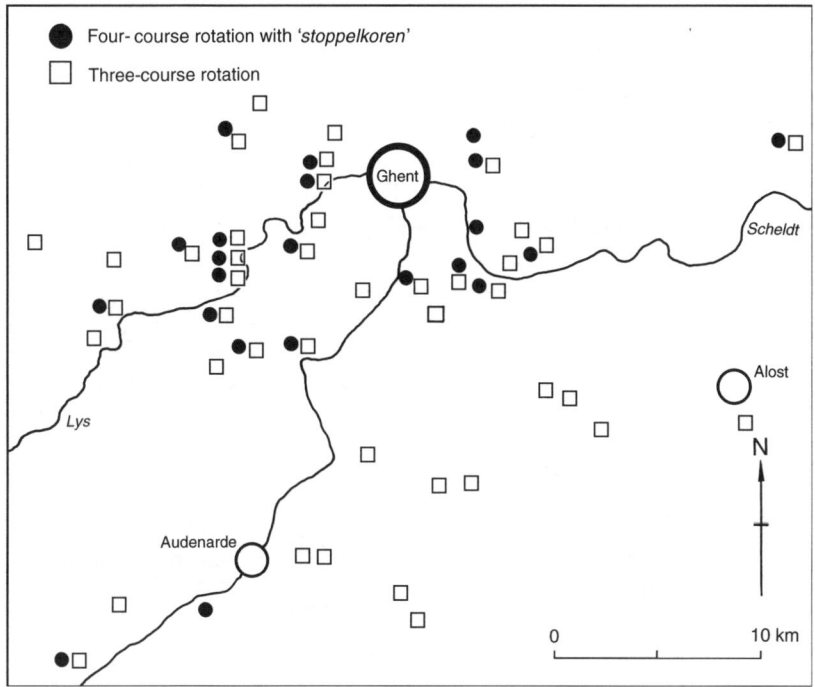

Figure 4.2. "*Stoppelkoren*" on large demesnes mentioned in leases copied in the Registers van de Keure' of Ghent (1400-9).

farms in, say, the second half of the thirteenth century, when population pressure was higher and the need to farm intensively even greater.

In conclusion, it appears that:

(1) crop rotations were extremely flexible. *Flurzwang* and other regulated field systems only partly restricted this freedom, especially after "the long thirteenth century". The large demesnes, smaller in number but possessing a disproportionately large amount of land, were probably more tied to regulated field systems than smaller holdings.

(2) Flemish farmers could more or less freely regulate the acreage of their arable land by simply reducing or expanding the amount of fallow and/or *dries*. By the late thirteenth century, for example, it appears that fallow was being steadily reduced on the land under permanent cultivation, while *dries*-land expanded on former pasture land and marginal land and probably acted as an intermediary for more permanent cultivation. It was in this way that farmers at the time adapted themselves to changing economic and agricultural conditions.

Crops

Flemish farmers also extended this flexibility and "progressiveness" to the sorts of crops they cultivated. In particular, the cultivation of legumes is often seen as being at the core of agricultural progress. And, indeed, there is no area in Europe where the cultivation of these crops was so important as in Flanders (figure 4.3).[32] Beans and peas were both used as fodder crops and for human consumption,

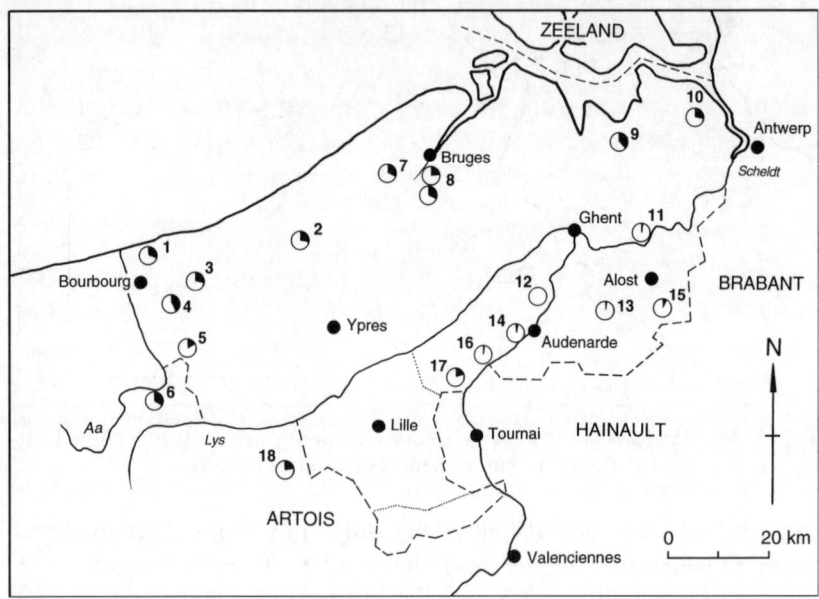

Figure 4.3. The importance of *Leguminosae* in the late middle ages. Percentage of land sown with *Leguminosae* in some areas and on some large demesnes: *1.* Saint-Georges 38% (demesne, 1315). *2.* Châtellenie of Furnes 27% (average, 1550-64). *3.* Bourbourg 34% (demesne, 1315). *4.* Saint-Pierre-Brouck 44% (demesne, 1315). *5.* Saint-Omer 21% (tithe, 1466). *6.* Zudausques 36% (tithe, 1399). *7.* Vlissegem 30% (demesne, 1333). *8.* Zuienkerke 23%, 35% (demesne, 1333-58). *9.* Otene 40% (demesne, 1295). *10.* Frankendijk 33% (demesne, 1295). *11.* Châtellenie of Ghent 2% (estimated average, fifteenth-century). *12.* Machelen 0% (demesne, 1317-25). *13.* Sint-Denijs-Boekel 3% (tithe, 1536). *14.* Châtellenie of Audenarde 4% (estimated average, fifteenth-century). *15.* Ninove 10% (demesne, 1454-75). *16.* Avelgem 1% (demesne, 1317-25). *17.* Mouscron 20% (demesne, 1450). *18.* Festubert 25% (tithe, 1324-28).

---

[32] Thoen 1992a, 52–4. For comparison, see Campbell 1988; Campbell and Overton, 1993, 54. In Norfolk, currently considered the most progressive area agriculturally in medieval England, legumes on average did not comprise more than 13.5 per cent of the total sown acreage.

while another more archaic type of *leguminosae*, vetches, was often cultivated exclusively as a fodder crop. The variety and the import-ance of the cultivation of these crops was rising in the thirteenth century and must certainly have some connection with the growing importance of stable-fed cattle and a more strict regime of manure conservation and allocation. In three-field systems legumes were sown with the spring grains or on part of the fallow.

Turnips were another important crop connected with the intensifi-cation of Flemish agriculture.[33] The cultivation of turnips as a green fodder crop (especially when densely sown on the fallow) is now thought unlikely before the end of the fifteenth century, when it became common practice on larger farms. However, from the thir-teenth century they may have been used as an industrial crop for the cultivation of rapeseed oil. Turnips at this time were mostly sown with the spring grains in three-course rotations.[34] From the begin-ning of the fifteenth century turnips were also sown for their tubers. This was a practice directed towards human consumption, which would continue to gain popularity among smallholders in the six-teenth century.

Among the crops grown for commercial purposes, plants servicing the burgeoning Flemish textile industry were particularly important. Dye plants, such as madder (for the colour red), woad (blue) or weld (brown, yellow), were particularly prevalent (see figure 4.4), and were especially cultivated on smaller holdings.

Another very important commercial crop was flax, the cultivation of which expanded considerably in the fourteenth century, especially in the area around Ghent, where it remained very important until the nineteenth century.[35] In my opinion, this first expansion was due to political factors. In 1314 Ghent received from the count the privi-lege of prohibiting rural woollen cloth-making as a cottage industry from an area of about 50 acres (20 ha) around the town, so that flax cultivation and linen production could be promoted within the pro-tected area and replaced in large part the production of woollen cloth.[36] Altogether, thanks to the intensity of cultivation, the yields for all these crops, as well as for the classical grains (wheat, rye,

---

[33] Slicher van Bath 1963, 122–5.
[34] Thoen 1988, II, 725–33, and 1992b, 54–5.
[35] Thoen 1988, II, 997–9 and Thoen 1992a.
[36] Thoen 1992a, 350.

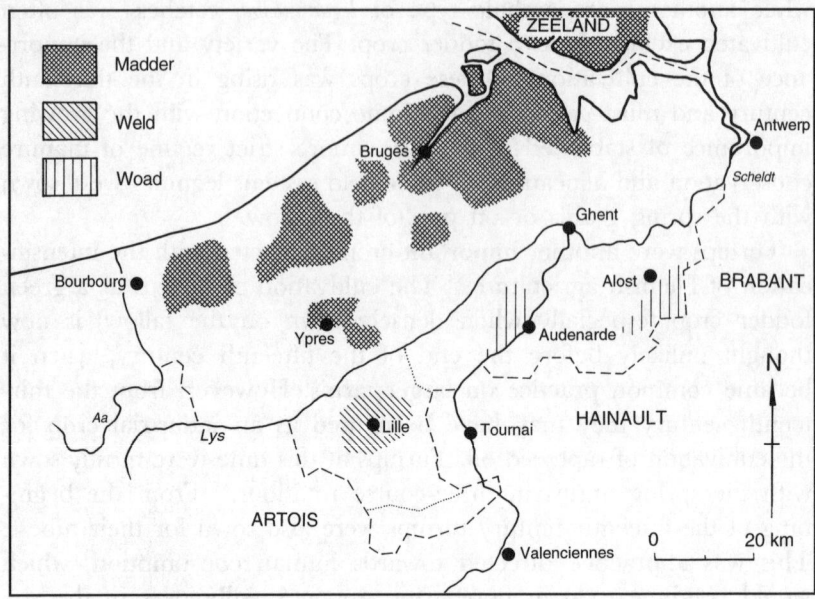

Figure 4.4. The cultivation of dye plants in the late middle ages.

barley, oats), were in general the highest in Europe at the time. Elsewhere, as in northern France, where measurable yields date back to 1187, yield-ratios were in the range of 1:3 to 1:10, the first extreme being more common than the second.[37] In contrast, the more intensive Flemish agriculture achieved yields of 1:20 to 1:30 for winter grains.[38] In areal terms, yields for winter grains were normally 2,000 to 3,000 litres per hectare (about 22 to 33 bushels per acre) for southern Flanders. In "central" Flanders yields of the same cereals based on data from smaller farms are slightly lower (about 1,700 litres per hectare—19 bus/ac—in the area around Audenarde), but this was probably the consequence of a lower proportion of fallow on these smaller holdings.[39] For the coastal area, the only available data from atypically large and extensively cultivated demesnes give lower yields of around 1,200 litres per hectare (13 bus/ac).[40] The yields from oats were even higher than those for winter grains, even though the land was seldom manured when oats were sown. Between

[37] Thoen 1994a, especially 168–9, note 14; Thoen forthcoming (a).
[38] Derville 1975/6 and 1987.
[39] Thoen 1988, I, 734–65.
[40] Mertens 1970, 75–91, 153–62; Mertens and Verhulst 1966.

1435 and 1540, for example, yields for oats from small holdings around Audenarde averaged 2935 litres per hectare (33 bus/ac). In the coastal areas, yields of oats of 2481 litres per hectare (28 bus/ac) were evident around 1400 and as much as 3067 litres per acre (34 bus/ac) around 1500.[41] Clearly, oats could still profit from the manuring of the autumn-sown grains, which did not deplete completely the fertilizing substances in the soil. In such a way, even marginal lands had become exceptional in Flanders.[42]

Tools and Tillage

In addition to the increasing intensity of cultivation, new implements and techniques were introduced to maintain labour productivity. Although further research is necessary, it seems that horses were already the common draught animals from at least the twelfth century, as Flemish farmers attempted to benefit from the higher productivity of horses compared to oxen.[43] This was a much more precocious use of horses compared to neighbouring areas such as Picardy and England.[44] Similarly, the Flemish hook or "pick" and the scythe superseded the sickle, increasing the labour productivity of harvesting several times. The first written sources for these implements date back to 1300.[45] The cultivation of cereals in narrow, high-backed ridges, thus facilitated sowing in lines, weeding, drainage and the use of the above-mentioned "pick" during the harvest.[46] Finally, the collection of manure from stables was an important feature of intensification of the Flemish rural economy.

Could the introduction of these "new" tools and techniques counterbalance declining labour productivity caused by the trend towards smaller holdings as agriculture intensified? Although such an idea is theoretically plausible, it seems unlikely to have been effective to such a degree. Indeed, sources seem to indicate that a shift from the plough to the more labour intensive spade was the most characteristic change caused by intensification of Flemish agriculture.[47]

---

[41] Thoen 1988, II, 1243; Vandewalle 1986, 205.
[42] Thoen 1988, II, 716–17, 1280–1.
[43] Thoen 1993a, 265–6.
[44] Fossier 1968, I, 377–81; Langdon 1986, especially 255 (fig. 42).
[45] Thoen 1988, II, 776–7.
[46] Thoen 1988, II, 779; Verhulst 1990b, 128–9.
[47] Thoen 1988, II, 781–4.

Elements to Explain the Intensification of "The Flemish Agriculture"

Thus, altogether, a more intensive, market-oriented agriculture resulted over "the long thirteenth century". New techniques and new crops
became general practice, on large and small farms alike, but in different forms according to the size of the farm. But why did this hap-
pen? Although this cannot be told in great detail here, it is clear that
explanations must be considered in relation to the global structural
changes of the Flemish rural economy. In a recent paper I explained
these changes in the context of increased population, growing urbanization, the changed surplus extraction relations within the classes
wielding authority and the changed balance of power between lords
and peasants.[48] In that study I also attempted to prove that urbanization and its increasing demands were not automatically and exclusively causing a new market structure in the countryside. In fact,
the extensive market-oriented and specialized agriculture of the eleventh and twelfth centuries could also have survived or even expanded
in an increasingly demand-oriented economy. But this did not hap-
pen. Indeed, in terms of the technical adjustments that such changes
brought about, each sector tended to find its own set of technological solutions. For our purposes here, it is most useful to analyse the
situation in the context of farm size: that is, large farms (demesnes
and the like) and small holdings (mostly cultivated by peasants).

Large farms, or what we consider as such in the Flemish context
(25–60 ha; 60–150 ac), probably continued to be market-oriented,
but it seems in many cases that their extensive and specialized character probably became less clear or changed from what it had been
in the eleventh and twelfth centuries (see above). Furthermore, a considerable number of "large" farms were added to the existing number. In the thirteenth and fourteenth centuries, both in the Flemish
interior and in the Flemish coastal region, a large number of *Einzelhöfe*
(isolated large farms, mostly on moated sites) were created, partly as
a result of reclamations and partly as a result of purchases by the
rising bourgeoisie. This does not mean that proportionally the number
of large holdings was increasing, because it is likely that the number
of smaller holdings was growing even faster in the same period.

Altogether, the impulse from the landowners to force their farmers to specialize was diminishing in the course of the twelfth and

---

[48] Thoen forthcoming (b).

later centuries because the landowners' administrative strategies were changing as a function of their changing needs. The expansion of the urban markets caused the collapse of the provisioning system of the large landowners described above. Rents in kind were more and more replaced by rents in money as a result of the greater circulation of coin in the countryside and the increased need for cash by lords who preferred buying more basic products on the markets as well as luxury products for which cash was more of a prerequisite. At the same time, because of the increasing size of the ruling classes, their numbers were being swelled (particularly by incoming bourgeoisie), the size of many larger farms was reduced, thus diminishing the economies of scale that extensive specialization on these farms had previously achieved. Moreover, from the fourteenth century until the end of the Old Regime, landowners with large farms increasingly used them for their own "survival" strategies: that is, to ensure food supplies in the event of the famines that began to occur more frequently after 1315.

The result of this was the important process of *Vergetreidung* (the extension of crop growing, often leading to an imbalance between cattle breeding and grain production) on the larger farms, which were now free to operate more flexibly. This phenomenon is related to the increased needs of city and country alike for cereals and to rising grain prices. Because real wages and transport costs were now lower, the production of cereals represented the most cost-effective and labour-productive policy for the majority of large farms. The generalization of the leasing system and the abandonment of direct exploitation led to a further move away from specialization. The transformation of demesnes into "independent" leasehold farms, beginning as early as the first half of the twelfth century on the count's estates and becoming predominant over the course of the twelfth and thirteenth centuries (except on those estates where direct consumption was felt to be more advantageous, as in the case of hospitals), also resulted in a corresponding change in the way these farms were managed. Provided he accepted certain conditions, the farmer obtained more freedom of action on his farm which resulted in a more flexible operation. Specialization was thus determined by market forces rather than by the consumption needs of a large landowner. On the other hand, leasehold farms were also characterized by a wider spread of risk, based on mixed crop and livestock production. These "mixed" farms, typical of intensive Flemish agriculture, were able to increase crop productivity through the use of farmyard dung, thus raising the farm's operation to a higher technical level and perhaps even boosting labour productivity. Finally, from the second half of the thirteenth century onwards, the more localized "inland" agroecosystem for provisioning towns, dating back to the previous period, was partly and progressively replaced by a larger system, integrating areas from outside Flanders, such as Picardy, Hainault, and even areas "overseas" (as in England: Langdon, this volume).

Similar sorts of changes were also occurring for small holdings, which were rapidly multiplying as a result of reclamations, the subdivision of peasant holdings, the fragmentation of large farms, and the (compared to cattle breeding) more labour intensive *Vergetreidung* on the larger farms which allowed for a combination of peasant farming and wage work. As in the case of larger farms, these small holdings slowly produced proportionally more for the market. Stimulated by urban demand, the expanding circulation of money, the growing necessity for supplementary incomes (as average holding size diminished), small farms in many regions indeed started to produce commercial crops. All of this was paralleled by a growing rural textile industry around cities such as Ypres, Bruges and Ghent, from the second half of the thirteenth century in particular.

Notwithstanding the overall reduction in farm size, the gap between rich and poor did not become smaller in the thirteenth century. Large farms, of about 25 to 80 ha (60–200 ac), controlled the largest part of the area of land used for agriculture. They could also boost their profits through intensive cultivation. In the twelfth century, this was initially through the general introduction of the three-course rotation combined with intensive fertilization with farmyard dung; while, during the thirteenth century, sowing on the fallow and more complex rotations of crops continued the process. Again, the coastal plain showed the way, the reason probably being the earlier introduction of leasing.

The small holdings also continued to be dependent on the large farms. The labour productivity of the rural textile industry and almost certainly also of the production of dye plants was particularly low, being supported generally by the growing pool of excess labour. Wage work on larger farms, which increasingly constituted the majority of land, similarly remained an essential part of lower-class incomes. This heavy reliance upon wage labour coupled with the farm size structure generally prevents designating medieval Flemish rural society as a true peasant economy (despite the growing desperation among the lower classes), but rather as a *mixed market-oriented survival economy*, a feature which would characterize rural Flanders until the nineteenth century. Large farms and small holdings thus became increasingly interrelated, being bound by the market and forced to adopt increasingly flexible strategies, albeit for different reasons.

The changed structure of farm holdings was also connected with changed social relations, especially among the upper classes. In the course of the twelfth and thirteenth centuries, Flanders became a principality with a power structure divided between the count, the lords and the bourgeoisie. These groups were more competitive than collaborative, which caused an early and intensive breakdown of the social structures. The increased split between the *seigneurie banale* (that is, power based upon the right to regulate the activities of people, which became more and more exclusively in the hands of the count) and the *seigneurie foncière* (that is, power based upon the possession of estates, through the right to exact rents and the like, in which the bourgeoisie played an increasingly important role), as well as many other factors, resulted in the more traditional seigneurial structures being superseded. Hence, there was more and earlier freedom among the lower classes, a lower level of surplus-extraction, fewer possibilities for seigneurial reactions, and an earlier relaxation of direct demesne farming in favour of leasing. These were all elements which favoured the fragmentation of holdings and the evolution towards a commercial-survival economy.

These changes held possibilities for both land and labour productivity. It is likely that consideration for the former predominated, however, particularly for the "survival" part of the economic structure, since labour was often sacrificed for improving land productivity on small holdings. Nevertheless, the economic implications of decreasing labour productivity were probably limited, since many holdings were so small that the additional labour input for survival simply absorbed hidden unemployment. As a result, labour-saving techniques were probably directed more towards the commercial part of the economy, such as in the development of horse-drawn transport for taking goods to market.

Conclusions

We have given a short survey of the current state of understanding and research concerning the evolution of Flemish agriculture from the extensive regime of the eleventh and twelfth centuries towards a much more intensive agricultural system. Crucial was the "long thirteenth century", during which most of the changes in agriculture took place.

It is likely that during this period the foundations were laid for the "Flemish husbandry" which became so important in later periods. For reasons which cannot be elaborated here, the system was not destroyed in the crisis period of the later middle ages.[49] Nor does this mean that agriculture became static. The dynamics of the commercial-survival system continued to promote changes (albeit at a slower rate), and further intensification also took place. The evolution of a "classic" three-course rotation into virtually permanent cultivation in the late eighteenth century, for example, is a much greater step than has often been supposed.[50]

Finally, it is ironic that the progressive nature of Flemish agriculture in the middle ages is exactly that which makes it difficult to discern, particularly in the leasing of demesnes which tended to cover up many of the details of farming. As a result, conclusions about the scale and rate of change in particular must remain hypothetical and conditional until further research can provide clearer answers.

Bibliography

Campbell, B. M. S. 1988, "The diffusion of vetches in medieval England", *Economic History Review* 41, 193–208.

Campbell, B. M. S. and Overton, M. 1993, "A new perspective in medieval and early modern agriculture: six centuries of Norfolk farming *c.* 1250– *c.* 1850", *Past and Present* 141, 38–105.

Derville, A. 1975/6, "Le rendement du blé dans la région Lilloise (1285– 1541)", *Bulletin de la Commission Historique du Départment du Nord* 40, 23–39.

---

[49] Thoen 1988, II, 1022–90.

[50] Contrary to what is generally written, averaged over the entire year, about half the land lies in fallow in a three-field system. In addition, the even more extensive "up-and-down" husbandry continued to play a role until at least the sixteenth century (see Thoen 1994b, 158). By the eighteenth century, however, the long fallow had disappeared in most areas of Flanders: Vanderpijpen 1983.

—— 1987, "Dîmes, rendement du blé et révolution agricole dans le nord de la France au Moyen Âge", *Annales E. S. C.* 42, 1411–32.

—— 1988, "L'assolement triennal dans la France du nord au Moyen Âge", *Revue Historique* 280, 337–76.

Fossier, R. 1968, *La terre et les hommes en Picardie jusqu'à la fin du XIIIe siècle*, 2 vols., Paris.

Langdon, J. 1986, *Horses, oxen and technical innovation*, Cambridge.

Lindemans, P. 1952, *Geschiedenis van de landbouw in België*, 2 vols., Antwerp.

Mertens, J. 1970, *De laatmiddeleeuwse landbouweconomie in enkele gemeenten van het Brugse Vrije*, Brussels.

Mertens, J. and Verhulst, A. 1966, "Yield ratios in Flanders in the fourteenth century", *Economic History Review* 19, 175–82.

Morimoto, Y. 1994, "L'assolement triennal au haut Moyen Âge". In *Économie rurale et économie urbaine au Moyen Âge*, eds Y. Morimoto and A. Verhulst, Kyushu/Ghent, 91–125.

Slicher van Bath, B. H. 1960, "The rise of intensive husbandry in the Low Countries". In *Britain and the Netherlands. Papers delivered to the Oxford-Netherlands Conference*, eds J. S. Bromley and E. H. Kossman, London, 130–53.

—— 1963, *The agrarian history of western Europe*, London.

De Smet, J. 1933, *Het memoriaal van Simon de Rikelike*, Brussels.

Thoen, E. 1988, *Landbouwekonomie en bevolking in Vlaanderen gedurende de late middeleeuwen en het begin van de moderne tijden. Testregio: de kasselrijen van Oudenaarde en Aalst*, 2 vols., Ghent.

—— 1992a, "Excursion to the region of Oudenaarde and Courtrai (valleys of the river Scheldt and the river Leie)". In *The transformation of the European rural landscape: methodological issues and agrarian change 1770–1914. Papers from the 1990 meeting of the standing European Conference for the Study of the Rural Landscape*, eds A. Verhoeve and J. A. J. Vervloet, Ghent/Wageningen, 338–51.

—— 1992b, "Technique agricole, cultures nouvelles et économie rurale en Flandre au bas Moyen Âge", *Douzièmes Journées Internationales d'Histoire, Septembre 1990, Centre Culturel de Flaran*, 51–67.

—— 1993a, "The count, the countryside and the economic development of the towns in Flanders from the eleventh to the thirteenth century. Some provisional remarks and hypotheses". In *Studia historica oeconomica. Liber amicorum Herman Van der Wee*, eds E. Aerts, B. Henau, P. Janssens and R. van Uytven, Louvain, 259–78.

—— 1993b, "Dries versus kouter. De wisselbouw in de Vlaamse landbouw van de middeleeuwen tot de zestiende eeuw", *Heemkring Scheldeveld* 22, 71–102.

—— 1994a, "Le démarrage economique de la Flandre au Moyen Âge: le rôle de la campagne et des structures politiques (XIe–XIIIe siècles). Hypothèses et voies de recherches". In *Économie rurale et économie urbaine au Moyen Âge*, eds Y. Morimoto and A. Verhulst, Ghent/Kyushu, 165–84.

—— 1994b, "Die Koppelwirtschaft im Flämischen Ackerbau vom Hochmittelalter bis zum 16. Jahrhundert". In *Économie rurale et économie urbaine au Moyen Âge*, eds Y. Morimoto and A. Verhulst, Ghent/Kyushu, 135–53.

—— 1995, "Précis d'histoire du seigle en Flandre du XII⁽e⁾ au XVIII⁽e⁾ siècles: culture et consommation". In *Le seigle. Histoire et ethnologie*, eds J-P. Devroey, J-J. Van Mol and C. Billen, Brussels, 101–16.

—— forthcoming (a), "L'exploitation de quelques domaines du comte de Flandre au nord de la France à la fin du 12⁽ème⁾ siécle".

—— forthcoming (b), "The medieval roots of capitalism in the former County of Flanders. *Status questionis* and hypotheses". In *The Netherlands in the Brenner debate*, eds J. L. Van Zanden and P. Hoppenbrouwers.

Tits-Dieuaide, M.-J. 1984, "Les campagnes flamandes du 13⁽e⁾ au 18⁽e⁾ siècle, ou le succès d'une agriculture traditionelle", *Annales E. S. C.* 39, 590–610.

Van Uytven, R. 1973, "De drankkultuur in de Zuidelijke Nederlanden". In *Drinken in het Verleden*, Louvain, 22–9.

Vanderpijpen, W. 1983, *De landbouw en de landbouwpolitiek in het Leie en het Scheldedepartement (1794–1814)*, unpublished doctoral thesis, University of Brussels VUB, 3 vols.

Vandewalle, P. 1986, *De geschiedenis van de landbouw in de Kasselrij Veurne (1550–1645)*, Brussels.

Verhulst, A. 1966, *Histoire du paysage rural en Flandre de l'époque romaine au XVIII⁽e⁾ siècle*, Brussels.

—— 1985, "L'intensification et la commercialisation de l'agriculture dans les Pays-Bas méridionaux au XIII⁽e⁾ siècle". In *La Belgique rurale du Moyen Âge à nos jours. Mélanges offerts à J.-J. Hoebanx*, Brussels, 89–100.

—— 1990a, "The 'agricultural revolution' of the middle ages reconsidered". In *Law, custom and the social fabric in medieval Europe. Essays in honor of Bryce Lyon*, eds B. S. Bachrach and D. Nicholas, Kalamazoo, 17–28.

—— 1990b, *Précis d'histoire rurale de la Belgique*, Brussels.

—— 1994, "Grundherrschaftliche Aspecte bei der Entstehung der Städte Flanderns". In *Économie rurale et économie urbaine au Moyen Âge*, eds Y. Morimoto and A. Verhulst, Ghent/Kyushi, 157–64.

—— 1995, *Landschap en landbouw in middeleeuws Vlaanderen*, Brussels.

Verhulst, A. and Gysseling, M. 1962, *Le compte général de 1187 connu sous le nom de "Gros brief" et les institutions financières du comte de Flandre au XII⁽e⁾ siècle*, Commission Royale d'Histoire, Brussels.

De Vos, M. 1991, *Bouwlandtermen in de Vlaamse dialecten. Spreidings- en betekenisgeschiedenis*, Tongeren.

Warlop, E. 1975, *The Flemish nobility before 1300*, Kortrijk.

5. AGRICULTURAL PRODUCTION AND TECHNOLOGY IN THE NETHERLANDS, C. 1000–1500

Peter Hoppenbrouwers

Attempting an overview of Dutch agricultural history during the period 1000–1500 is depressing for two reasons. Firstly, there is relatively little published work to survey, particularly on the history of agriculture in its narrower technical sense. Secondly, this is especially regrettable, since the half millennium before 1500 was crucial not only for the dramatic restructuring of rural society, but also for agricultural production and technology. At the start of the millennium, the northern (or present-day) Netherlands by all accounts had a backward, rather primitive peasant economy. Towns and markets were virtually non-existent, as was anything like centralized power. Five hundred years later, however, a not inconsiderable period, the coastal areas in particular were among the most highly urbanized and most densely populated of Europe, by which time a number of territorial principalities had formed (figure 5.1). By then, too, the region had developed all the basic features of a rural economy on the verge of commercialization and specialization.

The dearth of research has less to do with scholarly indifference than with a shortage of primary sources. When it comes to the history of agriculture proper, historians have relatively little of a documentary nature to work from during the first half of the period under consideration. Only from the second half of the thirteenth century onwards do relevant written sources begin to appear, and it is another century before they become numerous and detailed enough to provide hard evidence about the evolution of rural social structure and its agricultural potential. By the fourteenth and fifteenth centuries, however, several useful classes of documents had appeared. For the identification of individual farms and holdings, especially of the peasantry, these include lease and mortgage contracts, rent books, tax registers, and documents stemming from litigation, while for the elucidation and quantification of agricultural production and regional economic structure there are river toll registers, receipts of tithes, leases of land or rural grain mills in accounts, fiscal enquiries, and bylaws issued

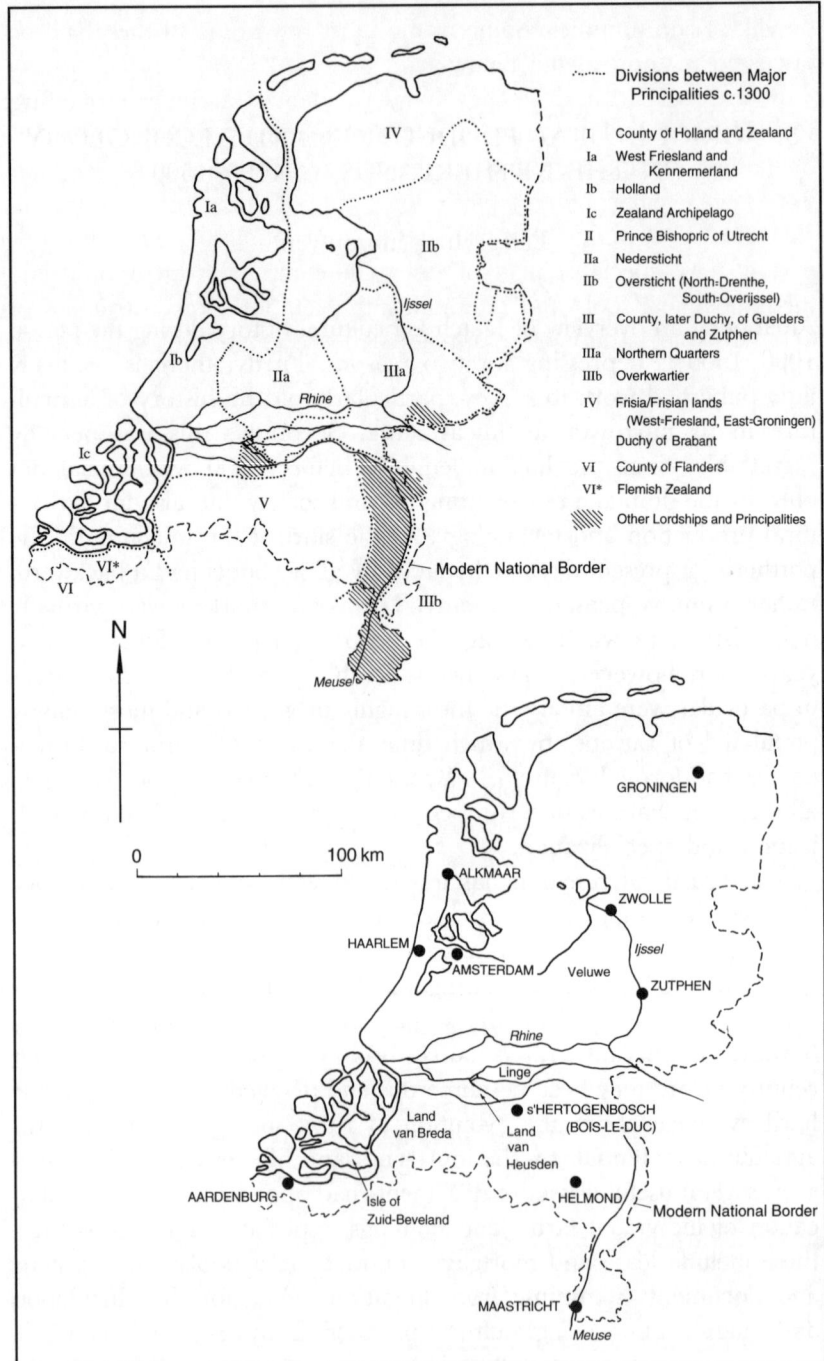

Figure 5.1. The territorial principalities and towns of the Netherlands.

by village communities. Some of the most rewarding of these will be discussed in more detail below.

Before 1250, on the other hand, evidence on how agriculture worked and what peasants did is overwhelmingly non-documentary in nature. Archaeology and its related auxiliary biological sciences, in particular palaeobotany, palynology and palaeozoology, are crucial. Historical geography also provides a conceptual umbrella for this period: as a specialist field of research, equipped with appropriate methodological tools and attempting the temporal reconstruction of a remarkable variety of man-made landscapes in the Netherlands, it has made impressive progress during recent decades.[1]

Issues and Debates: The Expansion Phase

It is usual to distinguish within the period 1000–1500 a succession of two long-term cyclical phases, one of expansion (c. 1000–1300) and one of contraction (c. 1300–1500). There is little reason to believe that the rural history of the region bounded by the present-day Netherlands has been at odds with the general European pattern, although, as we shall see, the latter period has some peculiarities.

The general picture of the expansion phase, though, is clear enough. The centuries between about 1000 and 1300 seem to have been very dynamic, showing everywhere a surge of human energy. This was realized in two main ways. First was the extension of existing settlements and fields. In early medieval Netherlands this was achieved over four main agricultural landscapes:

(1) the sandy soils of the east and south, spread over three plateaux, divided by the rivers IJssel and Rhine/Meuse respectively, to which can be added the narrow dune strip along the coastline in the west;

(2) the riverine areas of IJssel and Rhine/Meuse, consisting of alternating sandy levées and stream ridges, and swampy back-lands of river clay (to the east) and peat (to the west);

(3) the salt marshes of the coastal areas of Frisia in the north (present provinces of Groningen and Friesland) and Zealand in the southwest;

---

[1] Mainly thanks to the efforts of J. A. J. Vervloet of the Agricultural University of Wageningen and G. J. Borger of the University of Amsterdam.

(4) the fertile loess lands of South Limburg (mainly in the former Duchy of Brabant).

Second, in addition to the intensified exploitation of these agricultural zones, there was the reclamation and colonization of previously uninhabited lands. One should bear in mind that in the early medieval period more than half the total surface of the present-day Netherlands was covered with peat (figure 5.2). During the expansion phase the vast peat moors that stretched from north to south, like a broad ribbon, between the inland sand plateaux and the salt marshes or dune strips of the coastline were brought under cultivation, an unparallelled human achievement in Dutch history. In this way an entirely new, fifth, agrarian landscape was created.

In all five of these landscapes, agricultural expansion was accompanied by a number of well-known agrarian, technological, socio-political, institutional and ecological features:

(a) the extension of the arable and of the growing of bread grains (a trend German historical geographers have coined *Vergetreidung*);

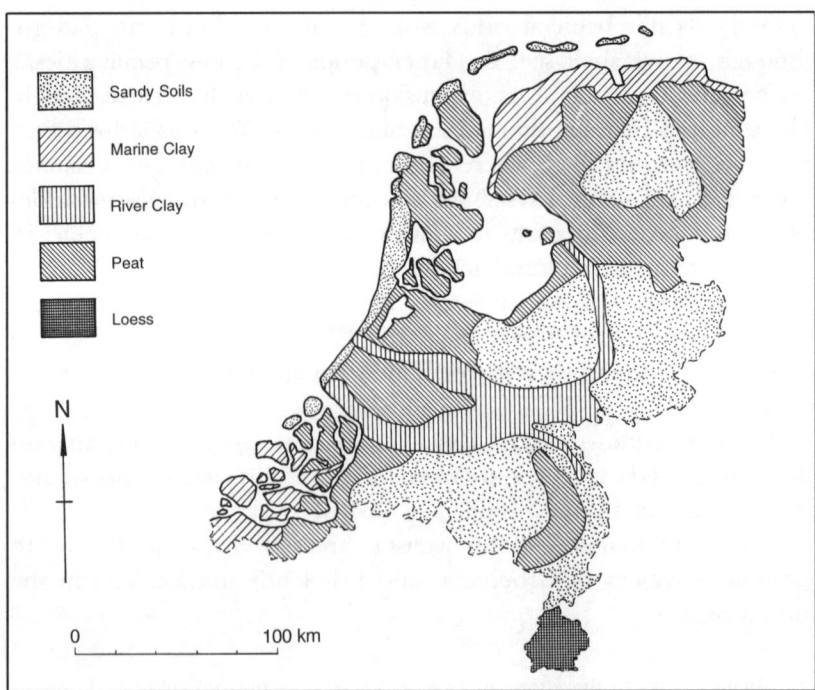

Figure 5.2. Dominant soil types in the Netherlands.

(b) a more intensive use of meadows, sometimes resulting in parcelling, along rivers and brooks;

(c) the creation of a physical infrastructure (dykes, sluices, and eventually wind pumpmills) indispensable to the exploitation of soils with a defective natural drainage;

(d) anthropogenic ecological changes: that is, signs of over-exploitation (dehydration of peat lands; degeneration of natural forest or deforestation; extension of heath moors; sand drifts);

(e) the "fixation" of settlement;

(f) radical changes in the exploitative relations between lords and peasants: concretely, the disappearance of manorial structures and the *seigneurie foncière* inherent to them (for example, labour services), and, in the opposite direction, the rise of "banal" lordship, based on the exercise of territorial rights of public justice and authority;

(g) as a counterweight to the latter, the formation of village communities, that is to say local associations of free peasants obtaining extensive authority, of a predominantly public nature, in the matter of local jurisdiction and administration;

(h) again, partly in competition with the latter, the formation of specialized local or supra-local organizations to control either common interests (waterboards) or the use of common resources (common waste lands); also, in general, the extension and formal establishment of religious and state networks (churches and parishes; judicial districts);

(i) finally, of course, the rise (or rather re-emergence) of markets and towns, which can at least be partly seen as a concomitant phenomenon of rural and agrarian expansion.

These issues have already led to a literature that is too extensive to treat in detail here. Therefore, only the two aspects most relevant to the theme of agro-technical development will be highlighted: that is, one, the extension of the arable and the *Vergetreidung*-process in existing settlements on the sandy soils and, two, the large-scale reclamations of salt marshes and peat moors in the coastal and river areas. In both cases, a considerable ingenuity was employed, especially in the matter of dyking. Altogether, it points to a rural society possessing significant technical skills and insights; having a superabundance of manpower; using both simple tools, like spades (usually with some iron parts), and more complicated ones like heavy ploughs provided with iron shares and coulters, as well as with mouldboards; and, last but not least, employing horse traction.

The Inland Areas: Agrarian Systems and the Extension of the Arable

For the sandy regions, discussions concentrate on the character of the transformation of early medieval agrarian systems that would have gone with the extension of the arable in existing settlements. Some historical geographers have interpreted the evidence in terms of a transformation of some form of agriculture using wood pasture (*Wald-viehbauerntum*) to a much more arable-oriented regime centred upon rye cultivation (*Heideviehbauerntum*). This transformation was probably accompanied by gradual changes in the composition of livestock, in particular a reduction in the number of pigs and also cattle in order to make room for growing flocks of sheep as the main suppliers of manure that would be mixed with sod cut from heath moors. Although it has been remarked that integrated agrarian systems such as *Wald-* or *Heideviehbauerntum* are very difficult to substantiate archaeologically,[2] since there is no way even to estimate the proportional importance of their arable and pastoral elements, there are clear signs of the growth of settlements and of a more intensified use of these sandy lands. For one thing, from the eleventh century onward, settlements—now soon to be called villages—began to be fixed in one place. Until then most must have been tiny, and farms were not renewed on the same site, but usually rebuilt at some distance elsewhere. This phenomenon of "wandering settlements" has been best researched for Drenthe, but established on various locations in North Brabant as well.[3]

In recent years, discussions about this phenomenon and the agrarian transformation accompanying it have been focused on two main issues: the formation of open fields, and sod-manuring. As to the first point, recent research has taught us not to exaggerate the speed with which things changed. Reconstructions of the evolution of open fields in Drenthe (where they were called *essen*) indicate that their medieval nuclei were actually rather small (even if their area may have been doubled during the eleventh to thirteenth centuries) in comparison to the total area known from nineteenth-century maps. By far the larger part of these nineteenth-century fields has thus been added since the sixteenth century. In addition, an infield-outfield system was associated with these fields until well into the seventeenth century:

---

[2] Bongers and van Vilsteren 1986.
[3] Bardet *et al.* 1983; Verhoeven and Vreenegoor 1991; Roymans and Kortland 1993.

while the central (open) fields were in permanent use for the growing of rye, the peripheral zones served most of the time as wild pasture,
only once in a while being sown with oats or buckwheat.[4]

This leads us to the second, and likewise hotly debated issue: the
age of sod- or *plaggen* manuring (see Widgren, this volume). It is focused
on the assertion made by the German palynologist Karl-Ernst Behre
that sod-manuring on the sandy soils of northern Germany was introduced during the tenth-eleventh centuries as a counterpart to the
efforts to increase substantially the production of bread grains and
winter-sown rye in particular. This whole idea has been challenged
by Bieleman and Spek of the Agricultural University at Wageningen.[5]
Both tend to see this whole process, at least in the northeast parts of
the Netherlands, as evolving at a much later date (see also below). It
should be mentioned, however, that recent palynological and pedological research in northeast Brabant by van Mourik[6] seems to support
Behre's dating of the introduction of sod-manuring. Van Mourik also
suggests that buckwheat (*fagopyrum*) was grown on the sandy soils from
about AD 1000, much earlier than is generally assumed,[7] and the
whole process points to an early depletion of natural forest, because
the extension of the arable seems to have taken place at the expense
of heath, not woodland. Initially, the heather was simply ploughed
under. Later sods consisting of both heather (*ericaceae*) and peat moss
(*sphagnum*)—mixed with animal manure—were brought on. After about
AD 1000 the pollen density decreased, indicating that cut sods gradually contained less organic, and more mineral (sand) material, probably as a result of over-exploitation. A more intensive land use in
east Brabant is also reflected in pollen samples from the village of
Dommelen. They show a rather consistent preference for rye, which
remained the dominant crop during the expansion phase. In Drenthe,
on the other hand, the clear preference of early medieval peasants
for growing rye shifted to one for more mixed crops later on, with
barley and oats, but also peas, beans, flax and "deder" seed (*camelina
sativa*; in English "gold-of-pleasure") being grown along with rye.
Only from the fourteenth century onwards did a tendency to rye

---

[4] For an overview: Bieleman 1992, 82–7.
[5] Bieleman 1994; Spek 1992.
[6] Van Mourik 1991 and 1993.
[7] Other evidence for the same early period, based upon pollen analyses, comes
from the Dommel valley between the present-day cities of Bois-le-Duc and Eindhoven:
Janssen 1972; Heesters 1973.

monoculture return.[8] In South Limburg, finally, expansion of the area under cultivation occurred relatively late, perhaps only from about 1200, with the transition from some form of wood pasture to a regular two-course system.

Salt Marshes and Peat Moors to Fields and Meadows

It is generally assumed that those parts of the coastal area that were not protected by dunes were dyked against the sea during the expansion phase. In the present provinces of Friesland and Groningen a closed system of sea dykes was probably not in place anywhere before the end of the thirteenth century; in the Zealand archipelago, however, this had already been achieved during the twelfth century.[9] Concerning the earliest endykements little can be established with certainty. The geneally accepted view is that local dykes along the salt marshes, mud-flats, tidal creeks and sea arms were gradually welded together. The oldest dykes were low, adjusted to the summer water level. Later on, they were heightened to above the winter water level. At the same time, some of the main sea arms were sealed. As a result, settlement spread out from the long-inhabited natural levées and sandy ridges along the coastline. Although possibilities for a more intensive use of the interior lands behind the coasts were created, nothing points to a sharply increased level of cultivation, perhaps because of a negative consequence of the construction of closed sea ramparts: at heightened tide levels draining problems were created for the lower lying basin clay deposits behind the fertile salt marsh ridges along the coastline. Thus, habitation mounds (*wierden*), so typical for protohistorical settlement in the northern coastal area, remained a necessity here. Altogether, it is puzzling that drainage techniques available since the late middle ages (especially pumping with windmills) were not applied here on any scale before the end of the eighteenth century.

In Zealand, then largely an archipelago of saltings that only gradually with human help grew into larger islands, "offensive" dyking (that is, aimed at taking land from the sea rather than protecting it from periodic inundation) went further back in time. From early on,

---

[8] Data summarized in Groenman-van Waateringe and van Wijngaarden-Bakker 1990, 290 (table 1).

[9] For the most recent dating, see van de Ven 1993, 78 and 75–7 respectively.

rich Flemish abbeys invested large amounts of capital in endykement projects. As dykes led to the accretion of new land, defensive dyking consolidated the formation of new marine clay polders. In Flemish Zealand, we know of such polders from the second half of the twelfth century thanks to much ado (in writing, luckily) about property rights and the status of settlers (*hospites*). The oldest sea dykes on the isle of Zuid-Beveland originated in the eleventh century, but only the disastrous storm of February 1134 led to a more systematic endykement, both of the defensive and offensive type, and to the closing of sea inlets with dams.

For Flemish Zealand, Gottschalk assumed that, after dyking, much of the older inland peat moors (*moeren*) that previously were used for extensive sheep grazing and peat digging were now turned into arable fields.[10] This process of intensification would have been accelerated by the remarkable prosperity of the small towns in the region, such as Aardenburg. In addition to bread grains, they created a demand for fruit and industrial crops, in particular madder. From the beginning of the fourteenth century there are also reports on the growing of lentils, rapeseed, leeks and onions near Aardenburg. By then, even fallowing had been sharply reduced, as can be shown from a lease from 1343 for lands held by the Ghent abbey of Saint Bavo in east Flemish Zealand. Of a total arable of about 25 ha (*c.* 60 ac) only 14 per cent lay fallow, while no less than 19 per cent was sown with various sorts of pulses.[11]

A very different type of endykement grew out of what is commonly—and rightly—regarded as the most conspicuous feature of the medieval expansion phase in the Low Countries: the reclamation and colonization of the vast peat moors behind the coast. The peat layers were drained by digging parallel ditches into them. After the scrub had been burnt, the peat could be tilled and settled upon. Although simple enough to do technically, basic requirements still had to be met. One necessity was the availability of equipment with iron parts (spades, axes, hoes, mouldboard ploughs) and probably also of horse power. Another was a high level of cooperation between the workers, because the subsidence that followed the lowering of the water table through drainage called for scrupulous water management. Initially, peasants could dodge the battle with the water by

---

[10] Gottschalk 1955, 31, 92.
[11] Gottschalk 1984, 255.

lengthening their ditches into new wilderness—if they were allowed. But this only meant delay. Inevitably, the peat reclamations set in motion a cat-and-mouse game between merciless nature and human inventiveness. It started with the deepening of ditches, and the construction of dykes and sluices. It continued beyond the medieval period, with the use of pumps and wind-watermills, culminating in today's sophisticated drainage schemes.

Much of the chronology and geography of the reclamations, the pedological processes they set in motion, the creation of newly inhabited and cultivated space, and the specific institutional arrangements that consequently took shape has been clarified in recent studies.[12] Despite this, little research has been dedicated to the actual exploitation of the reclaimed lowlands. It has always been assumed that the search for new farm land was the prime driving force behind the exploitation of the peat moors, and that farming on the reclaimed bogs involved both arable and animal husbandry.

Recently, Borger has questioned the relationship between population pressure and the initial phases of the peat reclamation, at least for the northern coastal areas. He argued that these lands "were sparsely populated" in the Carolingian period. "It was people rather than land which were scarce". What drew farmers to the peat moors was, according to Borger, the changing quality of the peat vegetation due to improved drainage by natural causes. The consequent breakdown of the peat was accompanied by rapid mineralization, while more effective aeration of the top soil favoured the nutrient uptake by plants, all of which helped to make peat soils viable for grazing sheep and growing grain. "So, a well-drained peatland will ensure high yields without fertilizing for a long time."[13] Only slowly will the natural supply of phosphorus and potassium diminish to the point that animal or mineral fertilizers have to be regularly applied. As for the latter, in West Friesland (province of North Holland) traces have been found of the use of calcareous clay that had been dug up and brought on the land as a mineral fertilizer. In any case, the extensive use of peatland as arable well into the fourteenth century is amply indicated by archaeological, documentary (for example, tithe payments) and toponymical evidence.

---

[12] The best surveys available in English are TeBrake 1985, chs. 6 and 7; Borger 1992; van de Ven 1993, especially 44–70.

[13] Borger 1992, 137–8.

The usual picture of a consistently applied pattern of exploitation in medieval times, especially in regard to the cultivation of the so-called *stroken* (strips of equal width), is probably too precise. The predominant historical view is one in which the arable did lie nearest to the farm (situated at the beginning of the strips), while cattle and sheep were kept on parts further away and not yet reclaimed. If peasants were forced to move their housesites to higher parts because of subsidence of the peat, the arable, if at all possible, would have "followed" the farm. This kind of presentation[14] does not agree with the seemingly haphazard use of reclaimed peat lands in the coastal regions as evident from post-medieval sources, where drier patches, regardless of their location, were used for growing oats or cash crops like hemp and flax, and lower and wetter parts were used as meadow or pasture.[15] Similarly, we should distinguish more than has as yet been done between peat reclamations in the middle of peat bogs on the one hand, and so-called fringe-reclamations (*randveenontginningen*) from the edges of the bogs on the other. Because the latter were located in transition zones of two (or even more) soil types, peasants had the option of concentrating specific agrarian activities on specific soil types.

Transition Zones: the Central River Area

The sort of "transition" agriculture just mentioned also characterized agrarian development in river areas, where features of both the inland and the coastal zones were mixed. Here many settlements were very old and had from early times developed types of open-field arable (*enken, ingen*). The inventory Pleijter and Vervloet have made of the remnants of curving ploughstrips and convex-shaped fields on levées in certain river areas indicates an early use of heavy ploughs with fixed mouldboards, drawn by a large number (possibly as many as eight) oxen.[16] These led to C- or S-shaped fields with ridges and furrows with measurements very much like examples known from England: about 675 m in length to about 15 m in width. As the topsoil layer of these ridges alternates humus material with river clay, the authors have speculated that after seasonal river flooding fresh

---

[14] For example, Bos 1988, 45.
[15] Personal communication from J. Bieleman.
[16] Pleijter and Vervloet 1986.

clay from the ditches between the ridges was brought up on the topsoil with the double advantage of adding fertility and maintaining drainage. They also argue that such a system predated the continuous dyking that began to take place here around 1100.

Behind the stream ridges along the river banks were the wet basin lands, consisting of heavy river clay, sometimes covered with peat. These wetlands were inundated practically every year during the winter season and can only have been used extensively as wild prairies and meadows. During the expansion phase they were drained on a large scale, in conjunction with dyking, and, if needed, the damming of rivers and tributaries. In the long run, the endykement of rivers had the same negative effects as described above for sea dykes: they exposed the backlands to unpredictable and disastrous flooding, in this case because river dykes, by narrowing the riverbeds, contributed to rapidly changing water levels, especially during wet seasons, and heightened pressure on the banks. Seepage also militated against arable farming in the basin lands, as they were eventually turned into permanent pasture and meadow. Not surprisingly, too, all known *wierden* (artificial habitation mounds for individual farmsteads) in the river areas date from after the endykement.[17]

During the initial stages of basin-land reclamation, however, this could not be foreseen, and new settlements did arise in these swampy areas, such as Babyloniënbroek in the Land van Heusden, the name suggesting that the first settlers must have felt like the Jews in exile. We know a considerable amount about the agrarian use of these lands around the middle of the thirteenth century thanks to the notebook of William of Rijkel, abbot of the Benedictine abbey of Saint Truiden, which had extensive possessions in this part of the river area.[18] The notebook reveals a number of important things. First, the arable extended from the slopes of the stream ridges into the lower lying swamp lands (*paludes*). Second, oats were by far the most important grain, comprising 80 per cent or so of the crop compared to about 15 per cent for barley, the rest being wheat (probably the archaic *triticum dicoccum*). Third, without any doubt horsepower was used for ploughing. The *curtis* of Saint Truiden at the Heusden village of Aalburg kept fourteen work-horses (against only one ox), while the arable held in demesne measured at least 30 ha (*c.* 75 ac). Fourth,

---

[17] Compare with Bosschaart and Driessen 1989.
[18] Hoppenbrouwers 1992, I, 255-7.

twenty head of cattle were kept both for producing dairy products and meat (and dung, of course). These characteristics have been confirmed for other parts of the river area, especially the disproportionate place of oats in arable production.

Issues and Debates: the Contraction Phase

The Development of Agrarian Production and Technique

Although mainly based on sources on river trade (especially toll registers), the survey of agrarian production during the later medieval period by Alberts and Jansen still remains the authoritative view.[19] The picture here is of a marked distinction between the agrarian economies of the coastal regions on the one hand, and the inland regions on the other. The former increasingly specialized in animal husbandry, an old tendency, now enhanced by both natural (peat subsidence) and commercial (expansion of urban markets) factors. It was first directed to more extensive agricultural activities, such as fattening, and haymaking, later on to intensive dairy farming as well. Before the end of the fourteenth century Holland, but also Frisia (the present provinces of Friesland and Groningen), exported both live cattle and butter and cheese to the Rhineland, Westphalia and the southern Low Countries.

The limited possibilities for arable agriculture were exploited for the production of cash crops for specific urban industries, such as hemp in the Gouda region and flax in the neighbourhood of Haarlem, an early indication that the rural economy of Holland would hinge more and more upon non-agrarian activities (see below). In the coastal regions only the Zealand archipelago developed a notable commercial arable agriculture. It was directed towards the production of high-value bread grains (wheat), madder as a main cash crop, and horticulture. At the same time the inland regions, according to Alberts and Jansen, displayed the more traditional pattern of mixed farming on poor soils, fixed first and foremost on subsistence. This meant a predominance of the growing of rye as the most important staple food. Of limited importance was the production of cash crops for smaller urban markets, such as hops in the villages southwest of the

---

[19] Alberts and Jansen 1964, chs. 6 and 8 (not revised in the second edition (1977)).

town of Groningen. Only in the lower lying regions (southwest Drenthe and the adjacent part of Overijssel) did the fattening of cattle and horsebreeding develop into fully commercialized sectors of agriculture, just as it did in the large river areas. Finally, the fertile loess area of South Limburg had an altogether exceptional position. Its fields produced a rich variety of crops that could be sold on the markets of Maastricht and Roermond: bread and brewing grains (rye, spelt (*triticum spelta*)), pulses, vegetables and fruit, as well as cash crops (flax, hemp, woad).

Alberts and Jansen had little to say about agrarian systems and techniques. Some details concerning intensive land use on the sandy soils of northern Brabant were filled in by Jansen in his PhD thesis of 1955 on short-term leases contracted before the aldermen of Bois-le-Duc from the second half of the fourteenth century.[20] Because open fields were absent in most of Bois-le-Duc's rural district, there was no need for strict rotations with extensive fallowing. Of the systems in use, Jansen mentioned two in particular. Both could be described as "improved" all-corn systems. In the first arable plots were intensively cultivated with rye, substituted every sixth year with oats or buckwheat. Fallowing seems to have been replaced at an early date by the growing of stubble crops, such as vetches, spurry and mangels. They could first serve as fodder and then be ploughed under for green manuring. In this system, the field was only cleared of crops once in six years for two to three months to allow repeated ploughing. The second rotation Jansen emphasized was one in which two years of rye were followed by one year of barley, for which there was a growing demand in the expanding Bois-le-Duc brewery business. Rotation systems such as these clearly required heavy manuring. For that purpose, cattle were kept in the byre during most of the year, and always fed there, while sheep, tended on the extensive common moorlands, were folded at night into pens, the floors of which were strewn with heather sods to be mixed with the manure. Reflecting the importance of this livestock, the lessor and tenant of the farm tended to have equal shares in the ownership—and, consequently, the produce—of all cattle and sheep present on the farm during the term; this implied that tenants had to return half of the livestock, but there are variations on this theme.

Although, even after thirty years, the picture of regional diversity

---

[20] Jansen 1955.

in later medieval agriculture portrayed by Alberts and Jansen still dominates, in various fields important details have been added and gaps filled in. I will treat these under four headings: (a) rotation systems and intensified use of the arable; (b) animal husbandry; (c) drainage techniques; and (d) farm buildings.

(a) Several authors have put forward evidence for a more intensified use of arable during the fourteenth and fifteenth centuries: first, by the sowing of vetches and/or pulses in stubble or fallow land (for example, in South Limburg, the large river areas and Flemish Zealand);[21] second, by the application of town waste and marl (the latter burnt to chalk) as manure (for example, in the neighbourhood of Maastricht and the Linge region in the area of large rivers);[22] and third, by the laying out of cultivation ridges (*bedden*) on plaggen soils (as in the Land van Breda, province of North Brabant).[23] All point to determined attempts at enhancing soil productivity.

In some cases, we have more specific details on the rotations proper. On two large farms of the Ghent abbey of St. Pieter in the Ossenisse *polder* a (fallowless) three-course system was applied in 1472. In it spring crops (barley, oats, pulses), winter crops (wheat, rye, barley again, rape-seed) and fallow were alternated. A three-course rotation is also known from South Limburg, but Renes has doubted the consistency of its application.[24] The avoidance of fallowing was made possible by several arrangements. For example, sometimes arable agriculture took place outside the open fields, on so-called *kampen*. They were used intensively without fallow for commerical horticulture or the growing of cash crops, and well marled for that purpose.

In the Land van Heusden a form of convertible husbandry was in use, as can be documented for two fields near the village of Hedikhuizen. One of them was outerdyke land or *waard* (that is, land between the river dyke and the river's low-water summer level).[25] On both, five to ten years (the term was quite irregular) of use as arable were alternated with three to four years as pasture. One other *waard* was said to be so "exhausted" after its continuous use as arable that the lessor was forced to lay the field in pasture for seven

---

[21] Jansen 1979, 51; Renes 1988, I, 125; van Bavel 1993, 437; Hoppenbrouwers 1992, I, 262; Gottschalk 1984, 505–6.
[22] Van Bavel 1993, 439; Jansen, 1979, 51; Renes, 1988, I, 122; van Bree 1972, 22.
[23] Bastiaens and van Mourik 1994.
[24] Renes 1988, I, 122–4.
[25] Hoppenbrouwers 1992, I, 261–2.

or eight years. Other sources from the same region speak of the incidental sowing on parts of common pastures; in one such field the rotation was three years' arable, then three years' pasture. A paradox, as yet unresolved, exists concerning the ingenuity with which seemingly marginal lands—for example, the shoals along the river Meuse—were tried out as arable during the fifteenth century, while at the same time continuously depressed rents on short-term leases suggest an agrarian slump.

(b) Three remarkable sources from the second quarter of the sixteenth century tell us about the number and composition of livestock kept on farms on the sandy soils of Guelders and east Brabant, as well as on the IJssel estuary. Most extensive are the so-called *Veetelling* of 1526, the records of a cattle census held in the Veluwe Quarter of the duchy of Guelders, and a census of livestock and cereal production held in the same year in the eastern part of the prince-bishopric of Utrecht (the Oversticht, or Overijssel).[26] From about the same period, but far less comprehensive, comes a list of livestock present on seventeen farms of the abbey of Binderen around the town of Helmond (province of North Brabant).[27] The most remarkable finding on all three lists is the astonishing number of livestock that was present on most farms. On the Veluwe, the numbers of horses and sheep easily eclipse those known from early nineteenth-century sources. The largest flock of sheep comprised 450 animals, but more than 20 per cent of all Veluwe farms had at least eighty sheep. Two-thirds kept at least three horses, one out of three at least five. By careful analysis of their geographical distribution and by comparison with the number of horses per farm, it appears that, of over 5,700 oxen enumerated in the census lists, most must have been raised for meat.[28] Consequently, horses were probably first and foremost kept as draught animals, but even then their large numbers remain a mystery, as the Veluwe was not known for its commercial horse-breeding and the sandy soil was too light to make large ploughteams necessary. The same goes for the Helmond region, where most farmers nevertheless held between two and six horses. Only in the neighbourhood of Zwolle might the high average number of horses be

---

[26] Roessingh 1979 analysed the Veluwe register, while van Zanden 1986 examined one of the Overijssel registers.
[27] Frenken 1995.
[28] Roessingh 1979, 15–16.

explained by the need to compose large plough-teams. This can be deduced from the fact that the average number of horses on farms situated on river clay was seven, whereas the overall average in the Zwolle region was "only" 4.4. For this reason, Van Zanden has assumed that on river clay fields ploughs were indeed drawn by teams of six horses.[29]

In the Veluwe area, the total number of adult cattle in 1526 almost equalled the number known from 1800. Close to 3,300 farms in possession of at least one horse held on average nine head of cattle. In the Zwolle area 202 of 212 households held cattle; about 20 per cent of these households had one or two head of adult cattle, another 20 per cent three or four head, about a quarter five or six head, more than 20 per cent seven to nine head, and the remaining 13 per cent ten or more head. Similarly, seventeen of the nineteen abbey of Binderen farms held cattle, ranging in number from nine to thirty head, the median being eighteen (both adult and juvenile).

The large numbers of animals on the farms of the inland regions clearly stand out against the more modest numbers from coastal areas, like Holland, despite the fact that, as mentioned above, such areas had long specialized in livestock raising. Still, numbers of horses and cattle of the order just mentioned were a rarity there. De Vries, after having surveyed the evidence (mainly from two extensive village-by-village surveys ordered by the Burgundian-Habsburg government in 1494 and 1514), assumed an average number of four to six cattle.[30] In contrast to the inland areas just described, the number of cattle must thus have quadrupled between c. 1500 and c. 1800.[31] Data on herds from Friesland are lacking before the middle of the sixteenth century, but we do know about farm sizes in the Frisian grassland area from a survey of 1511. The average was about 18 ha,[32] which would have been the equivalent of about nine head of adult cattle per farm, counting 2 ha for each head. Considerably rarer were horses, where there were only one or two per farm even around the middle of the sixteenth century.[33] It suggests ploughing in Friesland was performed by oxen rather than horses. On the other hand, the Groningen coastal regions were already renowned for the

---

[29] Van Zanden 1986, 96–7.
[30] De Vries 1974, 69–71.
[31] De Vries 1974, 138.
[32] Spahr van der Hoek 1952, 104; compare de Vries 1974, 58–9.
[33] De Vries 1974, 139, table 4.10, column 6.

breeding of horses by the later medieval period, a reputation they shared with the central river area (see above).

(c) Technologically, the greatest improvement in the rural economy of the time was undoubtedly in the area of drainage. The fifteenth century saw the introduction, in the western lowlands, of the wind-driven drainage mill and the laying out of so-called *boezems*, reservoirs for the temporary storage of polder water. The first known windmill for drainage was built at Alkmaar in Holland in 1407–8; in the century or so that followed over a hundred more appeared, mainly south of a line from Amsterdam to Haarlem.[34] Apart from endowing the landscape of Holland with its most valued man-made feature, drainage windmills must have re-increased the possibilities for arable agriculture from about the beginning of the sixteenth century. In the reclaimed peat lands around the town of Gouda the growing of oats, barley (even winter-sown) and all kinds of cash crops (hemp, flax, rape) made headway in the course of the fifteenth century. Horsebreeding was also of considerable importance. During the three annual fairs that were held in Gouda, up to 2,000 horses were bought, sold or traded at the beginning of the sixteenth century.[35]

(d) Van Oorschot has drawn attention to the improvement in living and working conditions of medieval peasants.[36] As a result of improvements in the construction of fireplaces sometime in the later medieval period, peasants and their livestock were able to live in separate areas within the farm, divided by a wall. In addition, in regions with mixed agriculture, such as Drenthe, the work area, usually combining cow shed and threshing floor, was improved by broadening the centre aisle and turning the cows with their backs to the side-walls, thus preventing them from fouling the threshing floor. At the same time, the use of the flail became more popular here.

Structural Changes in the Rural Economy and Society

Originally, Alberts and Jansen tried to couch their analysis of rural developments, and especially of town-country relationships, in terms of secular trend theory (that is, concerning the long-term cyclical

---

[34] As shown from the impressive inventory created by Bicker Caerten 1989; compare van de Ven 1993, 104–7.

[35] Ibelings forthcoming.

[36] Van Oorschot 1989, 42, relying here on the older work of Hekker 1955.

development of population and grain prices). They had to be particularly cautious on both demographic and real income development.[37] In the past few decades, however, much new research has been done on population history, and on variations in agrarian production, especially in terms of prices and wages.

As to general demographic development, there is no reason to doubt any more, as Alberts and Jansen still did,[38] the heavy toll taken by the Black Death of 1349–50 and its so-called echo-epidemics later on in the fourteenth and fifteenth centuries. What we still do not know is, leaving aside some localized exceptions, how large the population of any region was either before, or shortly after, 1349–50. It makes any assessment of long-term demographic development during the later medieval period far more a guess than an estimate—however creative some authors may have been.[39] Even very general suppositions, such as the total population figure around 1500 not being much higher than around 1300, should be taken with caution.[40]

The same goes for serial data on prices and wages. Important new series of grain prices have been made available for Maastricht, Liège and Zutphen,[41] while wages series have been created for Zutphen and Holland.[42] But, once again, most data are no earlier than the second quarter of the fifteenth century, so that there is too narrow a timespan to make a sufficiently comprehensive assessment of late medieval economic development. The only measurement of longer term fluctuations in the medieval period is the heroic attempt by J. C. G. M. Jansen to reconstruct the evolution of grain production in the Maastricht area from the mid-thirteenth to the beginning of the nineteenth century using tithe return series.[43] Rather boldly, considering a large lacuna in his data between 1389 and 1440, Jansen concluded that there could not have been any general or deep demographic and economic crisis in late medieval South Limburg.

All in all, much of the economic and social development in the "Netherlands" during the later medieval period still remains obscure. There is, however, every hope that our picture will gain in depth

---

[37] Alberts and Jansen 1964, especially 108, 124.
[38] Alberts and Jansen 1964, 105.
[39] Visser 1985; de Boer 1988.
[40] Compare the balanced overview in van Schaïk 1987, 183–92.
[41] Jansen 1979; Tijms 1983; Pieyns and Tijms 1993; Kuppers and van Schaïk 1981.
[42] Noordegraaf and Schoenmakers 1984; Noordegraaf 1985b.
[43] Jansen 1979.

and clarity with future research. Certainly, although the lack of good
serial data for periods earlier than the fifteenth century creates serious
difficulties, many other available sources have not yet been exploited
exhaustively, some scarcely at all. With regard to the directions that
research is taking, I will confine myself to highlighting two important
areas: first, the population shift between town and countryside and
its effects on the rural economy; and second, the question of whether
regional differences in medieval agrarian production sharpened due
to a restructuring of interregional product and factor markets.

Without any doubt, the population shift between town and countryside was most dramatic in Holland. The level of urbanization may
have doubled between *c.* 1340 (23 per cent) and 1514 (45 per cent),
whereas the total population figure would have been about the same
in both years.[44] Though less spectacular, the same process must have
taken place in a much more thinly populated area such as Guelders.
Even there urbanization reached a level of about 35 per cent around
1500.[45]

Rapid urbanization in the densely populated county of Holland
had tremendous effects on the structure of the rural economy and
society. New ideas on this subject have arisen from a critical discussion by Noordegraaf in 1985 of de Vries's *Dutch rural economy in the
Golden Age.*[46] Noordegraaf doubted de Vries's description of the rural
economy of Holland at the threshold of the early modern period as
fundamentally unspecialized and little commercialized. Recently the
debate has been reopened by van Zanden, who proposes an elegant
solution that does justice to both de Vries's opinion and Noordegraaf's
objections.[47] The core of van Zanden's thesis is that during the later
medieval period in Holland, and especially in the reclamation area,
a specific economic structure developed that could be characterized
as both very unspecialized and still highly commercialized. It is a
picture of peasant households making ends meet by combining small-
scale agriculture, geared to dairy farming and/or the growing of cash
crops, with a wide variety of non-agrarian activities, ranging from
fishing, ship building or working generally in the fastly expanding
transport business, to reed or peat cutting, salt or brick making, or

---

[44] Blockmans *et al.* 1980, 42–86, especially table 1 (44–5).
[45] Van Schaïk 1987, especially 183–92.
[46] De Vries 1974; Noordegraaf 1985a.
[47] Van Zanden 1988 and 1993 and forthcoming.

even just repairing the countless dykes and ditches. In van Zanden's view, these activities are comparable to the rural textile industries in Flanders for the same period, and he thus prefers to characterize late medieval Holland as undergoing "proto-industrialization". It would have meant, among other things, that, by 1500, wages would outweigh income from the sale of agrarian produce for the majority of peasant households—an observation van Zanden has extended to other areas at the time (for example, the IJssel delta).[48] Whether or not this meant that peasants increasingly loosened their relationship to their land and to subsistence agriculture, thus making way for the build-up of larger farms at the expense of smaller ones, depended upon the peculiarities of regional economic structures. In Holland, for instance, small farms were highly resilient for at least two reasons. First, agrarian specialization was directed to labour—rather than capital-intensive lines (dairy farming, the growing of cash crops, and all kinds of specialized horticulture). In this situation, larger farms did not lead to any significant economies of scale. In addition, large and small farms benefited equally from the continuous investments in dyking and draining technologies, because the costs were shared among landowners in proportion to holding size. In this way, smallholders profited to the same degree as more affluent landholders from substantial improvements in soil quality they could never have financed on their own.

The more Holland peasants were absorbed into product, service and labour markets (the first two often being extraregional or even international), the more their real income became dependent on the relative development of market prices of all kinds of agrarian and non-agrarian cost and income components. As we do not know the exact mixture of these components (which, in any case, could vary considerably over time and space), nor, often, the prices involved, it is almost impossible to appreciate the logic behind the choices that peasants made to earn a living, let alone to grasp general real income development in the countryside. All efforts in these directions must be regarded as the crudest approximations of economic and social reality. Such approximations have so far included (a) the calculation of real wage series over time by expressing nominal wages of (mostly urban) building-trade workers in quantities of bread grain (especially rye), (b) estimation of the preference for agrarian versus

---

[48] Van Zanden 1986, 104.

non-agrarian (the latter mainly wage-oriented) activities by comparing, over a long period of time but within the same geographical area, nominal wages and rents as recorded in short-term land leases, and (c) the reconstruction, on a regional level, of the entire surplus-extraction system. This last gives us an idea of the extent to which changes in the surplus-extraction process—especially growing state taxation—and the distribution of landed property over the main social classes (the "social-property system" in neo-Marxist terminology) stimulated or impeded productivity and technological growth.

My last point is about the ties that bound the rural economies of the inland regions to the urbanized one of the coastal area. Regrettably, explanatory theories of uneven regional development have till now largely concentrated on the post-medieval period. Nevertheless two of these theories dealing with later periods are worth mentioning, since both might be profitably applied to earlier times. One is developed in de Vries's seminal book of 1974.[49] It turns on the increasing disparity between the "progressive" coastal regions and the "backward" interior provinces under the impact of unequal development in terms of specialization and commercialization. Its applicability to the later medieval period has been briefly discussed above. The other, staunchly defended by J. Bieleman of Wageningen University, is aimed at proving the structural interrelatedness of the economies of the coastal and inland regions in terms of Thünensian "intensity circles".[50] Priester has expressed serious doubts on this effort.[51] In trying to project this model on, say, the fourteenth and fifteenth centuries, one has to realize that any Thünensian force field would have to be structured quite differently from the one Bieleman has presented for the early modern period, if only because, despite the high level of urbanization and high relative population density in the coastal areas, most individual towns were small. Under these circumstances, one wonders whether and, if so, on what level one should look for possible Thünensian constellations: that is, at the intra- or the interregional (in Bieleman's eyes even international) level? In a lucid comment on Bieleman's PhD thesis of 1987, Priester chose the first possibility and expressed doubts about the applicability of Thünensian economics generally for earlier periods.[52] Issues which Priester felt

---

[49] De Vries 1974.
[50] Especially Bieleman 1989; 1993.
[51] Priester 1988.
[52] Priester 1988.

disturbed the unimpeded working of Thünensian economics for the early modern period (and thus, even more so, for medieval times) included the unlikelihood of peasants being pure economic optimizers, in view of their strong desire to adopt risk reduction strategies; the unequal levying of taxes, which easily could be extended to other modes of surplus-extraction; and specific physical-geographical features, such as soil types and natural drainage conditions. These were all critical factors that distorted economic interaction between localities and regions.

All these factors reflected upon the economic and technological responses of farmers in the medieval Netherlands. The distinctions among regions or even social sectors still have to be worked out, but the variety of economic opportunities available to Dutch farmers and the difficulties encountered in realizing them were perhaps more pronounced in the Netherlands than anywhere else in medieval Europe. The principles underlying economic and technological development thus appear in sharper relief than normal here. It is, however, the particular misfortune of Dutch agricultural history of the period that historical resources, especially documentary ones, are so meagre—the absence of reliable crop yield data, for instance, will not have escaped the observant reader. But diligence and ingenuity have in the past gone far in elucidating some of the central issues and problems and will undoubtedly continue to do so in the future.

Bibliography

Alberts, W. J. and Jansen, H. P. H. (with the collaboration of J. F. Niermeyer) 1964, *Welvaart in wording. Sociaal-economische geschiedenis van Nederland van de vroegste tijden tot het einde van de middeleeuwen*, The Hague.

Bardet, A. C., Kooi, P. B., Waterbolk, H. T. *et al.* 1983, *Peelo, historisch-geografisch en archaeolgisch onderzoek naar de ouderdom van een Drents dorp*, Amsterdam.

Bastiaens, J. and van Mourik, J. M. 1994, "Bodemsporen van beddenbouw in het zuidelijk deel van het plaggenlandbouwareaal", *Historisch-Geografisch Tijdschrift* 12, 81–90.

van Bavel, B. J. P. 1993, *Goederenverwerving and goederenbeheer van de abdij Mariënweerd (1129–1592)*, Hilversum.

Bicker Caerten, A. 1989, *Middeleeuwse watermolens in Hollands polderland, 1407–8–rondom 1500*, Wormerveer.

Bieleman, J. 1989, "Die Verscheidenartigkeit der Landwirtschaftssysteme in den Sandgebieten der Niederlande in der frühen neuzeit", *Siedlungsforschung. Archäologie-Geschichte-Geographie* 7, 119–30.

—— 1992, *Geschiedenis van de landbouw in Nederland 1500–1950*, Meppel/ Amsterdam.

—— 1993, "Dutch agriculture in the Golden Age, 1570–1660". In *The Dutch economy in the Golden Age. Nine studies*, eds K. Davids and L. Noordegraaf, Amsterdam, 159–83.

—— 1994, "Plaggenbemesting in Drenthe; oud fenomeen in nieuw perspectief", *Historisch-Geografisch Tijdschrift* 12, 1–12.

Blockmans, W. P. *et al.* 1980, "Tussen crisis en welvaart: sociale veranderingen 1300–1500", *Algemene Geschiedenis der Nederlanden*, vol. 4, Haarlem, 42–86.

de Boer, D. E. H. 1988, "Op weg naar volwassenheid. De ontwikkeling van produktie en consumptie in de Hollandse en Zeeuwse steden in de dertiende eeuw". In *De Hollandse stad in de dertiende eeuw*, ed. E. H. P. Cordfunke, Zutphen, 28–43.

Bongers, M. J. W. M. and van Vilsteren, V. T. 1986, "Waldviehbauerntum: een wildwestverhaal?", *Historisch-Geografisch Tijdschrift* 4, 76–81.

Borger, G. J. 1992, "Draining-digging-dredging. The creation of a new landscape in the peat areas of the Low Countries". In *Fens and bogs in the Netherlands*, ed. J. T. A. Verhoeven, Deventer, 131–71.

Bos, J. M. 1988, *Landinrichting en archeologie: het bodemarchief van Waterland*, Amersfoort.

Bosschaart, A. M. W. and Driessen, P. M. M. 1989, "Terpen in de Nederbetuwe en de Tielerwaard", *Historisch-Geografisch Tijdschrift* 7, 10–17.

van Bree, G. W. G. 1972, *Steenbergen in de middeleeuwen. Landbouw, veeteelt en productie*, Zevenbergen.

Frenken, A. M. 1955, "Pachtopbrengst en veestapel van de hoevenaars der abdij Binderen in 1532", *Brabants Heem* 7, 67–71.

Gottschalk, M. K. E. 1955, *Historische geografie van Westelijk Zeeuws-Vlaanderen. I. Tot de St-Elisabethsvloed van 1404*, Assen.

—— 1984, *De vier ambachten en het Land van Saaftinge in de Middeleeuwen*, Assen.

Groenman-van Waateringe, W. and van Wijngaarden-Bakker, L. H. 1990, "Medieval archaeology and environmental research in the Netherlands". In *Medieval archaeology in the Netherlands. Studies presented to H. H. van Regteren Altena*, eds J. C. Besteman, J. M. Bos and H. A. Heidinga, Assen/ Maastricht, 283–97.

Heesters, W. 1973, "Uit de voorgeschiedenis van een Brabants dorp", *Brabants Heem* 25, 125–49.

Hekker, R. C. 1955, "De voorgeschiedenis van de boerderij in Oust-Nederland", *Driemaandelijkse Bladen* new series 7, 81–97.

Hoppenbrouwers, P. C. M. 1992, *Een middeleeuwse samenleving. Het Land van Heusden (ca. 1360–ca. 1515)*, 2 vols., Wageningen/Groningen.

Ibelings, B. J. forthcoming, "Aspects of an uneasy relationship. Gouda and its countryside (15th–16th centuries)".

Jansen, H. P. H. 1955, *Landbouwpacht in Brabant in de veertiende en vijftiende eeuw*, Assen.

Jansen, J. C. G. M. 1979, *Landbouw en economische golfbeweging in Zuid-Limburg 1250–1800*, Assen.

Janssen, C. R. 1972, "The palaeoecology of plant communities in the Dommel valley", *Journal of Ecology* 60, 411–37.

Kuppers, W. and van Schaïk, R. 1981. "Levensstandaard en stedelijke economie te Zutphen in de 15de en 16de eeuw", *Bijdragen en Mededelingen van de Vereniging Gelre* 72, 1–45.

van Mourik, J. M. 1991, "Zandverstuivingen en plaggenlandbouw; het bodemarchief van de Peelterbaan", *Historisch-Geografisch Tijdschrift* 9, 88–95.

——— 1993, "Zandverstuivingen en plaggenlandbouw; het bodemarchief van Tungelroy", *Historisch-Geografisch Tijdschrift* 11, 14–27.

Noordegraaf, L. 1985a, "Het platteland van Holland in de zestiende eeuw. Anachronismen, modelgebruik en traditionele bronnenkritiek", *Economischen Sociaalhistorisch Jaarboek* 48, 8–18.

——— 1985b, *Hollands welvaren? Levensstandaard in Holland, 1450–1650*, Bergen.

Noordegraaf, L. and Schoenmakers, J. T. 1984, *Daglonen in Holland, 1450–1600*, Amsterdam.

van Oorschot, A. C. 1989, "Het zandgebied tot 1850". In *Geschiedenis van Emmen en Zuidoost-Drenthe*, ed. M. A. W. Gerding, Meppel/Amsterdam.

Pieyns, J. and Tijms, W. 1993, "De graanprijzen van Luik (1400–1940). De effracties van de schepenen (1400) 1409–1792, de clericale effracties (1528–1793) en de effracties van het département de l'Ourthe en de provincie Luik sedert 1794", *Studies over de sociaal-economische geschiedenis van Limburg* 38, 67–154.

Pleijter, G. and Vervloet, J. A. J. 1986, "Kromakkers en bolliggende percelen. Enige opmerkingen over opbouw en ouderdom van een aantal akker-meten bij Tull (prov. Utrecht)", *Historisch-Geografisch Tijdschrift* 4, 13–21.

Priester, P. 1988, "Het model van J. H. von Thünen en het "geval" Drenthe", *Ons Waardeel* 8, 197–205.

Renes, J. 1988, *De geschiedenis van het Zuidlimburgse cultuurlandschap*, 2 vols., Assen/Maastricht.

Roessingh, H. K. 1979, "De veetelling van 1526 in het kwartier van de Veluwe", *A. A. G. Bijdragen* 22, 3–57.

Roymans, N. and Kortland, F. 1993, "Bewoningsgeschiedenis van een dekzandlandschap langs de Aa te Someren". In *Een en al zand. Twee jaar graven naar het Brabantse verleden*, eds N. Roymans and F. Theuws, 's-Hertogenbosch, 22–41.

van Schaïk, R. 1987, *Belasting, bevolking en bezit in Gelre en Zutphen (1350–1550)*, Hilversum.

Spahr van der Hoek, J. J. [with the collaboration of O. Postma] 1952, *Geschiedenis van de Friese landbouw. Deel I*, Drachten.

Spek, T. 1992, "The age of plaggen soils. An evaluation of dating methods for plaggen soils in the Netherlands and northern Germany". In *The transformation of the European rural landscape: methodological issues and agrarian change 1770–1914*, eds A. Verhoeve and J. A. J. Vervloet, Brussels, 72–91.

TeBrake, W. H. 1985, *Medieval frontier. Culture and ecology in Rijnland*, College Station.

Tijms, W. 1983, *Prijzen van granen en peulvruchten te Arnhem. Deel 2 [Koevorden (1639–1909), Maastricht (1342–1914), Nijmegen (1558–1916)]*, Groningen (*Historia Agriculturae*, XI–2).

van de Ven, G. P. (ed.) 1993, *Man-made lowlands. History of water management and land reclamation in the Netherlands*, Utrecht.

Verhoeven, A. and Vreenegoor, E. 1991, "Middeleeuwse nederzettingen op de zandgronden in Noord-Brabant". In *Middeleeuwen in beweging. Bewoning en samenleving in het middeleeuwse Noord-Brabant. Bronne, methodiek, nieuwe resultaten*, eds A.-J. Bijsterveld, B. van der Dennen and A. van der Veen, 's-Hertogenbosch, 59–76.

Visser, J. C. 1985, "Dichtheid van de bevolking in de laat-middeleeuwse stad", *Historisch-Geografisch Tijdschrift* 3, 10–21.

de Vries, J. 1974, *The Dutch rural economy in the Golden Age, 1500–1700*, New Haven/London.

van Zanden, J. L. 1986, "De telling van de veestapel en de graanproduktie in Zwollerkerspel in 1526. Een stukje van een legpuzzel", *A. A. G. Bijdragen* 28, 93–107.

—— 1988, "Op zoek naar de 'missing link'. Hypothesen over de opkomst van Holland in de late Middeleeuwen en de vroeg-moderne tijd", *Tijdschrift voor Sociale Geschiedenis* 14, 359–86.

—— 1993, *The rise and decline of Holland's economy*, Manchester.

—— forthcoming, "A third road to capitalism? Proto-industrialization and the moderate nature of the later medieval crisis in Flanders and Holland, 1350–1550".

6. AGRICULTURAL TECHNOLOGY IN MEDIEVAL DENMARK

Bjørn Poulsen*

This survey of Danish agricultural technology history covers the period from *c.* 1000 to *c.* 1540. The "Denmark" we are discussing here is somewhat larger than its modern version, comprising not only Jutland and the eastern islands (for example Zealand), but also parts of northern Germany and southern Sweden (figure 6.1). It was a border region between the classical western and central European regions of predominantly arable agriculture and the more pastoral areas of Sweden

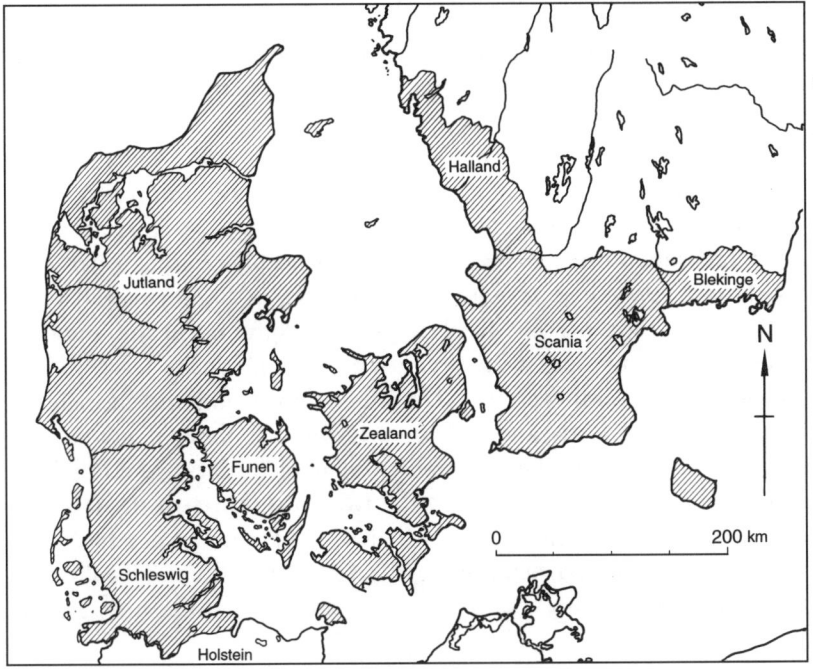

Figure 6.1. Medieval Denmark

* I would like to thank Erland Porsmose for useful comments and Per Ingesman for letting me use his unpublished edition of the accounts of the archbishop of Lund. Niels Hybel and Niels Engberg also kindly gave me access to articles in press.

and Norway to the north. The geography of this north European border zone was and still is characterized by both plains and hilly landscapes with generally good soils, although there are also widespread instances of sandy moors and, on the Jutish west coast, marshes.

In broad outline, important technological change can be seen over time. An increased use of iron during the third and fourth centuries AD led to longer scythes, longer sickles for harvesting, and the development of wagons for work in the fields.[1] The mouldboard plough spread much more slowly. Initially used in the North Sea marshes from the first century AD, it only made its way into Denmark and other areas of northern Germany as part of the settlement expansion of the ninth and tenth centuries.[2] The decisive introduction of a whole new technological complex, however, only occurred from the eleventh to the thirteenth centuries, as land was cleared for cultivation, livestock numbers increased, mills were built, and a much more pronounced material culture developed, not least in the form of improved tools of all types.[3] This was accompanied by a new mood of energy and industriousness, as typified by the twelfth-century Holstein peasant, Gottschalk, who without rest cleared the wilderness of beech, oak and other trees and thus enlarged his arable.[4]

It is, as elsewhere in Europe, possible to connect these occurrences with the collapse of a war and slave economy and the emergence of "feudal society", but in Denmark and the other Nordic countries it was more specifically associated with the shift from a Viking economy based on raids and tributes to a more peaceful regime, where the leaders of society had to earn their fortunes at home instead of in foreign adventures.[5] This was clearly the background of the founding of new settlements and widespread land reclamation.

From the eleventh century, Danish society could increasingly be seen in terms of lords and peasants. From that time, if not before, large estates owned by kings, princes, religious institutions and nobles existed. Slavery still existed up to the twelfth century on the farms of both lords and peasants. There was also undoubtedly a large number of freeholders, particularly as the rapid expansion of cultivated land from the eleventh century onwards encouraged the immigration

---

[1] Hedeager and Kristiansen 1988, 148.
[2] Compare Myrdal, this volume.
[3] On the concept of "technological complex" see Myrdal, this volume.
[4] Assmann 1979, 48.
[5] Compare Bois 1992.

of peasants from other areas. These were joined by a growing number of semi-free tenants (the Danish *landboer* class), who would in time comprise the most numerous group in rural society.

Initially, the estates of the wealthy tended to be composed of large manors, which until about 1200 were directly managed under the supervision of bailiffs. Hereafter these manors were increasingly farmed out to tenant farmers.[6] Similarly, a phasing out of large-scale farming occurred step by step, marked, for example, by the disappearance of the grange system *c.* 1250, but first and foremost by a general spread of peasant farming. It is likely that the peasantry already in the years 1100–1350 constituted the most important farming sector, but when large-scale seigneurial production effectively ceased in the fourteenth century their efforts became decisive. Technological innovations hereafter came from peasants, who often owned large holdings that were only lightly burdened by seigneurial exactions. By the fifteenth century these peasant holdings constituted effective, if largely family run, units of production; on most farms, it was only in times of emergency that servants were hired.[7] This rule was only broken by the emerging group of very wealthy peasants.

The demographic crisis of the fourteenth century struck Denmark, like the rest of Europe, and was followed by stagnation in grain production, at least in the western part of the country.[8] Rather than being a social or ecological crisis, however, it was more a time of fruitful transformation, as both lords and peasants were drawn into a larger commercialized economy. The western part of the region turned to the raising of cattle for the export market to northern Germany, the Rhine area and the Netherlands. From the second half of the fourteenth century to the sixteenth century, lords and peasants alike shipped both cattle and grain to the Dutch and north German Hansa towns. In terms of agricultural production, though, lords only exercized minor influence on their peasants. Such demesne lands as were left to them were aimed primarily at self-provision. The considerable grain and cattle exports from Denmark to northern Germany and to the Low Countries of the fourteenth and fifteenth centuries thus fell largely into the hands of Danish peasants. Not until the sixteenth century would lords show renewed interest in agriculture, by enlarging demesnes and involving themselves more

---

[6] Hybel 1996.
[7] Poulsen 1993.
[8] Gissel *et al.* 1981.

directly in farming. Until then, their influence was limited to the working capital they supplied to their tenants (usually in the form of livestock, grain and working tools provided at the start of the lease). Otherwise, apart from groups such as the Cistercians, interest by lords in the practical side of farming (including technological innovations) seems to have been weak. The tradition of agricultural literature in northern Europe during the middle ages, for example, was virtually non-existent.

Sources

The evidence for detailing technological innovation in medieval Denmark is fragmentary. Accounts from medieval Danish demesnes are few.[9] This to some degree reflects the fact that book-keeping and other aspects of documentation were slower to develop in Denmark. Accounts from ordinary peasant farms are not found before the sixteenth century,[10] while cadastral surveys and wills from the medieval period rarely reveal much with respect to agricultural technology. Some late medieval inventories are more rewarding, as are thirteenth-century laws. It must be said, too, that the archaeological material is fairly good and well studied. A number of excavations of rural settlements have been published, and the Danish finds of parts of medieval ploughs are remarkably good for Europe at this time.[11] In several instances, too, traces of ploughing and field systems have been discovered.[12] Accompanying all this is a fairly rich set of iconographic sources, primarily surviving frescoes from village churches. This pictorial evidence, of course, should be treated with the usual caution, but we have here some very useful depictions of a number of agricultural work situations,[13] which can be supplemented with iconographic evidence from the north German area—for example, illustrations from the early fourteenth-century manuscripts of the *Sachsenspiegel*.[14] Taking

---

[9] For an example of an account, see Poulsen and Pedersen 1993.
[10] Lorenzen-Schmidt and Poulsen 1992.
[11] For example, Steensberg and Østergaard Christensen 1974; Engberg 1994.
[12] For a recent survey see Møller 1991.
[13] Compare Götlind 1993, 185–244.
[14] A large number of German medieval illustrations has been presented by Bentzien 1990. On *Sachsenspiegel* see Bergman 1995.

all these together, it is possible to sketch in the main lines of technological development in the various aspects of medieval Danish agriculture.

Villages and Fields

Up to AD 1000, Danish villages tended to change sites periodically. Although the process is not fully understood, the emergence of a more stable village site during the period 900–1100 was probably connected to a number of factors. These include the intensification of feudal rights, the creation of a parish system and the building of churches; also, a technological explanation connecting the immobilization of villages to the mouldboard plough and the introduction of the open-field system has been advanced.[15]

In any case, there were marked regional differences in field systems in medieval Denmark, although this is only fully delineated in seventeenth-century data. At that time, the country was roughly divided into two parts.[16] In the east, on the islands and in east Jutland open-field systems similar to that of the English midlands predominated, with the large open fields being fenced in a permanent fashion. On the other hand, the sandy soils of western Jutland were generally dominated by small, unfenced fields operating under a long fallow period of three to six years. In this system (called *græsmarksbrug* in Danish), commons, pastures and moors were at least as important as the arable fields. Overall, farming was much more extensive and regulations for grazing and cropping much less regulated than in east Denmark.

The introduction of the open-field system seems to have occurred between 1000 and 1200. This is reflected in Danish provincial laws from around 1200, showing a gradual movement towards collectively regulated systems, perhaps initially with small, block-shaped fields.[17] A paragraph of the Jutland *Jyske Lov* of 1241, for instance, indicates the existence of fields cultivated in common and with strict rules for cropping, while in 1313 a cadaster clearly states that two-field systems existed in a number of villages around the east Jutland town

---

[15] Porsmose 1988, 220–33.
[16] Frandsen 1983b and 1983a.
[17] Hoff 1984 and 1990.

of Aarhus.[18] Fences, winter and spring crops and commons are
also mentioned. No detailed chronology can be fixed, but it also
appears as if fully evolved three-field systems were known in parts of
Denmark around 1200. This agrees with German experience, where
mature three-field systems can be dated to the twelfth and thirteenth
centuries and where it was certainly a feature of German eastern
colonization.[19]

In areas of *græsmarksbrug* where moorland predominated, peat began to be used for manure.[20] Accompanied or not by cattle manure,
turf from the large outfields guaranteed that "infields" could be cultivated for long periods without fallow, a system that was particularly associated with the growing of rye. In Schleswig turf manuring
allegedly dates from the ninth century, and it might also have been
used in the Danish town of Ribe at the same time.[21]

In the western areas of marshland, farming was from early times
very individualistic, and the only regulations found here deal with
the fight against the sea. Marsh settlements expanded along the
German North Sea coast and up to Ribe from the eighth to the
tenth centuries.[22] When the sea began to rise around 1000 so-called
"summer dykes" were built.[23] They were about 1.60 m high and
protected grain during growing periods, while the settlements were
placed on raised platforms. By 1100 the "summer dykes" began to
merge together, forming longer distance coastal dykes, while in the
twelfth century higher "winter dykes" of 2 to 2.80 m were being
built. The Dutch, who were already colonizing the deltas of the Weser
and Elbe, undoubtedly led this spread of drainage technology. In all
regions, the height of dykes grew through the middle ages, and despite
large losses in land due to flooding in the later fourteenth century
steady reclamation went on. In the fifteenth and sixteenth centuries
when reclamations of land reached a peak in Nordfriesland dykes
were built higher than ever (*c.* 3.40 m) and, where they were particularly exposed, protected by a fence of wooden stakes sunk into
the sea-facing base of the dykes.

---

[18] *Jyske Lov*, III, 59 (Brøndum-Nielsen and Jørgensen 1932, 480); Frandsen 1983b, 7.
[19] Rösener 1986, 130; Krenzlin 1952.
[20] Behre 1980; Stoklund 1990.
[21] Van Mourik 1990, 169–76; Lerche and Jensen 1985.
[22] For the reclamation in the south see Hofmeister 1975.
[23] Kühn and Panten 1989; Kühn 1992; Reinhardt 1984.

Grain and Livestock

Not surprisingly, the areas of open fields were dominated by grain production while the western *græsmarksbrug* and the marshes concentrated on cattle breeding. But generally both grain production and cattle breeding existed in close symbiosis. This was especially the case in the areas of open fields where manure from the animals was crucial for maintaining the fertility of the arable land.

Likewise the various agricultural activities interlocked closely with each other. Some sense of this can be gained by looking at the manor of Aahus (Scania) in 1532, where the cycle of agricultural work is revealed through the recorded boon works of the manor's 270 tenants.[24] From April 28 to May 18 the agricultural year started with ploughing with the light ard plough,[25] whereafter barley (and probably oats, although it was not recorded in the source) was sown from May 19–25. From June 23–29 dung was spread. This was followed by haymowing in the meadows, beginning June 30 and lasting to August 24. In the last phase of haymaking the hay was raked together, turned over and carted to the manor. From July 28 to August 10 the rye sown in the previous year was harvested, immediately followed by the harvest of barley from August 11 and the harvest of oats from August 25. By September 7 the grain harvest was finished, and the sheafs were bound and carted to the barns. Autumn ploughing with a heavy plough began on September 15 and ended on October 13, followed by a week of harrowing the fields and sowing rye. Threshing in the barn began on October 27 and lasted until the end of January the following year. The agricultural year on Aahus thus appears in a rhythmical movement which, in fact, was not untypical for large parts of medieval Denmark, even if the rhythm varied from region to region and over time. The activities took place in a well adjusted sequence, and this was even more pronounced on peasant farms where family labour had to be rationed carefully.

The grain cultivated on Aahus was typical of that grown in Denmark generally. Barley was commonly grown since prehistoric times and constituted the most important crop around 900–1000.

---

[24] Rigsarkivet, Copenhagen, Reg 108 A, ny pk. 36.
[25] Ploughing in the spring was done by *aarmend*; their work was called *ærie*. This must indicate that their work was done with an ard plough; compare Steensberg 1981b, 341.

The cultivation of oats also dated back to the Bronze Age, and its role in the medieval agricultural expansion was mostly as horse fodder, especially with the steady reduction of grazing areas and the rising number of horses (see below).[26] Exceptions in this respect were the moor regions which were unsuitable for oats. A new crop was found here in rye which became widespread in Denmark during the period 900–1100. Its introduction, as we shall see later, was probably connected to the new technology of the mouldboard plough with its emphasis on ridge and furrow cultivation. Ridge and furrow helped drainage and was seemingly a precondition for winter crops such as rye. Except for some areas, such as parts of the east Jutland coast, rye cultivation spread quickly and eventually became the dominant winter crop.[27] The cultivation of wheat, on the other hand, declined. Only on the southern banks of the Baltic, on Fehmarn, and on the southernmost of the Danish islands, Lolland-Falster, was it grown in sizable quantities. Otherwise, it was barley and rye that formed the main grain components of the new technological complex gaining force from around the year 1000.

Such fragmentary evidence as we have suggests yields were low. Data from the castle of Tranekær on Langeland in 1511 seem to indicate that the barley harvest only gave a return of 2.6:1.[28] A 1388 account of a manor in southwest Jutland shows, probably more typically, that the returns on rye were 3.6:1 and barley 3.4:1.[29] The manor of Aahus in Scania in 1532 mentioned above had a similar return of 3.8:1 on barley, and on a ducal manor in Haderslev in south Jutland a three-fold return on rye, barley and oats was recorded in 1544.[30] The returns on barley on the nearby ducal manor at Sønderborg moved between 4:1 and 6:1 during 1518–37, while the manor's rye cultivation in the same years peaked with a nine-fold return.[31] These last mentioned returns were probably optimal results in one of Denmark's most fertile areas. Altogether, the evidence for an indisputable increase in yields is lacking.

---

[26] Porsmose 1988, 290.

[27] Sometimes it could be mixed with barley, as indicated by a piece of barley-rye bread found on the fifteenth-century site at Store Valby in Zealand: Steensberg and Østergaard Christensen 1974, 351.

[28] Lütken 1909, 75.

[29] Poulsen 1990.

[30] Ingesman 1990, 129; Falkenstjerne and Hude 1895 and 1899, 2.

[31] Rigsarkivet, Copenhagen, Slesvig og holstenske regnskaber før 1580, Sønderborg amts naturalieregnskaber 1518–37.

Such increases in crop productivity as there were may have been connected to cattle raising. Cattle not only supplied their dung while grazing on fallow but also in the form of manure carried from the stables. The spreading of dung from stalls was in use since the Iron Age, but it undoubtedly became more widespread in the middle ages, especially around the growing number of towns and cities, where the practice can be documented from the thirteenth century onwards.[32] It is probably no coincidence that the earliest Danish dung fork (thirteenth-century) was found in the town of Lund.[33] The introduction of the mouldboard plough was similarly important as it allowed a much more effective ploughing of manure into the soil. Finally, the increased numbers of cattle in late medieval agriculture resulted in a growing supply of manure, and it is possible that this could have contributed to higher yield ratios in certain regions.

It is difficult to measure long-term changes in cattle numbers, as considerable variations occurred, including the early large-scale cattle raising in the marshes. It also depends on which aspect of the cattle "industry" one is considering. In the case of dairy herds, it seems that numbers were modest, with herds of more than four to five cows being rare in thirteenth- and fourteenth-century Denmark. For example, six quite large tenant farms in west Jutland, in a noted cattle-raising area, only had stocks of three, four, four, five, six and twelve cows respectively c. 1290.[34] Similar levels of dairy animals are evident on a number of Schleswig farms around 1410, where individual peasants normally possessed four to five cows. Two members of the lower nobility in the same area and time, on the other hand, had somewhat larger stocks, namely 10 and 30 cows. Although, in the first decades of the fifteenth centuries, large manors with only 12 to 14 cows could still be found, the normal size soon rose to twenty or more. The north Jutland manor of Aagård, for example, had 52 cows in its stalls in 1494, while 60 cows were evident on the Scania manor of Flyinge in 1536. Peasants with larger numbers of dairy animals were also evident at this time; for example, a farmer on Lolland had 15 cows in 1505.

To this fifteenth- and early sixteenth-century growth in dairy animal numbers can be added the even more important increase in

---

[32] Andrén 1986.
[33] Myrdal 1985, 59; compare Steensberg 1981a.
[34] Nielsen 1869; Poulsen 1985.

meat cattle.[35] This was especially evident in areas with plenty of
meadows, especially Jutland. A north Jutland manor like Aagård in
1494 had 65 beef cattle, while peasants in these regions were also
increasingly engaging in the breeding of oxen. This was soon reflected in impressive levels of exports. The first quantification of this
is from 1485, when duty was paid on 13,020 oxen at Gottorp, the
customs post where nearly all Danish and most Schleswig cattle passed
on their way to the south. As the importance of the Dutch market
grew, so did the number of cattle exported: in 1501 28,300 oxen
were registered at customs points leaving Denmark, and in the 1540s
the export had reached about 35,000 animals.[36]

Although it would become secondary to cattle raising in the later
middle ages, horse breeding experienced great success during the
twelfth and thirteenth centuries, where much grazing was reserved
for horse studs. In the thirteenth century horses were the most important Danish article of export next to the herring of Scania: for
instance, 8,400 horses were exported abroad via Ribe around 1230.
The sources tell us not only about the studs of kings, bishops and
lords, but also of peasants: for example, the case of a Jutland free-
holder, Torkil Skarpe, who around 1300 acquired a pasture for his
wild stud and built his herdsman a house on the site.[37]

Since the eleventh century pig keeping must be termed an essential part of the agrarian economy of medieval Denmark, but with
the regeneration of woods after the demographic crisis of the fourteenth century its importance increased markedly.[38] For example, some
4,000 pigs were fed or "masted" upon acorns and other forest fruits
on the manor of Boller (Jutland) in 1492, while mast-lists for the
county of Schleswig in the first half of the sixteenth century show
peasants taking herds of up to 40 pigs to the woods for feeding.[39]

Wood and Iron

The woods were not only important for animal grazing, but also as
suppliers of firewood and timber. Timber in particular was needed for

---

[35] Enemark 1983, 16–18.
[36] Blanchard 1986; Enemark 1983.
[37] Christensen *et al.* 1938–94, series 3, vol. 4, no. 230; compare no. 231.
[38] Fritzbøger 1994.
[39] Christensen 1928–39, vol. 4, no. 7105; Rigsarkivet, Copenhagen, Slesvig og
Holstenske regnskaber før 1580, Gottorp amts pengeregnskaber 1541-2.

the extensive fencing around open fields, and in all areas wood was essential for tools. In the eleventh century all free men had unrestricted access to the woods for timber and other purposes. Increasingly, however, this was limited as lords in the twelfth and thirteenth centuries extended exclusive property rights over many woodland areas, and throughout the later middle ages there was constant friction over forest privileges.[40] In the fifteenth century peasants in densely populated Zealand and the treeless marshes of Jutland had to buy wood from Norway or Halland. Usually, however, enough wood was available to give everyone the necessary material for tools, although it was clearly a limited resource, as, for instance, shown by mouldboard ploughs with inserted pebbles to guard against wear.[41]

The iron supply was more problematic. Admittedly, since the Iron Age, the extraction of iron from bog was rather common in Denmark, and furnaces for iron smelting from c. 1100–1300 are known. This domestic iron-making, although generally small-scale in terms of individual operations, was overall quite significant. Thus, for example, a constant production of iron from peasant hearths in Jutland can be documented until the end of the sixteenth century, and the east Danish production in Scania became very considerable, at least from the twelfth century.[42] In Schleswig-Holstein a rise of 15 per cent in the number of medieval furnaces compared to those of the preceding two centuries can be documented, and the Halland production was also on the increase. But supply generally lagged behind demand until German and Swedish iron production from the thirteenth century onward decisively influenced the availability and price of iron. An account from the early years of the sixteenth century tells us for instance how Lübeck merchants at that time shipped Swedish iron (*osemund*) to the Schleswig town of Flensburg where a local merchant distributed it to village smiths and peasants all over the region.[43]

The close analysis of finds can reveal the origins of the iron found in tools. A plough share from the Jutland island of Alrø, possibly sixteenth-century, was made of bog iron.[44] On the other hand, a complete twelfth-century "tool-chest" found near the village of Veksø

---

40 Holm 1988.
41 Lerche 1994, 199–205.
42 Steensberg 1952; Jöns 1992–3; Olsson 1995.
43 Stadarchiv Flensburg, Altes Archiv. B. Königliches Gymnasium, no. 565.
44 Buchwald 1992.

on Zealand comprised four iron tool pieces from Zealand or Jutland, five from southern Norway and six from southern Halland.[45] This latter find, possibly hidden because of war, dramatically illustrates the early availability of iron tool parts for agriculture. These included a heavy coulter for a plough (564 mm long and weighing 4.9 kg), four solid hoe pieces, nine sickle blades of the angled type measuring up to 275 mm, and two scythe blades (for short-handled scythes) 270 mm and 355 mm respectively. A 312 mm knife, probably for cutting leaves for cattle was also found, as well as a little bell for one of the animals and four whetstones for sharpening the tools. Finally, non-agrarian tools in the find included chisels, a plane, augers, nails, keys, handles, hooks and plates for a chest.

The Veksø hoard indicates the value attached to such tools in the twelfth century, as do lists of rural goods made in later times. In the early war-torn decades of the fifteenth century, for instance, Schleswig peasants complained that, as well as cattle, money and cloth, they had lost scythes, axes and augers.[46] Altogether, judged in terms of investment and the high profile given to them by farmers, iron-strengthened tools clearly played an important part in the agriculture of the time.

Ploughing, Harrowing and Hauling

The most important tool was probably the plough, particularly in its mouldboard form, which was perhaps the most important element of medieval agricultural development in the North European low-land zone.

From the Bronze Age and through the Iron Age the light ard plough was used in north Europe. Only in the north German marshes were implements capable of substantially turning the soil found already in the first centuries AD, as at Feddersen Wierde near Bremen and in Ostermoor in the Elbe marshlands, where turned-over furrows have been excavated.[47] It is thus likely that here a mouldboard plough with an asymmetrical share was being used (although all plough shares heretofore found in the region from this time period have in

---

[45] Engberg and Buchwald 1995, 70.
[46] Poulsen 1988, 92.
[47] Steensberg 1977.

fact been symmetrical).[48] The diffusion north to Denmark and to the rest of northern Germany at any rate clearly took place much later. From radiocarbon dates of parts of Danish ploughs found in moors, the earliest is the Navndrup beam from Jutland with a calibrated dating of 1285.[49] Ploughing traces, however, predate this somewhat. Already in the tenth to eleventh century, long flat fields are found, suggesting the use of a mouldboard plough; and the first Danish cases of high-backed ridges, that only a mouldboard plough could have created, date from the late twelfth century.[50]

Probably around 900 to 1100, then, the mouldboard plough was introduced into Denmark, gradually diffusing from southern areas towards Scania in the northeast.[51] Variations soon appeared, the most impressive being the ploughs with wheeled fore-carriages that appeared in Denmark and northern Germany in the thirteenth century. The Navndrup beam was a wheel plough, and a nice example of the type is depicted on the county seal of the Schleswig Froesherred c. 1300 (figure 6.2).[52] Later written sources also record plough wheels, as in 1505 when a peasant from the island of Lolland allegedly had three *plow redhe* (that is, three fore-carriages for wheel ploughs).[53] Ploughs without wheels, the so-called swing ploughs, were, however, also frequent during the middle ages. Most of the fifteenth-century Danish pictures of ploughs in fact shows wheel-less examples.

As already indicated, ploughs needed iron for shares, coulters and draught chains. Thus, a number of iron shares have been found in Denmark and Germany, both symmetrical (generally for ards) and assymmetrical (generally for mouldboard ploughs). Both the ard and the mouldboard plough could be worked with arrow-shaped shares. The two heaviest and longest tanged shares found in Denmark, datable from the thirteenth-fourteenth century, were both arrow-shaped; from Zealand, their weights were 3.0 and 2.1 kg. Grith Lerche's conclusion is that they were probably from some robust type of wheel ard or plough used for clearing land.[54] Asymmetrical triangular shares

---

[48] Felgenhauer-Schmiedt 1993, 167.
[49] Vensild 1970; Lerche 1994, 22.
[50] Vejbæk 1984; Born 1979, 310ff.
[51] Myrdal 1985, 76.
[52] Grandjean 1953, 18.
[53] Zangenberg 1928, 43–4.
[54] Lerche 1994, 213.

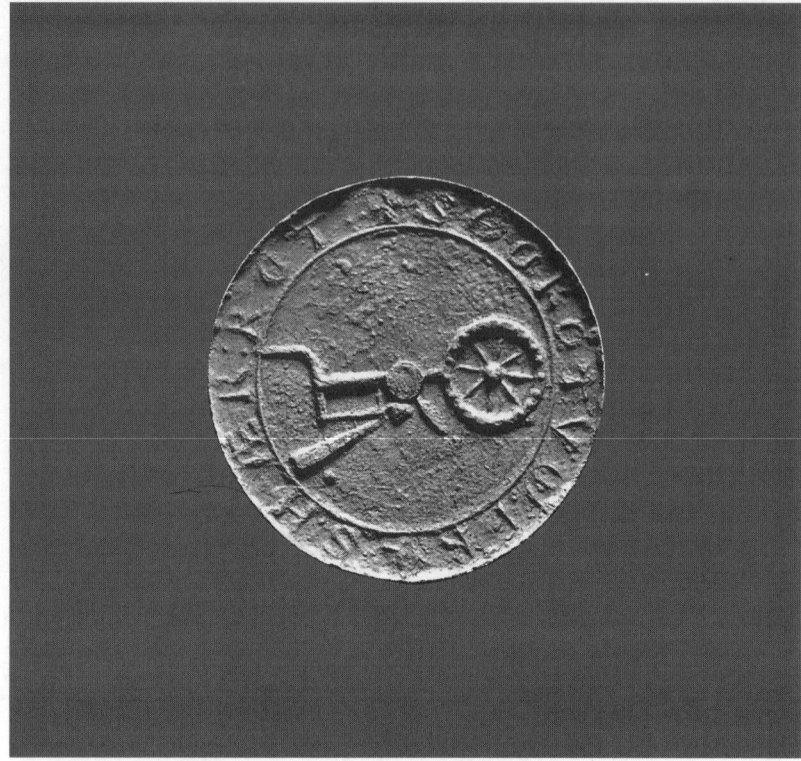

Figure 6.2. Seal of the south Jutland county (*herred*) of Froes, *c.* 1300, showing a
realistic depiction of a plough with a mouldboard.

with no left wing could also be very heavy. A specimen found on the
south Jutland castle of Nørrevold from the mid-fourteenth century in
its present state weighs 3.9 kg.[55] Shares of this type were seemingly
standard on late medieval mouldboard ploughs, as evident from
German finds (for example, at Mecklenburg). Similarly, it was this
share type that peasants of Schleswig often chose for representation
on seals.[56] The 1488 seal of the wealthy peasant, Hans Brodersen,
shows such a share with a coulter. The coulters themselves were of
considerable weight. Three Danish examples from the twelfth and
thirteenth centuries weigh 3.6, 4.9, and 5.1 kg.[57]

---

[55] Lerche 1994, 216.
[56] Poulsen 1991.
[57] Lerche 1994, 226ff.

Such ploughs were obviously very substantial pieces of equipment, which required considerable animal power. Deciding how many animals should pull such ploughs and what animals they should be (for example, horse or ox) were important issues for farmers of the time. The development of more effective harnessing for, as well as shoeing of, horses in the early middle ages made it possible to use these animals more effectively in agricultural work,[58] but oxen generally remained predominant, especially on large farms, as shown in table 6.1.

From table 6.1 demesnes obviously preferred oxen for ploughing, and it looks as if ploughteams of eight animals were normal. The only case where horses apparently had replaced oxen was at Nykøbing in 1531. Overall, there seems to have been a shift towards horses in the late fifteenth century. Aagård in Jutland, for example, had 39 horses, and some of these might have been used for ploughing, since this was by no means a new thing; three "plough horses", for instance, were stalled at the bishop of Roskilde's demesne at Østergaard, outside Copenhagen, in 1328.[59] On the whole, however, the horses had other functions on the demesnes; as indicated in table 6.1, many horses were termed as "wagonhorses" (vognheste), and, of course, they were widely used for harrowing (see below).

Generally, peasants also used oxen for ploughing, especially in earlier centuries. In regard to ploughtaxes levied from the 1230s, a law from c. 1260 stipulated that anyone not paying the ploughtax should lose the best ox from their team.[60] Some rich peasants had levels of oxen similar to demesnes. For example, a peasant at Bregninge on Lolland in 1505 had eight oxen and five horses, along with three ploughs (presumably for the oxen) and four ards (for the horses?; see below).[61] In general, however, peasants had smaller ploughteams. Of eleven tenant holdings supplied with livestock from their lords during the period 1290–1339, four were given two oxen apiece, another four were given four oxen each, two were given seven oxen each, and one was given eight.[62] If these numbers can be assumed to represent the total complement of draught animals on these farms, as

---

[58] Hagar 1982.
[59] Christensen *et al.* 1938–94, series 2, vol. 10, no. 53.
[60] Wegener 1871, 14: *si quis non soluerit . . . amittet meliorem bouem de aratro . . .*
[61] Zangenberg 1928, 43–4.
[62] Nielsen 1869, 61ff. Christensen *et al.* 1938–94, series 2, vol. 4, no. 163; vol. 5, no. 158; vol. 12, no. 167.

Table 6.1. Number of plough oxen, ploughs, workhorses and wagonhorses on a number of Danish demesnes, 1290–1536.

Manor	Year	Oxen	Ploughs	Workhorses*	Wagonhorses
Simmersted	1290	8	1		
Højstrup	1406	19	?		
Møgeltønder	1417	16	2		2
Saltø	1486	5	2		
Kokkedal	1492	16	2	1	4
Bregentved	1494	22	3	13	4
Ågård	1495	22	3	39	
Saltø	1501	12	2		
Næsbyhoved	1509	16	2		2
Tranekær	1510	10	?	?	
Gisselfeldt	1511	26	?	7	
Nykøbing	1531		1	14	
Saltø	1536	8	?	10	
Borrebygård	1536	19	?	9	4

* Mainly given as "øg" in the sources (from the old nordic *ekyr*—beast of burden, horse—itself related to the latin *jugere, jugum* (yoke)).
Sources: Nielsen 1869, 81; Molbech and Petersen 1858, 297, 300; *Danske Magazin* 1851, 3rd series, vol. 3, 272, 285; 1860, 3rd series, vol. 6, 93; 1873, 4th series, vol. 2, 3–16; Enemark 1983, 16; Christensen 1956, 315–18; Zangenberg 1928, 36–50; Christensen 1828–39, vol. 6, no. 11338; Lütken 1909, 75; Barner 1882, 76; Rigsarkivet København, Registratur 108A, gl. pk. 50.

seems likely, it would appear that most peasant farms probably had ploughteams of two to four oxen only. Sometimes the livestock supplied was given to several farms in common. Thus, three farms in south Jutland in the 1290s had to share six oxen and two horses, while four others shared four oxen and two horses and two farms shared eight oxen; similarly two farms on Funen at the same time were sharing two oxen and four horses.[63] This suggests that smaller peasant holdings might have routinely combined their animals to achieve larger teams, but this was certainly not necessarily the rule. Even larger farms often seemingly had limited numbers of draught animals. In several complaints from the Duchy of Schleswig during the years 1409–21, seven peasants stated that they each had been robbed of two oxen, while complaints from members of the lower

---

[63] Nielsen 1869, 61; Christensen *et al.*, series 2, vol. 4, no. 163.

gentry show losses of one, four, six, eighteen and twenty-four oxen.[64]
In any case, small numbers of oxen on peasant holdings were in line
with popular perception, as typified by the dismissive remark attributed to the evil thirteenth-century Queen Berengeria in a sixteenth-
century popular song:

> What shall a peasant have more in his home
> Than two oxen and a cow?[65]

All this, however, did not mean that peasants generally had no horses.
Already in the early years of the twelfth century even poor Holstein
peasants had a horse; similarly, in the thirteenth century horses were
extremely common on Mecklenburg farms, and one can assume that
the general progress of workhorses as documented for medieval
England was also mirrored to some degree in Denmark and northern Germany.[66] The Schleswig complaints mentioned above also
indicate that most peasants had two horses, while some had as many
as five to seven horses. One peasant claimed that his horses were
taken from him while at work in the fields.[67] Iconographic evidence,
however, provides the strongest evidence that horses were becoming
prominent in late medieval Danish agriculture. The great majority
of the relatively numerous fifteenth-century frescoes in Danish village
churches show ploughing being performed by two horses. Only the
paintings of two east Jutland churches from about 1500 show different arrangements, where two oxen are combined with two horses
(figure 6.3).[68] Further, the few depictions of wagons also only show
teams of horses, as, for instance, in a late fifteenth-century fresco
from the church of Rinkaby (Scania; figure 6.6). Overall, a shift from
oxen to horses can clearly be seen on peasant farms over the later
medieval period. The one clear exception to this was the moor regions where the soil did not favour oats and where oxen consequently
remained the most important draught animal to this century.

   As stressed by John Langdon, one of the reasons why the peasants
in general changed to horses was that they were more suited to small
peasant operations.[69] Admittedly, they cost more individually to feed

[64] *Danske Magazin* 1889–92, 5th series, vol. 2, 92, 99, 100, 101, 104.
[65] Steenstrup 1873, 254.
[66] Assmann 1979, 150; Bentzien 1980, 85.
[67] *Danske Magazin* 1889–92, 5th series, vol. 2, 90.
[68] Church of Gjerrild, Djurs Nørre herred (county), Randers amt (province); church
of Hyllested, Djurs Sønder herred, Randers amt.
[69] Langdon 1986.

Figure 6.3. Fresco from the east Jutland church of Hyllested, *c.* 1475-1525, showing a peasant, helped by a boy, pushing what may be a wheel-less ard, hauled by two oxen and two horses (photograph: Danish National Museum, Copenhagen).

than oxen, but they had the great advantage of being multifunctional; they could be used in ploughing, harrowing, hauling and riding, and, as in the last three functions, were particularly useful when speed was desired. The peasant accordingly could do with fewer draught animals if he concentrated on horses. Horses also benefited from the fact that ploughing was not done by the heavy mouldboard plough solely. In Denmark, the lighter ard did not disappear from the agricultural scene completely with the introduction of the mouldboard plough. Traces of ard ploughing from the thirteenth century have been uncovered in Ribe, and pictures as well as written sources tell us of their continued existence during the middle ages.[70] In fact it remained in use for a very long time even in regions of very developed grain production like Mecklenburg and Zealand. On Zealand the ard was still being employed in the nineteenth century, where it was called a *krog*. Similarly, in the above-mentioned Lolland inven-

[70] Madsen 1980.

tory from 1505 four small ard ploughs were listed alongside the three mouldboard ploughs. Evidently the ard complemented the heavy mouldboard plough in the areas of open field. The account of Aahus of 1532 mentioned above records ards being used, even on the largest farms, for the light spring ploughing before the sowing, and that this was a common practice is also indicated by a 1513 source from east Jutland.[71] In the fifteenth-century church frescoes both ards and mouldboard ploughs are shown. Later ethnographic surveys in Scania, in fact, show the ard dominating, just as the frescoes from the same area show only ards in the medieval context.[72] Significantly, the frescoes only show horses pulling ards.

Harrowing, of course, was also an area where horses excelled. The harrow followed the plough, and on late medieval demesnes inventories normally record the same number of harrows as ploughs. Harrows break heavy clods, remove weeds and spread manure. All known medieval specimens from Denmark and northern Germany are rectangular in shape with tines of wood,[73] but other forms may well have existed. Like the plough the harrow spread north and eastward from the west coast marshes, where it is documented in Holland in Roman times. A German find from Hattersum (Kreis Wittmund in Ostfrisland) contains parts of harrows from the ninth century and is contemporary with a harrow find from Viborg in Jutland.[74] Only a little later is a specimen from Mecklenburg, dated to the late tenth century.[75] The earliest example of harrow traces crossing ridged strips has been uncovered at Lindholm in a mid-eleventh-century context.[76] It seems probable, then, that the harrow was introduced in parallel with the mouldboard plough, but in contrast to the plough must have been mainly horse-drawn, thus contributing significantly to increased use of horses in agriculture.

Finally, the last major area for the use of animals in medieval agriculture was in hauling. Both grain and hay were transported from fields or meadows on vehicles. In very wet meadows, where wheeled vehicles could not go, it was necessary to transport the hay on horse-drawn stretchers (høbårer), which were often just made from a couple

---

[71] Hübertz 1845, 104, no. 94.
[72] Myrdal 1985, 30; Myrdal, this volume.
[73] As an improvement, its tines were often made of iron from about 1500.
[74] Bärenfänger 1993; Lerche 1982.
[75] Schuldt 1980.
[76] Lerche 1981.

of poles; for instance, they are mentioned in connection with transport of hay at the manor of Aahus in December 1532.[77] Otherwise,
wheeled vehicles comprised the predominant form of carriage in
agriculture. Both carts and wagons, drawn by two horses, are shown
as heavy work vehicles with insertible sides. Increasingly, four-wheeled
wagons seem to have become dominant, even if two-wheeled carts
still remained in use, thanks to the introduction of the revolving front
axle, an important innovation of the thirteenth and fourteenth century; such an axle appears, for instance, in an illustration of a heavy
horse-drawn, work wagon from a *Sachsenspiegel* manuscript of 1336.[78]
Transport of grain and hay from the fields to the farm and also of
manure from the farm to the fields made peasant investment in vehicles a virtual necessity, as did the trips to the newly developing
market towns of the twelfth and thirteenth centuries. Carrying services
for lords added to this tendency; in 1258, for instance, the bishop of
Ribe decided that every fifth *otting* (an *otting* was approximately one
farm) should send him a wagon in case of war.[79] From the thirteenth
century no peasant farm can be assumed to have been without a
wagon or cart.

Hoe and Spade

Animal power in cultivation was supplemented with significant amounts
of human muscle power. There are indications, for instance, that the
plough did not fully dominate as a cultivation tool before the fourteenth century. Where Adam is normally depicted as working with a
plough on fifteenth- and sixteenth-century frescoes, in earlier pictures
he works with a hoe, as in the twelfth-century fresco in Todbjerg
church, near Aarhus (figure 6.4).[80] Similarly, archaeological finds from
the eleventh century onwards indicate two new forms of tools: iron-
shod spades and iron hoes.[81] It is likely that some fields in eleventh-

----

[77] Rigsarkivet, Copenhagen, Reg 108 a, ny pk. 36, fol. 39r: *aage hø pa stængene.*
[78] Haupt 1995.
[79] Christensen *et al.* 1938–94, series 2, vol. 1, no. 157.
[80] Other early examples of Adam with hoe being Kippinge (1300–1325), Bregninge
(1275–1300), Kirkerup (*c.* 1325).
[81] Iron hoes and an iron spade shoe have been found in the excavated village of
Sædding near Ribe, dating from tenth to eleventh centuries: Stoumann 1978, 54.
See also the hoes of the tool find from Veksø discussed above. Iron-shod spades
were already present in Carolingian Germany: Roth and Wamers 1984, 146. A

Figure 6.4. Fresco from the east Jutland church of Todbjerg, *c.* 1125-50, showing Adam working with a hoe and Eve spinning (photograph: Danish National Museum, Copenhagen).

to twelfth-century Denmark were tilled with hoes.[82] This might have been a preferable option for the generally smaller fields of the early middle ages, many of which were seemingly only around 2 ha (*c.* 5 ac). But, even later, hoes and spades existed alongside ploughs in lists of farmers' tools and helped them in the tillage of the larger fields evolving in the twelfth and thirteenth centuries—on a rough estimate, fields of the high and late middle ages averaged about 15 ha (37 ac). The combined use of hoe and plough is evident from the above-mentioned twelfth-century find from Veksø, containing four hoes and a plough share, and from *Sachsenspiegel* scenes showing peasants working with both plough and spade in their fields.[83] Hoes, on

small hoe (*rastellum*) is mentioned in an early twelfth-century agricultural setting in Holstein: Assmann 1979, 102. For the use of spades for turf digging see Lerche and Jensen 1985.

[82] Compare Myrdal 1985, 116.
[83] Both 1995, 147.

the other hand, were a natural accompaniment to all who partici-
pated in the eleventh- to thirteenth-century land reclamations of forest
areas, and they also found full use in the dyke construction along
the west coast. Indeed, these sorts of tools were absolutely necessary
for the many very small farms in the thirteenth and fourteenth centuries, whose owners could not afford ploughs or draught animals.
This is, among other things, indicated by a cadaster of *c.* 1370 stat-
ing that these smallholders should not do their boon work with a
plough but *in rastrando* (that is, with a hoe).[84]

Working with Grain

Harvesting was the greatest user of manual labour, for which the
sickle dominated as a tool. Its blade was made of iron and existed in
two main forms: first, the angled sickle, where the blade curved directly forward from the handle; second, the longer bow-sickle or
"balanced sickle", where the blade initially angled backward from
the handle and then forward to give a rather more pronounced cur-
vature. The cutting edge of the sickles could either be smooth or
toothed.

The balanced or bow-sickle, which was already used in South
Germany in Merovingian times and for instance is known from a
eleventh-century find at Ditmarschen, is believed to have been introduced to Denmark around 1200.[85] A fresco from the church of Soender
Naerå on Funen, dated 1220–50, shows a bow sickle in use (figure
6.5). The bow sickle soon became dominant in Denmark, as seen in
fifteenth-century representations, except one from northern Scania,
where angled sickles remained in use to the nineteenth century.[86]
Given the tougher straw of winter rye, there may have been a connection between the dissemination of rye and the need for a sickle
that could deliver a more powerful stroke.

The scythe as a grain-reaping implement was not introduced to
Denmark before the late middle ages. In the Aahus account from
1532 the oats and barley are said to have been cut (*skaeres*), while the
rye was cut and mowed (*mejes*), the latter clearly indicating the use of

---

[84] Christensen 1956, 15.
[85] Steensberg 1943, 117–19; la Cour 1961, 202–3; Bentzien 1980, 41–2.
[86] Steensberg 1943, 218ff.; Myrdal 1985, 125.

Figure 6.5. Fresco from the Funen church of Soender Nærå, *c.* 1220-50, showing a man harvesting with a bow or balanced sickle (photograph: Danish National Museum, Copenhagen).

a scythe. Scythes were much more effective for harvesting when they had a cradle to catch the straws, and significantly the Aahus account records the purchase of a cradle-sned (*mejered*) for four shillings in May 1532.[87] In parts of Germany, the scythe was in use as a tool for grain harvesting around 1300, and by the fifteenth century it was widespread in the Mecklenburg area. From there, it presumably spread to Denmark, probably in the last part of the fifteenth century.[88] The first documented instance of scythes clearly intended for harvesting in Denmark comes from the inventory of a Lolland farm in 1505, containing "three scythes with their cradles" (*meyeiærn med there redhe*).[89] The use of scythes for reaping was probably a specialization for

---

[87] Rigsarkivet, Copenhagen, Reg 108 A, ny pk. 36; Regnskab for Åhus slot 1532–33, fol. 52v.

[88] Bentzien 1980, 74–6, 116–17.

[89] Zangenberg 1928, 46.

areas concentrating on grain production during the late middle ages and using larger fields than normal. The special, and perhaps more prestigious, nature of the activity was also reflected in its monopoly by men, whereas the sickle continued to be used by both men and women.

The threshing of grain by flail was always hard work. The word "thresh" itself is Germanic (Danish: *tærske*, German: *dreschen*) and means "to trample to pieces with the feet", indicating the original process. The history of the flail itself is not very well known; in parts of Germany it dates back to the early middle ages, but it appears not to have been introduced to Scandinavia before around 1000–1200.[90]

The threshed grain was ground in mills. Watermills spread quickly from the late eleventh century onwards.[91] In the twelfth century they can be documented at many places: for example, the five watermills in the Scanian town of Tommarp in 1161.[92] Hand querns, on the other hand, gradually disappeared, among other things as a consequence of the mill privileges of lords, even if they could still be found in villages.[93] In the thirteenth century horse mills and windmills appeared. Horse mills, for instance, are mentioned in 1247 at the monastery of Sorø.[94] Windmills, which are recorded in England and Flanders from the 1180s and are documented on the Lower Rhine in 1222, appear for the first time in Danish sources in 1259 and 1261; in the latter case, four windmills were recorded for the small town of St. Heddinge.[95] Practically every parish had its mill by the fourteenth century, most of them as part of demesnes.

Hay Harvest

The hay harvest, which was a precondition of cattle stalling, was from the Iron Age performed using a short-handled scythe. The short-

---

[90] Porsmose 1988, 286; Myrdal 1985, 136; Bentzien 1980, 45, 78–9.

[91] Nielsen 1980; 1986; Steensberg 1971; compare Roesdahl 1980, 123. Christian Fischer of the Silkeborg Museum doubts the existence of watermills in Denmark before about 1100 (personal communication).

[92] Christensen *et al.*, series 1, vol. 2, no. 143.

[93] Madsen 1988.

[94] Langebek 1776, 453. A potentially earlier example appears in a tale of a boy miraculously saved from being crushed in a horse mill at Æbelholt kloster: Olrik 1968, 283.

[95] Christensen *et al.* 1938–94, series 2, vol. 1, no. 274; no. 333. For the invention of windmills generally see Lohrmann 1995.

handled scythe continued in use throughout the middle ages and is depicted on a fifteenth-century fresco in Fanefjord Church.[96] Gradually, however, it was superseded by the long-handled scythe, which was probably introduced to Denmark in the thirteenth century. As described above, the long-handled scythe, often with two short bars on the shaft to aid handling, also gradually gained acceptance as a grain reaping instrument, and accordingly two types of scythes began to appear in the written sources. On the above-mentioned peasant farm on Lolland in 1505, there were besides the three scythes with their cradles also two "grass" scythes (*gresleer*). On the demesnes of Bregentved in 1494 we find four "hay scythes", and at Nykøbing castle there was a single "grass scythe" in 1531.

The mowed hay was gathered together, stacked and turned using wooden rakes. A rake of traditional type is depicted around 1300 in the north German manuscript *Sachsenspiegel*, but in the middle ages a technological improvement appeared with the introduction of the bow rake (where the handle and supporting pins were attached to the rake-head at three or more points) in the fifteenth century.[97] Another tool which could be used both to handle grain and hay on wagons and for the feeding of the cattle was the two- or three-pronged fork made of iron. In a fifteenth-century fresco in Rinkaby church mentioned above a woman is shown using a fork for moving a grain sheaf from a cart (figure 6.6), and finds from several excavations in Germany and Denmark indicate that forks were in use from the thirteenth century onwards.[98]

Conclusion

As part of the new social configuration, established 1000–1200, a different technological complex was introduced into Danish agriculture. It is difficult to prioritize the various factors, since they must be considered as interdependent. However, increasing iron supplies, heavy mouldboard ploughs, the introduction of winter rye, land reclamation, and the establishment of the open field system all contributed

---

[96] Steensberg 1943, 116 (short-handled scythe from Borgø, thirteenth century), 200 (Fanefjord Church fresco).

[97] As depicted on an early sixteenth-century painting in the church of Skarhult in Scania; see also Myrdal 1984.

[98] Trier 1993–4; Steensberg 1981a.

Figure 6.6. Fresco of the Scanian church of Rinkaby, *c.* 1450-1500, showing a peasant family harvesting corn and carting it from the field with a wagon (photograph: Danish National Museum, Copenhagen).

to both increased overall agricultural production and the strengthening of rural communities.

The various improvements could appear in differing combinations, depending upon the region. In the north German and west Jutland moor regions, for instance, the increased cultivation of rye was followed by reclamations made possible by the use of turf as manure. In general, the growth of grain production encouraged both the introduction of new tools and methods and the revitalization of old ones. A flail for more effective threshing was now used. The balanced sickle cutting the tough straws of winter rye became common. Horses, which could be fed from the increased production of oats, were now employed far more as hauling animals, thanks to new forms of harness, and gained a solid footing on peasant farms in particular. Finally the processing of the increased grain production was facilitated by the construction of mills, many of them built by lords trying to augment their share of the surplus.

In the late middle ages and the sixteenth century another phase of innovation occurred when generally larger farm holdings adjusted to new needs and circumstances. At this time the scythe as a grain harvesting instrument was introduced, making harvesting on larger fields more effective and permitting certain regions to stand up to a growing competition from east European grain zones. The marketing of grain from the thirteenth century onwards went by wagon to town markets, while from the fifteenth century peasants established contacts with cities in the south, either through merchants or using their own ships. At the same time cattle production was restructured, concentrating on long drives of oxen for sale on foreign markets. This would turn into a dominating source of revenue in coming centuries.

Bibliography

Andrén, A. 1986, "I städernas undre värld". In *Medeltiden och arkeologin: festskrift till Erik Cinthio*, eds A. Andrén, M. Anglert, V. Ekstrand, L. Ersgaard, H. Klackenberg, B. Sundnér, and J. Wienberg, Lund, 259–69.

Assmann, E. (ed.) 1979, *Godeschalcus und Visio Godeschalci* (Quellen und Forschungen zur Geschichte Schleswig-Holstein 74), Neumünster.

Bärenfänger, R. 1993, "Frühmittelalterliche Eggenbalken und weitere Holzfunde aus Hattersum, Kr. Wittmund/Ostfriesland", *Archäologisches Korrespondenzblatt* 23, 127–39.

Barner, K. 1882, *Familien Rosenkrantz's historie* 2, Copenhagen.

Behre, K.-E. 1980, "Zur mittelalterlichen Plaggenwirtschaft in Nordwestdeutschland und angrenzenden Gebieten nach botanischen Untersuchungen". In *Untersuchungen zur eisenzeitlichen und frühmittelalterlichen Flur in Mitteleuropa und ihrer Nutzung* 2, ed. H. Beck, Göttingen, 30–44.

Bentzien, U. 1980, *Bauernarbeit im Feudalismus*, Berlin.

Bergman, R. 1995, "Die Miniaturen des Sachsenspiegels und archäologischen Realien als Sachquellen zur ländlichen Alltagskultur Westfalens im Mittelalter". In *Beiträge und Katalog zur Ausstellung aus dem Leben gegriffen: ein Rechtsbuch spiegelt seine Zeit* 2, ed. M. Fansa, Oldenburg, 173–87.

Blanchard, I. 1986, "The continental European cattle trades: 1400–1600", *Economic History Review* 39, 427–60.

Bois, G. 1992, *The transformation of the year one thousand: the village of Lournand from antiquity to feudalism*, Manchester.

Born, M. 1979, "Acker- und Flurformen des Mittelalters nach Untersuchungen von Flurwüstungen". In *Untersuchungen zur eisenseitlichen und frühmittelalterlichen Flur in Mitteleuropa und ihrer Nutzung* 1, ed. H. Beck, Göttingen, 310–37.

Both, F. 1995, "Landwirtschaftsgeräte im Zeitalter des Sachsenspiegels". In

Beiträge und Katalog zur Ausstellung: aus dem Leben gegriffen: ein Rechtsbuch spiegelt seine Zeit 2, ed. M. Fansa, Oldenburg, 143–53.

Brøndum-Nielsen, J. and Jørgensen, P. J. 1932, *Danmarks gamle landskabslove med kirkelovene* 2 (*Jyske Lov*), Copenhagen.

Buchwald, V. G. 1992, "Jernfremstilling i Danmark i middelalderen—lidt om bondeovne og kloder", *Aarbøger for Nordisk Oldkyndighed og Historie, 1992*, 265–286.

Christensen, C. A. (ed.) 1956, *Roskildebispens jordebøger og regnskaber* (Danske middelalderlige regnskaber, 3rd series, vol. 1), Copenhagen.

Christensen, C. A., Nielsen, H., Skyum-Nielsen and Blatt, F. (eds) 1938–94, *Diplomatarium Danicum*, series 1–4, Copenhagen.

Christensen, W. (ed.) 1928–39, *Repertorium diplomaticum regni Danici medievalis* 2nd series, vols. 1–9, Copenhagen.

la Cour, V. 1961, *Næsholm*, Copenhagen.

Danske Magazin, series 1–8, 1745–present, Copenhagen.

Enemark, P. 1983, "Oksehandelens historie *ca.* 1300–1700". In *Sortbroget kvæg*, eds A. Pedersen, P. Enemark, E. J. Ipsen, and V. Bro, Århus, 9–87.

Engberg, N. 1994, "Resultater og tendenser i dansk landsbyarkæologi". In *Hikuin 21* (Landbebyggelse i middelalderen: huse og gårde), ed. N. Engberg, Højbjerg, 11–20.

Engberg, N. and Buchwald, V. F. 1995, "Værktøjskisten fra Veksø". In *Nationalmuseets Arbejdsmark*, ed. O. Olsen, Herning, 62–75.

Falkenstjerne, F. and Hude, A. 1895 and 1899, *Sønderjyske Skatte og Jordebøger*, Copenhagen.

Felgenhauer-Schmiedt, S. 1993, *Die Sachkultur des Mittelalters im Lichte der Archäologischen Funde* (Europäische Hochschulschriften, series 28, vol. 42), Frankfurt am Main.

Frandsen, K-E. 1983a, "Danish field systems in the seventeenth century", *Scandinavian Journal of History* 8, 293–317.

—— 1983b, *Vang og tægt: studier over dyrkningssystemer og agrarstrukturer i Danmarks landsbyer 1682–83*, Esbjerg.

Fritzbøger, B. 1994, *Kulturskoven: Dansk skovbrug fra oldtid til nutid*, Copenhagen.

Gissel, S., Jutikkala, E., Österberg, E., Sandnes, J. and Teitsson, B. 1981 (eds), *Desertion and land colonization in the Nordic countries c. 1300–1600*, Stockholm.

Götlind, A. 1993, *Technology and religion in medieval Sweden*, Falun.

Grandjean, P. B. 1953, *Slesvigske købstæders og herreders segl indtil 1660*, Copenhagen.

Hagar, H. 1982, "Seldon". In *Kulturhistorisk leksikon for Nordisk middelalder 15*, eds A. Karker, H. Pohjolan-Pirhonen, J. Benediktsson, M. Larusson, F. Hødnebø, and J. Granlund, Copenhagen, 106–17.

Haupt, H. 1995, "Der Wagen im 14. Jahrhundert". In *Beiträge und Katalog zur Ausstellung: aus dem Leben gegriffen: ein Rechtsbuch spiegelt seine Zeit* 2, ed. M. Fansa, Oldenburg, 155–61.

Hedeager, L. and Kristiansen, K. 1988, "Oldtid o. 4000 F. Kr–1000 E. Kr." In *Det Danske landbrugs historie*, vol. 1, *Oldtid og middelalder*, ed. C. Bjørn, Odense, 11–203.

Hoff, A. 1984, "Middelalderlige gærder og hegn: ældre og yngre dyrknings-systemer i Jyske Lov", *Fortid og nutid* 31, 85–102.

—— 1990, "På sporet af vikingetidens landbrug?", *Bol og By: Landbohistorisk Tidsskrift, 1990*, 7–49.

Hofmeister, A. E. 1975, "Die Organisation de hochmittelalterlichen Bin-nenkolonisation in den Marschhufensiedlungsgebieten an Weser und Elbe unter besonderer Berücksichtigung der Stader Elbmarschen", *Berichte zur Deutschen Landeskunde* 49, 107–20.

Holm, P. 1988, "Kampen om det som ingen ejer: Om rettighederne til den øde jord indtil 1241 som baggrund for den tidlige middelalders bondeuro". In *Til kamp for friheden: sociale oprør i nordisk middelalder*, eds A. Bøgh, J. Würtz Sørensen, and L. Tvede-Jensen, Århus, 90–108.

Hübertz, J. R. 1845, *Aktstykker vedkommende staden og stiftet Aarhus*, vol. 1, Copenhagen.

Hybel, N. 1996, "The creation of large-scale production in Denmark *c*. 1100–1300", *Scandinavian Journal of History* 20, 259–80.

Ingesman, P. 1990, *Ærkesædets godsadministration i senmiddelalderen*, Lund.

Jöns, H. 1992–3, "Zur Eisenverhüttung in Schleswig-Holstein in vor- und frühgeschichtlicher Zeit", *Offa: Berichte und Mitteilungen zur Urgeschichte, Früh-geschichte und Mittelalterchäologie* 49–50, 41–55.

Krenzlin, A. 1952, *Dorf, Feld und Wirtschaft der grossen Täler und Platten östlich der Elbe*, Remagen.

Kühn, H. J. 1992, *Die Anfänge des Deichbaus in Schleswig-Holstein*, Neumünster.

Kühn, H. J. and Panten, A. 1989, *Der frühe Deichbau in Nordfriesland. Archäo-logisch-historische Untersuchungen*, Bredstedt.

Langdon, J. 1986, *Horses, oxen and technological innovation: the use of draught animals in English farming from 1066 to 1500*, Cambridge.

Langebek, J. (ed.) 1776, *Scriptores rerum Danicarum medii aevi* 4, Copenhagen.

Lerche, G. 1981, "Additional comments on the Lindholm Høje field", *Tools and Tillage* 4, 110–16.

—— 1982, "A Viking harrow down a well", *Tools and Tillage* 4, 183–91.

—— 1994, *Ploughing implements and tillage practices in Denmark from the Viking period to about 1800 experimentally substantiated*, Herning.

Lerche, G. and Jensen, S. 1985, "A note on farming practice in the Viking period: sod manure (*træk*)", *Tools and Tillage* 5, 122–5.

Lohrmann, D. 1995, "Von der östlichen zur westlichen Windmühle: Beitrag zu einer ungelösten Frage", *Archiv für Kulturgeschichte* 77, 1–30.

Lorenzen-Schmidt, K.-J. and Poulsen, B. (eds) 1992, *Bäuerliche Anschreibebücher als Quellen zur Wirtschaftsgeschichte* (Studien zur Wirtschaft- und Sozial-geschichte Schleswig-Holsteins 21), Neumünster.

Lütken, V. 1909, *Bidrag til Langelands historie*, Rudkøbing.

Madsen, P. K. 1980, "Medieval ploughing marks in Ribe", *Tools and Tillage* 4, 36–45.

—— 1988, "Mølledrift og mølletvang i den tidlige middelalders Odense", *Land og by i middelalderen* 4, 45–58.

Molbech, C. and Petersen, N. M. (eds) 1858, *Udvalg af danske diplomer og breve*, Copenhagen.

Møller, P. G. 1990, "Højryggede agre: forskning og bevaring", *Bol og By: Landbohistorisk Tidsskrift, 1990*, 90–118.

Van Mourik, J. M., 1990, "Spuren von Plaggenlandbau im Gebiet der Schleswiger Landenge", *Offa: Berichte und Mitteilungen zur Urgeschichte, Frühgeschichte and Mittelalterarchäologie* 47, 169–76.

Myrdal, J. 1984, "The hayrake", *Ethnologia Scandinavica*, 25–33.

—— 1985, *Medeltidens Åkerbruk: agrarteknik i Sverige ca. 1000 till 1500*, Stockholm.

Nielsen, L. C. 1980, "Omgård: a settlement from the late Iron Age and Viking period in west Jutland", *Acta Archaeologica* 50, 173–208.

—— 1986, "Omgård: the Viking age watermill complex: a provisional report on the excavations in 1986", *Acta Archaelogica* 57, 177–204.

Nielsen, O. (ed.) 1869, *Indtaegtsangivelser og kirkelige vedtæger for Ribe domkapittel og bispestol kaldet "Oldemoder"* (Samling af Adkomster), Copenhagen.

Olrik, H. (ed.) 1968, *Danske helgeners levned*, vol. 2, Copenhagen.

Olsson, S.-O. 1995, *Medeltida danskt järn: framställning av och handel med järn i Skåneland och Småland under medeltiden*, Halmstad.

Porsmose, E. 1988, "Middelalder: o. 1000–1536". *Det danske landbrugs historie*, vol. 1, *Oldtid og middelalder*, ed. C. Bjørn, Odense, 207–417.

Poulsen, B. 1985, "Korn eller kvæg: landbrugets specialisering i senmiddelalderen belyst ved studier på Stevns og i Odsherred", *Bol og By: Landbohistorisk Tidsskrift, 1985*, 7–20.

—— 1988, *Land By Marked: to økonomiske landskaber i 1400–tallets Slesvig*, Flensburg.

—— 1990, *Bondens penge: studier i Slesvigske regnskaber 1400–1650*, Odense.

—— 1991, "Den sønderjyske herredsfoged i senmiddelalderen: et herredsfogedregnskab fra Sønder Gos herred 1474–75", *Sønderjyske Årbøger 1991*, 73–86.

—— 1993, Tjenestefolk på landet i reformationstidens Sønderjylland, *Bol og By: Landbohistorisk Tidsskrift, 1993*, 7–37.

Poulsen, B. and Pedersen F. S. (eds) 1993, "Regnskabet for Ribebispens gård Brink 1388–89", *Danske Magazin* 6th series, vol. 3, 316–36.

Reinhardt, W. 1984, "Zum frühen Deichbau im niedersächsichen Küstengebiet", *Probleme der Küstenforschung im Südlichen Nordseegebiet* 15, 29–40.

Roesdahl, E. 1980, *Danmarks Vikingetid*, Copenhagen.

Rösener, W. 1986, *Bauern im Mittelalter*, München.

Roth, H. and Wamers, E. 1984, *Hessen in Frühmittelalter. Archaeologie und Kunst*, Sigmaringen.

Schuldt, E. 1980, "Eine Egge des 10. Jahrhunderts aus der slawischen. Siedlung von Großraden, Kreis Sternberg", *Bodendenkmalpflege in Mecklenburg, 1980*, 203–7.

Steensberg, A. 1943, *Ancient harvesting implements: a study in archaeology and human geography*, Copenhagen.

—— 1952, *Bondehuse og vandmøller i Danmark gennem 2000 år*, Copenhagen.

—— 1971, "En skvatmølle i Ljørring", *Kuml*, 130–45.

—— 1977, "*Sula*: an ancient term for the wheel plough in northern Europe?", *Tools and Tillage* 3, 91–8.

—— 1981a, "Greb". In *Kulturhistorisk leksikon for Nordisk middelalder* 5, eds A. Karker, H. Pohjolan-Pirhonen, J. Benediktsson, M. Larusson, F. Hødnebø, and J. Granlund, Copenhagen, 452–4.

—— 1981b, "Plov". In *Kulturhistorisk leksikon for Nordisk middelalder* 13, eds A. Karker, H. Pohjolan-Pirhonen, J. Benediktsson, M. Larusson, F. Hødnebø and, J. Granlund, Copenhagen, 330–42.

Steenstrup, J. C. H. R. 1873, *Studier i Kong Valdemars Jordebog* 1, Copenhagen.

Stoklund, B. 1990, "Tørvegødning: en vigtig side af hedebondens driftssystem", *Bol og By: Landbohistorisk Tidsskrift, 1990*, 47–72.

Stoumann, I. 1978, *De der blev hjemme: en vikingelandsby ved Esbjerg*, Esbjerg.

Trier, B. (ed.) 1993–4, *Zwischen Pflug und Fessel: mittelalterliches Landleben in Spiegel der Wüstungsforschung: Begleitbände zur Wanderausstellung des Westphälischen Museums für Archäologie*, Münster.

Vejbæk, O. 1984, "Hus og ager: højryggede under en bebyggelse fra 1100-årene syd for Filsø i ål sogn", *Mark og Montre 1984*, 49–58.

Vensild, H. 1970, "Navndrupåsen: et bidrag til hjulplovens historie", *Bidrag til Viborgegnens topografi og historie* 1, Viborg, 52–70.

Wegener, C. F. (ed.) 1871, *Aarsberetninger fra det kongelige Geheimearchiv* vol. 5, Copenhagen.

Zangenberg, H. 1928, "En lollandsk Gårds Inventar i 1505", *Lolland-Falsters historiske Samfund, Aarbog 16*, 36–50.

7. THE AGRICULTURAL TRANSFORMATION OF SWEDEN, 1000–1300

Janken Myrdal

The Problem

Ever since Georges Duby and Lynn White Jr. formulated the concept of a medieval "agricultural revolution" in the early 1960s, it has been widely accepted that a profound agricultural change occurred between 1000 and 1300. Today we realize that the change in many respects started well before 1000 (see for example Raepsaet, this volume), but nonetheless the period from the tenth to the thirteenth centuries is still regarded as a remarkable period of growth, not only in terms of population and economic activity, but also in the development and spread of agricultural techniques.[1] In Sweden the pattern applied, but on a smaller scale and with a certain time lag. Towns, castles and churches increased in number during the twelfth and thirteenth centuries, and Stockholm rose slowly as a major town in the thirteenth century; although its size (c. 4000 in the late fifteenth century) was always small compared to other European capitals, it far outstripped other towns in the country. In terms of the level of material culture stone cathedrals (as at Uppsala) and churches began to make their appearance throughout the thirteenth century.

A stronger central authority was established with a decisive civil war ending in 1247, after which taxation of the peasantry in particular was introduced. Nevertheless the peasantry managed to maintain a certain degree of freedom and social status. At the end of the middle ages, half of the farming acreage was held by peasants, and serfdom was never introduced into the country.

To the end of the thirteenth century much of this economic change in Sweden was underpinned by an intensification of productivity in the countryside through an increase in total food production. But was this associated with a change in agricultural technology? For

---

[1] The early diffusion of watermills is emphasized as an important invention which, however, did not improve cereal yields.

example, elements of the so-called agricultural revolution in other parts of Europe at the time, such as the heavy wheel plough and the three-field rotation, scarcely spread to Sweden at all.

Methodology

Written evidence is sparse in Sweden. Medieval accounts are rare and mainly survive from the end of the middle ages. The eight regional law codes and one national law code are important sources for the thirteenth and fourteenth centuries. There are in addition about twenty thousand diplomas from the late twelfth to early fifteenth century, but only a few hundred of them (mainly inventories) directly deal with agrarian techniques. Most surviving examples of Swedish medieval literature are translations of texts from other countries, but in those collections of miracles which concern local saints some concrete details are mentioned in passing. Still more rewarding are the few treatises on agriculture, but the earliest in Sweden date from the early sixteenth century.

Without extensive archives, it is necessary to exploit a combination of different sources. Research on medieval Swedish agriculture has to lean heavily on archaeological evidence, especially for the period before the thirteenth century. This is not necessarily a weakness, because archaeology gives information of a different character from the written sources. Implements, for example, can be studied in detail. However, most of the preserved finds are of iron, whereas the majority of the implements were made of wood. While excavations in towns have yielded substantial numbers of wooden artefacts, those in the countryside have produced very few. Thus, in order to collect enough evidence to discuss wooden implements, it is necessary to draw on examples from a wide geographical area, which will blur regional differences. The identification of broken pieces of wood is also difficult as the material culture has immense variety. Such identification therefore requires an intimate knowledge of the whole range of wooden artefacts which have so far come to light, as documented, for example, in nineteenth-century ethnological sources.[2]

Used with caution, illustrations are an additional source. One of the main advantages of illustrations is that they depict whole imple-

---

[2] See introduction in Szabó et al. 1985, 2–6.

ments, in contrast to the partial survival in the archaeological record.[3] Hundreds of churches in Sweden contain medieval wall paintings, especially from the fifteenth century. Biblical and other religious scenes were depicted in a contemporary setting and the painters were often artisans with a knowledge of the region. However, the painters are known to have used European prototypes for their pictures, such as *Biblia Pauperum*. It is possible to study this material in two different ways. One approach is to compare the paintings with European examples to detect possible prototypes. The second approach is to relate the paintings to the later local material culture in order to identify local details. In figure 7.1, for example, I have compared

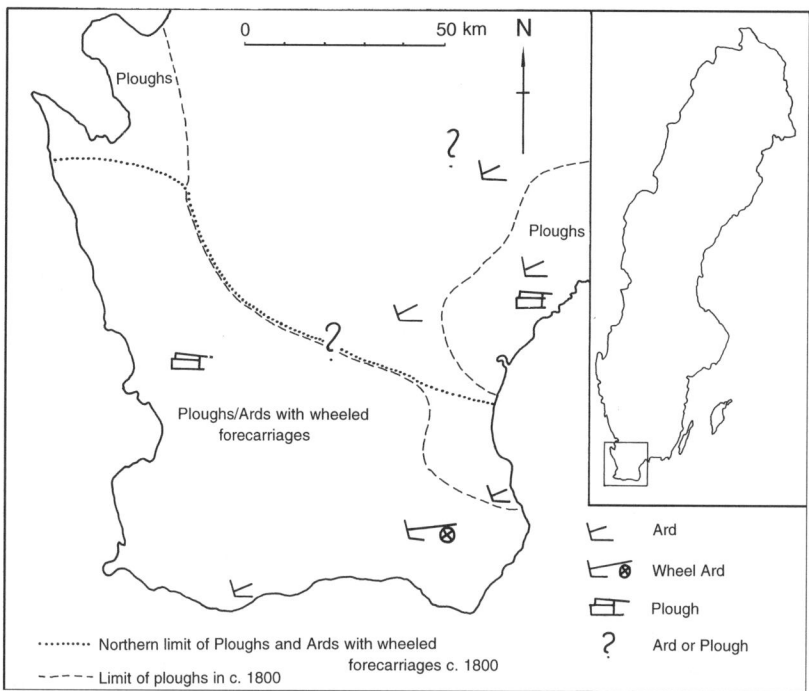

Figure 7.1. Plough and ard as shown in wall paintings in Scania from the late fifteenth to sixteenth centuries compared to the spread of the wheel ard and wheel plough in *c.* 1800. The ard without wheels was used in most of Scania, alongside the other two implements, in *c.* 1800 (Myrdal 1985).

---

[3] The Institut für mittelalterliche Realienkunde, in Krems, Austria, has collected and registered a huge amount of material culture depicted in medieval illustrations. The work of the institute is regularly presented in their publication series Medium Aevum Quotidianum.

different plough types in wall paintings from the late fifteenth and
early sixteenth century in Scania (nowadays in southern Sweden),
with the real distribution of the same types in the eighteenth century. The two distributions correspond; in those areas where wheel
ards were depicted, the implement was also used three hundred years
later, and ploughs were not represented in those areas where ards
later dominated.

After the Reformation wall paintings were executed less frequently,
and they become less useful because the subjects were derived from
prototypes which became more accessible with the wider distribution
of prints. There are, however, other types of representation, such as
those found on seals and gravestones.[4] In the sixteenth century every
county in Sweden was required to have a seal in order to underwrite
decrees from the state. Many counties chose as a motif something
that was typical of the region: coastal counties had fishes or ships,
while inland counties on the plains used ears of corn or implements.

The paucity of the written documents makes any discussion of the
relative role of different social classes in technological development
extremely difficult.[5] Perhaps in future a comparison of excavated
assemblages from monasteries, castles and villages could help such
an enquiry, but to come to some reliable conclusions the excavations
should preferably be from the same region and same period.

Even with the combination of different sources the data base remains limited, especially as it is necessary to concentrate on high-
quality evidence. Well-dated archaeological implements, for example,
run into the tens rather than hundreds. Imprecisely dated archaeological finds are of little use for a discussion of technological change,
and of course it is possible that an artefact could have existed centuries before its first documented occurrence. It is also important to
emphasize that the first appearance of a new technique is of less
importance than its general acceptance, which means that it is crucial, for example, to indicate whether an artefact is scarce or plentiful in archaeological assemblages. Given all these limitations, it is
often difficult to measure technological diffusion within time periods
of less than two or three hundred years.

---

[4] Alexander Fenton has used illustrations on Scottish medieval gravestones to discuss
ploughs, but few tombstones in Sweden depict agricultural implements, an exception
being the fifteenth-century stone from Vadstena, Östergötland, which depicts an
ard. Fenton 1970; Myrdal 1985, 85.
[5] Götlind 1993, 62 who argues that the role of the Cistercians as innovators in
agriculture has been overemphasized.

Technological Complex

Nonetheless, some reconstruction of the history of Swedish agricultural technology between the Birth of Christ and the eighteenth century is possible. Technological development rarely stopped completely, but was characterized by alternating periods of slow and rapid change.[6] It is also clear that there was an interdependence between several innovations which were introduced during periods of profound change.[7] The major transformations were thus characterized not only by a faster adoption of particular innovations, but also by the introduction of a new technological complex.

Part of this reflects the complicated nature of activities like farming, where the work process is broken down into several interdependent stages. Harnessing, for example, influences ploughing, which is associated with fallow-breaking which in its turn is connected with manuring. Thus, a single innovation makes little sense unless viewed in the context of the whole operation or as a part of a whole complex of technology.

In technological complexes, then, different techniques support the existence of others so a whole cluster of directly or indirectly interdependent techniques can be recognized. The complex will favour technological innovations that suit, or can be accommodated into, the system, and reject others. An innovation can thus exist for a long time before it becomes widely adopted, and the establishment of a new complex is preceded by the existence of under-used techniques which will have their day later. When a new technological complex breaks through, innovations of a different kind will also spread in a rather short time: for example, the diffusion of existing but unexploited techniques; or the introduction of techniques from other regions; or the invention of a new technique; or more often a combination of these alternatives.

The concept of a technological complex also assumes that there is a technological progress in pre-industrial agriculture.[8] Agrarian technology follows a broad path of necessity, where some innovations have to develop before others can get a chance to spread. The limited number of solutions to technical problems in pre-industrial production

---

[6] Myrdal 1985, 156–8; Myrdal and Söderberg 1991, 430–2.
[7] Myrdal 1988a.
[8] Persson 1988, 1–13.

has been described by André Leroi-Gourhan in his encyclopaedic examination of technology.[9] It also implies that the number of combinations is limited, which is one of the basic ideas behind the concept of a technological complex.

There are, however, no absolute connections between the different components of the complex. Different innovations came together in new combinations fairly frequently. Thus, in order to understand a technological complex it is necessary to bear in mind both its internal coherence and its fluctuating boundaries. This meant that the development of pre-industrial agrarian technique was not characterized by a smooth advance, where one innovation followed on from another, but more often by a change in tempo. The process can partly be understood as the result of the introduction and subsequent stabilization of technological complexes.

The concept is neither new nor, of course, restricted to the history of agrarian technique. In his book about horses and oxen in the middle ages, John Langdon discusses a "technological package" in much the same sense as I use "technological complex".[10] Paul David used a similar idea, of "technological interrelatedness", in his discussion of the introduction of the harvester in nineteenth-century England.[11]

Historians of modern technology often refer to "technological systems" which is a slightly different concept because they often combine social structure with technology, and as a synonym "socio-technological systems" is sometimes used. Another difference is that a "technological system" is a more or less physically integrated system working as a whole, as for example the electrification of a country.[12] In pre-industrial societies the farm, and in certain respects the village community, could be described as a technological system. The technological complex, however, consists of small units in which more or less the same technological connections are repeated. The "technological system" is perhaps a concept that is more appropriate for the analysis of a modern industrialized society with a high division of labour and where different branches develop, which is combined with a national tech-

---

[9] Leroi-Gourhan 1971, 1973. Correlations with later research are presented in Lemonnier 1992.
[10] Langdon 1986, 4, 16.
[11] David 1971, 155–6.
[12] Hughes 1987; Joerges and Braun 1994, 7–29.

nical integration in every branch. Therefore different branches must be described separately and be treated as national or supra-national organizations.

One Element in a Complex: the Plough

The technological complex is a useful way to conceptualize technological change. The complex is often composed of several elements, and it is these, no matter how apparently insignificant, which go to make up the whole complex. As an example of how different sources can be used to analyse the change of one important part of a technological complex, the history of medieval Scandinavian ploughing implements will be presented. The change actually involved different components; one was the plough replacing the ard, others were changes in the details of plough and ard. The plough had a mould-board to break and turn the soil, the ard merely tore up the earth without turning it. Research about the plough has been immense, and much of it has concentrated on the spread of the plough and how it replaced the ard (see Raepsaet in this volume for the early introduction of the plough in the northern Roman provinces). I will mainly concentrate on the east-central parts of the country (from Uppland to Östergötland), the region for which we have the best sources (see figure 7.2).

The evidence from excavated fields shows that the plough was introduced into southern Jutland in about AD 800–900, and then it spread to the whole of Denmark including Scania (then a part of Denmark, nowadays southern Sweden) by the twelfth century. Four preserved parts of the wooden frames from ploughing implements survive from the thirteenth to the sixteenth centuries and all were from ploughs of a rather heavy construction. A plough beam from a wheel plough, dated to the thirteenth century, also survives: it is probably this type of plough which became dominant in the region and which also spread to Denmark. It was certainly introduced from Germany.[13]

---

[13] Lerche 1994, 21–2, 149.

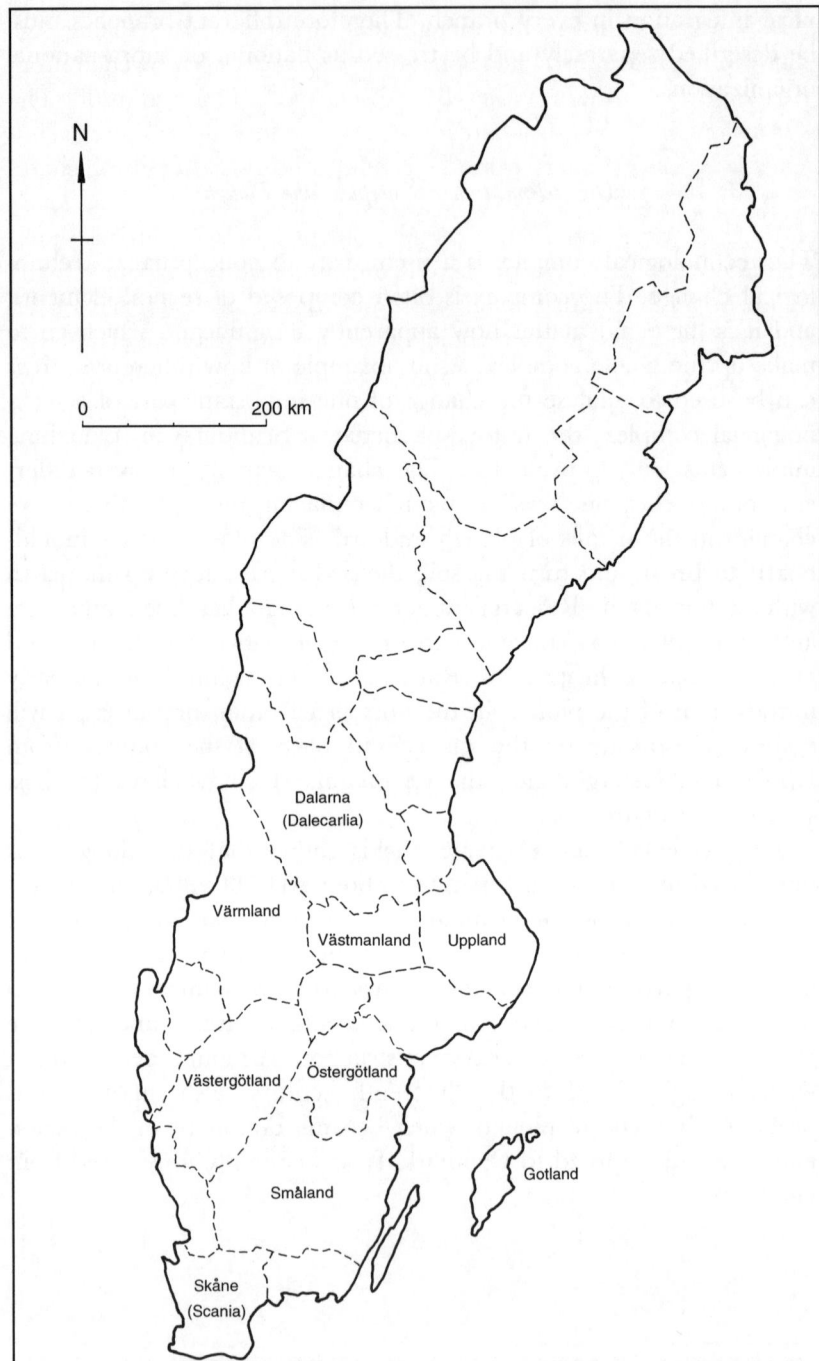

Figure 7.2. The provinces, "*landskap*", of Sweden discussed in the text.

In southern Norway the plough was adopted in the Viking Age (ninth to tenth centuries) and spread to parts of western Sweden (from Värmland to northern Västergötland) between 1000 and 1200. The word plough, "plog", was already in use in Norway by the eleventh century and in Denmark from the twelfth century. Several iron shares from Norway have survived from before 1000, and all are socketed and symmetrical, but even a plough can have a symmetrical share. A coulter has also been preserved, but it is insecurely dated to the eighth or ninth century, although some think it need not have been attached to a plough, but could be a separate implement (a "rist" or "ristle"), used to cut the turf before ploughing.[14] However, furrows made by a plough with a mouldboard have been identified in an excavation of a Viking-Age field in north Norway.[15] In western and central Norway written documents indicate that the ard was the dominant implement, but a thorough history of the plough in Norway has yet to be written.

The typology of the western Scandinavian plough is first identified on fifteenth- and sixteenth-century seals, and it is clear that foot ploughs were used instead of wheel ploughs; indeed it is doubtful if the heavy wheel plough ever existed in these regions. The ethnologist Ragnar Jirlow noted several similarities of detail between the Norwegian and the western Swedish ploughs in the eighteenth and nineteenth centuries and argued that these indicated a long period of interaction.[16] The early Norse plough probably did not have a wheeled forecarriage. The preeminence of the plough without a wheeled forecarriage is seen elsewhere, especially in northern England from at least the thirteenth century.[17]

On the basis of linguistic evidence Lynn White suggested that the wheel plough was introduced from Denmark to England in the late ninth and early tenth centuries.[18] However, the diffusion could have been in the opposite direction. From a typological point of view, the connection would have been between northern England and Norway.

---

[14] Veibaek 1974, 35–6; Søvberg 1976, 91–2.
[15] Pers. comm. Girth Lerche.
[16] Jirlow 1970, 50–2.
[17] Langdon 1986, 128–41; Leser 1931, 159.
[18] White 1962, 51–4. Puhvel 1964, 180–1 gives some support to White, but this field of research is extremely difficult.

In both Denmark and southern Norway the ard continued to be used alongside the plough until the late middle ages, and in certain regions beyond.

In the rest of Scandinavia the ard was the only ploughing implement. It is the ard and not the plough which is mentioned in regional laws and diplomas of northern Sweden until the fourteenth century, and this is confirmed by three preserved wooden frames of ards of *c.* 1300 from Dalarna. However, the plough is mentioned in late medieval documents, especially the inventories, and on the seals of these regions the plough is also depicted from the fifteenth and sixteenth centuries. The plough had a second phase of adoption in the fifteenth and sixteenth centuries when it spread into and then became the dominant implement in northern Sweden (from Dalarna (Dalecarlia) northwards) and also perhaps in central Norway (the Tröndelag). By contrast the ard continued to be the most important implement in a considerable part of eastern and southern Sweden and in Finland until the early nineteenth century (figure 7.3).

Archaeological finds and illustrations are the major sources for the types of ards and ploughs because the written sources do not normally distinguish between different types. The six surviving wooden plough frames from Sweden and a few illustrations indicate that for the twelfth to the fourteenth centuries the dominant types were the bow ard and the quadrangular ard or plough.[19]

A new construction was introduced in the fourteenth and fifteenth centuries, the "high ard" and "high plough". The plough frame was constructed in the same way for both ard and plough. Both types had a crossbar going from the handle to the front brace (figure 7.3) on which the ploughman could lean in order to press the implement into the soil. The high plough is peculiar to northern and central Scandinavia, and was an indigenous invention. Several medieval wall paintings and early modern seals also show this peculiar type of implement, which demonstrates the local adaption of illustrations.

The high plough could be used with only one draught animal, and could thus be described as a light plough. An interesting fact is that both the plough and the high-construction were introduced and spread at the same period in northern Scandinavia, and it was the high plough and not the quadrangular plough which replaced the ard in the north; and this may be explained by the way draught

---

[19] Myrdal 1985, 76–87.

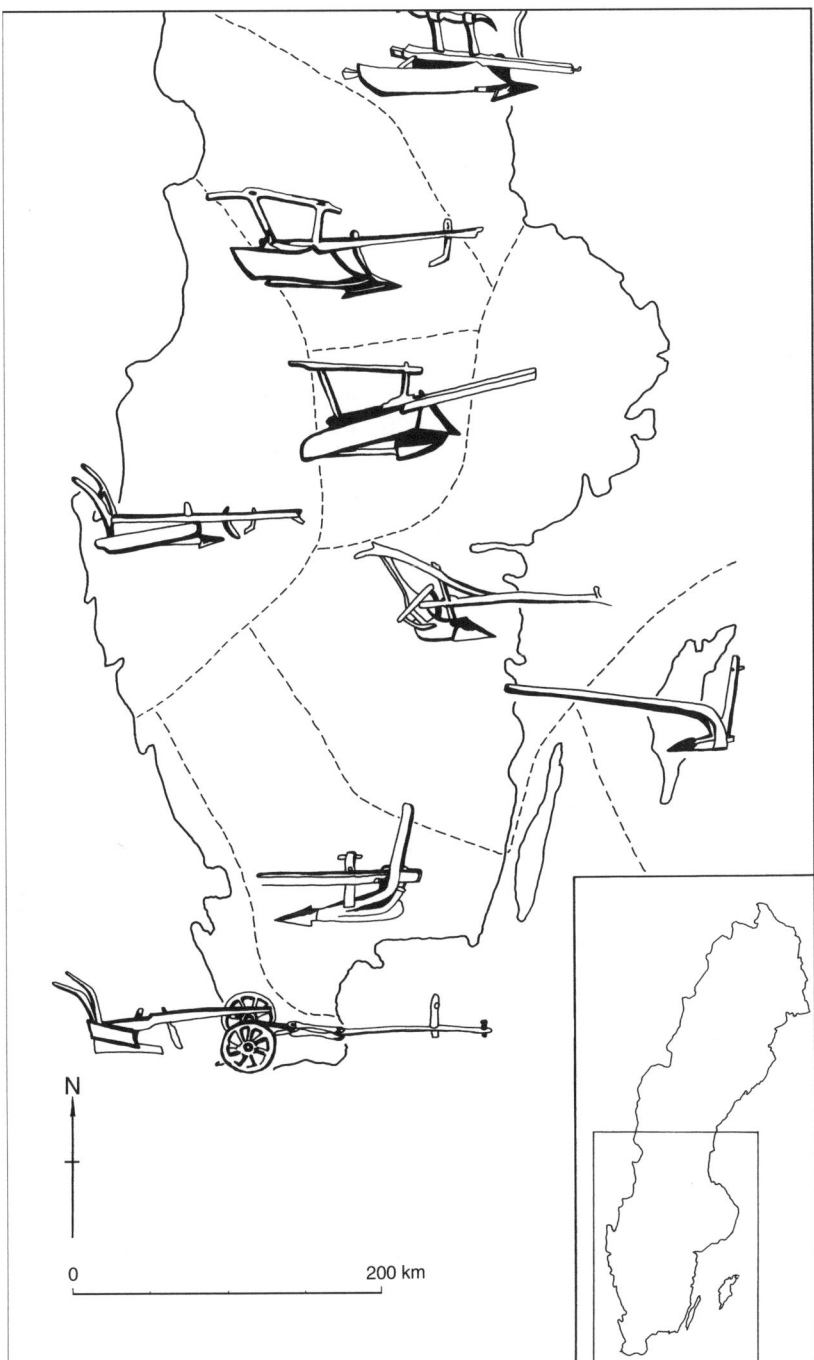

Figure 7.3. A simplified map of plough- and ard-types in eighteenth-century Sweden, with the three different types of high plough in the north, and with the high ard in the east (except on the island of Gotland, where the bow ard was common). The quadrangular plough was common in the west, and in the south the quadrangular ard. In the most southern parts of Sweden the wheel plough and wheel ard were used.

animals were used. In northern Sweden where small farms dominated, the peasants normally had only one horse for draught purposes.
The harnessing technique had also been adapted to this one-horse
system, and thus a plough that required two draught animals would
not have been adopted. Not until the invention of the high plough,
which only required one draught animal, did the plough spread in
this area. In eastern Sweden the ard of high type became dominant,
but this high ard normally had two draught animals.

One of the most important parts of the implement was the share.
We have nineteen well-dated iron shares from the second half of the
first millennium until about 1500 in Sweden (including Scania). The
shares are from different regions in southern and central Sweden
and regional differences have not been identified.

Further information comes from sixteenth-century accounts which
record work done in the smithies and give details about the amount
of iron used to forge shares. A third source is the illustrations of
shares, especially on seals. All the surviving shares are socketed, and
the illustrations also show socketed shares.[20]

The earliest iron shares, from the fifth to tenth centuries, have a
weight of about 0.3 kg (five examples). By the eleventh and twelfth
centuries there were two types, the early small share continued but
there was also a larger type with a weight of about 1 kg (seven examples). From the thirteenth to the fifteenth centuries there are still
a few rather small shares with a weight of about 0.5 kg, but most
of the shares are 1–1.5 kg (seven examples). The individual weight
of shares recorded in the sixteenth-century accounts can be estimated
to have been over 2 kg.[21]

Although the sample is small, there was nevertheless an increase
in the weight of shares which can be explained mainly by a change
in the shape of the share as indicated by the seals and archaeological finds. After AD 1000 shares were lengthened and thus increased
in weight. The earliest iron shares were small tips of iron, barely
100 mm long. The heavier, 1 kg-type, was normally 200–300 mm
long. Most of the archaeological finds are worn and thus have been
shortened by use, but as the shares from both before and after 1000
are all rather heavily worn the difference in length cannot be explained
this way. A second increase in the weight of shares occurred in the

---

[20] Myrdal 1985, 76–98; Myrdal and Söderberg 1991, 381–401.
[21] Myrdal 1985, 89–91; Myrdal and Söderberg 1991, 396–401.

sixteenth century when the shares were made broader to produce the "winged share".

The sources are generally silent about the methods of ploughing, but there are a few hints. The earliest written sources from the thirteenth and fourteenth centuries distinguish between shares to break the fallow and shares to cover the seed. This certainly reflects a difference in size: the larger ones were used to break the fallow, which implies that a separate fallow-breaking ard was introduced with the longer shares.

The number of ploughings has been considered an important aspect of medieval farming practice.[22] Written documents are the main source for such information, but only for the thirteenth century and later in Sweden. Two thirteenth-century law codes from central Sweden (the laws of Uppland and Västmanland) stipulated how a tenant should hand over his farm to the next tenant or to the landowner and indicate that it was normal to plough the fallow twice in the autumn before the spring sowing. In the sixteenth century, according to the agricultural treatises and some manorial accounts, three fallow ploughings seem to have been common in Uppland. Increased ploughing was also associated with a change from autumn to summer ploughing of the fallow. This was probably connected with a growing importance placed on winter rye at the expense of barley, which was previously the main crop. A change from spring to early autumn sowing is facilitated by a clean fallow in the late summer, which allows an earlier sowing and gives the winter seed a higher yield.

The New Technological Complex 1000–1300

The colonization of land was at the heart of the agrarian transformation. Pollen analysis combined with archaeological remains indicate that land reclamation started, at least in Scania and Uppland, as early as the eighth and ninth centuries.[23] These results accord with recent research outside Scandinavia which emphasizes a similarly early

---

[22] Parain 1966, 151–3; Duby 1968, 22–3, 104–5. Dyer notes that the thirteenth-century agricultural treatises recommended more ploughing of the fallow than is actually recorded in manorial accounts (1980, 126–7).

[23] Berglund 1990, 112, 172, 224, 248, 431; Broberg 1990, 84–97.

start for the expansion.[24] Although regional differences would be expected, the evidence for a detailed regional and chronological study of the process does not exist. Expansion of cultivation continued until the thirteenth century, and this late phase can be traced all over the country.[25]

An important component in the new technological complex was the iron-shod spade. Excavations of sites from the first millennium AD have produced many iron implements, but no spade socket of iron. The first iron-shod spades in Scandinavia occur in southern Denmark and date from the tenth to eleventh centuries. In Sweden such spades are both depicted in illustrations and found in archaeological excavations from the twelfth to thirteenth centuries. The iron spade is well represented in all source materials from the fourteenth and fifteenth centuries.[26] The iron-shod spade made it easier to grub up compacted soil with many roots, and the spread of this implement was connected with the continuing reclamation of land.

The new spade type facilitated ditch digging, and the implement's adoption facilitated the creation of ditch systems in and around fields. The regional laws mention ditches and those from east-central Sweden specifically mention that cooperation was necessary to dig and maintain the ditch systems in the fields. Ditched fields have not been detected in the archaeological record before the eleventh century, although field systems dated between AD 500 and 1000 are rare. Ditch digging enabled both a better preparation of the soil and also lower and wetter land to be reclaimed and cultivated.

On the plains in western Sweden the spread of ploughs and ridged fields facilitated drainage, and so made drainage ditches less necessary. They existed in this region, but did not become common before the agrarian revolution in the nineteenth century. In the southern woodlands, in Småland, the light and sandy soils did not need ditching.

An important part of agrarian technology is the axe. A thorough investigation of the history of this implement in Sweden has yet to be made, but it seems that a specialized felling axe to cut down trees came into use in Scandinavia after the tenth century; a more precise dating is not possible.[27]

---

[24] Randsborg 1991, 179–83; Bois 1992, 95–116.
[25] Myrdal 1985, 52–3. In northernmost Sweden the colonization continued well into the first half of the fourteenth century: Wallerström 1995.
[26] Myrdal 1985, 108–12. In England and France the iron-shod spade had been in use since the Roman period: see Raepsaet, this volume.
[27] Schovsbo 1987, 53; Thålin-Bergman 1976.

If one part of the new technological complex was geared towards the clearance of land, another part was centred around ploughing implements and the cropping systems (see figure 7.4). The infield system, with annual cropping, was dominant during the first millennium AD to 800–1000. (For differences and connections between two- and three-course rotation and two- and three-field systems, see Widgren in this volume.) The two-field system, with fallow every second year, was gradually introduced into eastern Sweden in about 1000. Regional laws from Uppland and Östergötland indicate that this cropping pattern was the dominant system in about 1300. In large parts of the woodland regions of southern (Småland and southern Västergötland) and northern Sweden the infield system continued to be the main system during and after the middle ages, although in parts of Västergötland the three-field system spread from the thirteenth century.[28]

Annual cropping did not exclude fallow. Fields were left fallow for long periods before they were taken up again. The breaking of long-established fallow cannot be achieved by the use of ards with a wooden share, and probably not by ards with a share tipped with iron. The fields had to be broken by hand with hoes.[29] Where the two-field system was introduced, the area lying fallow increased as half was fallowed every year. A specialized fallow-breaking share, the longer

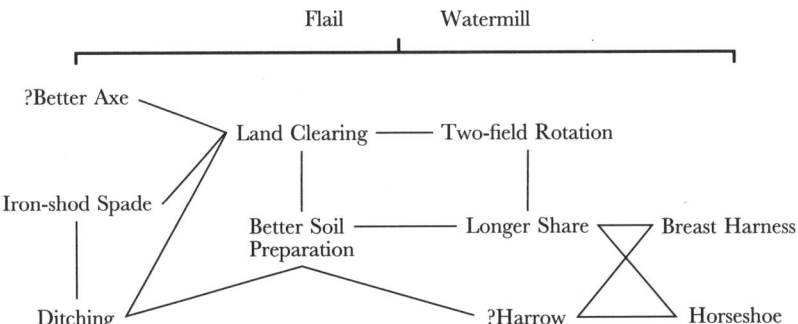

Figure 7.4. Some components in the technological complex used between 1000 and 1300 in east-central Sweden.

[28] This is similar to Denmark, including Scania, where it had spread from about the thirteenth century, but a break is more apparent during the fifteenth and sixteenth centuries: Frandsen 1983, 255–9, and English summary 260–8.
[29] Hansen 1969, 78–9, and Lerche 1994, 23. Experiments with small tips of iron have not yet been conducted.

iron share, was developed at about the same time. The long ard
share facilitated the introduction of the two-field system with regular
fallow. A similar connection between a change of cropping system
and a change of ploughing implement is also valid for Denmark.[30]

A regular fallow also had to be combined with a massive reclamation of land. The transition from an infield system to a two-field
system meant that, if the sown area was not to be reduced, then the
arable area had to be doubled. The change of course was not so
dramatic, but the introduction of the two-field system occurred con-
temporaneously with an expansion of the cultivated area. The two
main sectors of the new technological complex—clearance and cultivation—thus worked together.

It is even possible that the adoption of the harrow occurred at the
same time, but there is insufficient evidence.[31] In the northern provinces of the Roman Empire harrows dating to the first centuries AD
have been found. From about the same time large rakes found in
bogs in Jutland (in Thorsberg) have been interpreted as cultivation
implements, the forerunners of the harrow. Wooden artefact assemblages from bogs in Jutland, from marsh sites along the North Sea
coast (such as Feddersen Wierde and Elisenhof) and from castles in
eastern Germany (for example Gross Raden and Tornow) have to
my knowledge not yet produced any evidence of harrows from the
first millennium (but several ards or parts of ards have been found).
A closer survey of the actual finds could change that picture. The
earliest datable harrow in Scandinavia is Viking-Age and from southern
Denmark. From about the same time a proto-harrow, consisting of
a log with twigs, was found in the Ladoga region (in Aldeigjuborg).
The Scandinavian written sources often mention harrows in the twelfth
and thirteenth centuries.

If the harrow was adopted in Scandinavia around 1000, this innovation may have been connected with better horse management, as
horses were normally used for pulling harrows; but, as Georges Raepsaet
remarks in this volume, harrows could also be drawn by oxen.

An important element of agrarian change between 1000 and 1300
was the major improvement in horse harnessing. In the ninth to
tenth centuries the throat harness was replaced by the breast har-

---

[30] Lerche 1994, 295.
[31] Bentzien 1980, 35–8; Granlund 1961; Myrdal 1985, 99–102.

ness.[32] The archaeological record clearly shows that the iron horse-shoe was adopted in Scandinavia at the same time and this had an impact on the efficiency of farm horses as well as on transport and warfare.[33] The medieval development of yokes for oxen in Scandinavia is still uncertain as we lack finds.

As far as we can tell in Dalarna and the regions further north, horses were the sole draught animal (a famous example from northern Norway dates from the ninth century). In central and southern Sweden oxen were more common as draught animals. Early documents indicate an increased use of work horses in Uppland from the thirteenth century. By the sixteenth century, when more detailed information is available, horses had replaced oxen as the main draught animal as far south as Östergötland. The sixteenth century is also the first time when it is possible to compare peasant holdings with the demesnes, thereby giving the earliest indication of the social dimension to technology. Accounts from demesne agriculture can be compared with several wealth-tax registers, where the numbers of animals owned by the peasants were recorded. The small farmers often preferred horses, whereas the larger farms and especially the demesnes had more oxen. It is also significant that in the sixteenth century the ox was increasingly used in east-central Sweden; this trend is first recorded on the demesne estates.[34] John Langdon has shown a similar development in medieval England where the small farmers preferred horses, no doubt because the horse was more versatile which suited the small farmers who could not afford more than one or two draught animals.[35]

A recent Danish investigation of the archeological evidence for vehicles shows that wagons used for transportation changed to service an emerging market economy, while little happened to the farm wagons.[36]

Clearing or assarting, the better preparation of the soil by ditching and improved ploughing enabled a more efficient exploitation of the natural fertility of the soil. The loss of nutrients, especially nitrogen, was, however, replenished by the new fallowing arrangements

---

[32] Hagar 1970.
[33] Liestol et al. 1961.
[34] Myrdal and Söderberg 1991, 192, 197, 268.
[35] Langdon 1986, 48, 176–94.
[36] Schovsbo 1987, 142, 157.

which had a similar effect to manuring.[37] There was thus an important connection between better soil preparation and the change of
crop rotation. The whole complex of technological innovations could
thus cause a major increase in cereal yields.

We also find improvements in the processing of cereals. Most
important was the introduction of the watermill, from about the tenth
century in Denmark and slightly later in Sweden, followed by the
windmill in the thirteenth century in both countries.[38] Hand milling
was extremely labour-consuming, and the watermill made it possible
to cope with the increased grain production.

The flail, for threshing, also probably came into use at this period.
As it was wholly made of wood the evidence is slender, and one has
to look at northern Europe as a whole to get sufficient data. Excavated wood assemblages from northern Europe do not contain flails
before the tenth century, but between about 1000 and 1200 illustrations from all over northern Europe show flails. In Sweden archaeological finds, illustrations and written documents from the thirteenth
to fourteenth centuries all prove that flails were in use, but of a
more rudimentary type than those used later.[39]

One area of agrarian technology in Sweden which did not go
through a major change between 1000 and 1300 was the harvest.
During the second half of the first millennium the angled sickle was
in use everywhere in Sweden and still was common in the eleventh
and twelfth centuries. The bow sickle had spread at least to Gotland
during the thirteenth century, but the angled sickle was still used in
western Sweden and Östergötland. The bow sickle was not in general use over most of the country until the fifteenth century.[40]

The harvesting of cereals with a scythe did not become customary
until the sixteenth century, and then only in parts of eastern Sweden.
The general spread of this harvesting technique came much later, in
the eighteenth century. The scythe had been used for cutting hay
since the Birth of Christ.

---

[37] Shiel 1991, 63–4.

[38] See Ek 1963 for the written sources. In Denmark the watermill seems to have
been introduced earlier: Steensberg 1983, 101–9. Archaeological research has shown
that watermills were in use in Scania from the twelfth century: Andersson 1989,
283, 287. For a twelfth-century watermill used in iron production, see Magnusson
1995b, 57–61.

[39] Myrdal 1985, 136–8.

[40] Myrdal 1985, 120–8.

It is very difficult to analyse changes in the management of live-stock because it often involved a change in the way animals were handled rather than the employment of new implements. However, animal husbandry appears to have remained unchanged throughout the transformation of agriculture in Sweden; the transformation be-tween 1000 and 1300 primarily affected grain production. Cow-sheds and manuring had been in use for centuries. On the processing side the introduction of the plunge churn for butter making in northern Europe, including Sweden, between the ninth and the twelfth cen-turies was a major innovation.[41]

As is to be expected, these individual improvements tended to ar-rive at different times and to be adopted with varying speeds, while their penetration could vary tremendously from region to region. Some improvements had begun well before the period, and the tenth-century technological changes were preceded by land clearance, much of it simply reclaiming previously cleared land abandoned around the middle of the first millennium. From the tenth to the thirteenth centuries the new innovations began to arrive more quickly. Thus, for example, the introduction of the iron-shod spade and the longer ard share can be dated to the eleventh and twelfth centuries. The introduction of the two-field system to Sweden seems to have been about the same time, although less securely dated, and was more restricted, mostly to eastern-central Sweden. The introduction of the watermill and the improvements to horse harnessing seem to have come earlier, belonging to the tenth and eleventh centuries.

The period around the Birth of Christ experienced another tech-nological burst in Scandinavia, and also to a lesser degree in the six-teenth century, but these should be regarded as other technological complexes, with different components.[42]

Finally, a distinction should be made here between land productivity and labour productivity. The first was connected to the growth of population and the second to the growth of the non-agrarian sector of that population. Intensification of landuse in general also had an early beginning and provided the basis for improved food production and demographic expansion. Reclamation from forest, for example, or the conversion of pasture to arable or meadow should be seen as making non- or low-productive land more fruitful. Similarly, the

---

[41] Myrdal 1988b, 126–33.
[42] Myrdal 1988a.

improvements to ploughs and hand tools, such as the longer ard-share and the iron-shod spade, also contributed to land productivity.

On the other hand, improvement to labour productivity was just as important. The spread of horses as draught animals led to quicker ploughing and the introduction of water- and wind-powered milling facilitated much faster food processing. Thus the interlocking between land productivity and improving labour productivity could be a feature of the technological complex, even if a more highly labour-intensive agriculture and an increasing land productivity dominated the change. Regional differences should not be forgotten, and this change of tech-nology was especially valid for the plains in east-middle Sweden around the political centre of the country. Probably it implied a growing differentiation within this region, with a change to a more labour-intensive agriculture, while in other regions agriculture remained on a more extensive level.[43]

Swedish agrarian technology, then, went through a major change between the eleventh and the thirteenth centuries, reflecting the establishment of a powerful new technological complex. But this complex differed from others being established elsewhere. In northern France, for example, the heavy plough became the dominant imple-ment, while in Sweden the newly introduced plough only partially displaced the ard, which remained competitive due to such improve-ments as longer shares. In northern France, too, changes in ploughing implements were accompanied by the use of the three-field system, whereas in eastern Sweden the two-field system was preferred.

The merging of new and old techniques should be emphasized. Many of the new innovations spreading to Sweden between 1000 and 1300 had existed for centuries in southern parts of Europe, but other innovations had a particularly early start, the most important being the stabling of cattle which had been a fundamental technique of livestock breeding in the Low Countries and Scandinavia a mil-lennium before it spread to England and France around AD 1000.[44]

---

[43] For regional differences in England, see Campbell this volume.

[44] See, for instance, Chapelot and Fossier 1985, 104–5. Grenville Astill has re-marked that the first spread of byres and stabling of cattle in medieval England could have been restricted to plough animals (1988, 47, 55). The scythe had been in existence in England and France since the Roman period, and the cattle must have been kept outdoors but were given hay during the winter. However, stabling of cattle appears to be an underestimated element in agrarian change in England and northern France between 1000 and 1300. It probably led to more efficient manure collection which matched an improved ploughing and harrowing technique.

By widening the geographical perspective we can see many other technological complexes that emerged simultaneously to the ones in Sweden and France. In northeastern Europe, for example, the sokha, or "socha", a type of plough with two shares, was a central element in a technological complex in this part of Europe during the ninth century.[45] The drainage campaigns which were conducted along the southern North Sea coast from the Netherlands to Jutland formed another technological complex.[46] An important explanation for these variations was of course the different environmental conditions between Scandinavia in the north, the marshes along the North Sea coast and the plains in northern France, but this does not explain why the waves of technological change in Europe were contemporaneous.

To explain this correspondence in time but not in all the constituents of the different technological complexes, it is important to remember the floating borders of these complexes where there were opportunities to combine the different elements. But it was the central elements of the technical complexes that differed, and this has another important theoretical implication. It casts doubt on the technological complex as the sole cause of the transformation because it seems strange that several technological complexes, which differ in crucial aspects, appeared in different regions of Europe at the same time. It is tempting to look for general causes behind the shape of the different technological complexes, which means applying a wider perspective to the problem.

My conclusion is thus that the technological complex is an important explanation for agricultural transformation, but also that it must be seen in a wider context. The technological complexes in Europe differed partly because they were adapted to different ecotypes, and thus made possible a more intense exploitation of nature, which put medieval society as a whole on a larger scale. But the complexes were evolving within a new framework of economic and social conditions which were common to Europe as a whole.

I shall briefly touch on two examples from this wider context, as they affect Sweden. Change of social structure and change of non-agrarian production both had their own dynamic, and the interconnections with technological change should not be over-simplified.

---

[45] Smith 1977, 15–23; Chernetsov 1972, 41–3. Another eastern European complex introduced in this period in Bohemia outside the sokha region is described by Beranova 1984, 20–2.

[46] Bantelmann 1967; Besteman 1990, 111–17.

Different aspects, such as technology or the social structure, of a context have to be analysed separately to understand their inherent development (or continuity) and as a part of the whole to understand how they interact.

The first example concerns the rising iron production in Sweden during the eleventh to fourteenth centuries, which involved a change from the collection of bog iron to mining, and also the introduction of developed smelting techniques such as the blast furnace.[47] The enlarged production interacted with the introduction of new implements which required more iron. Iron shares, for example, became worn after a few seasons' use, and consequently their replacement must have consumed large quantities of iron in the agrarian society. The connection between iron production and a new agrarian technology was also a part of a more developed symbiosis between agrarian and non-agrarian production.

The other example is the village community, which became a more developed form of settlement from about AD 1000. Villages (and thus the related communal cooperation) had then existed for a long time, but are legislated for, with apparently new rules, in the regional laws from the twelfth and thirteenth centuries.[48] At least in parts of southern Scandinavia, the settlement form of the village became more stable during the same period.[49] The development of such communities was partly related to an intensification of landuse which included reclamations and the transformation of common land to private fields; it could also include the laying out of a network of ditches around the village and the creation of common fencing systems to define an enlarged arable area. These and other changes in agriculture demanded more well-defined property rights and new rules for cooperation.[50] The development of technological complexes cannot be divorced from these issues.

---

[47] Magnusson 1995a, 33.
[48] Hoff 1990, 42–7.
[49] For southern Scandinavia see Porsmose 1987, 224. For central Sweden see Windelhed 1995.
[50] Myrdal 1989.

Bibliography

Andersson, H. 1989, "Byn, huvudgården och kyrkan". In *By, huvudgård och kyrka*, eds H. Andersson and M. Anglert, Lund, 281–8.

Astill, G. 1988, "Rural settlement: the toft and the croft". In *The countryside of medieval England*, eds G. Astill and A. Grant, Oxford, 36–61.

Bantelmann, A. 1967, *Die Landschaftsentwicklung an der schleswig-holsteinischen Westküste dargestellt am Beispiel Nordfriesland: Eine Funktionskronik durch fünf Jahrtausende*, Neumünster.

Bentzien, U. 1980, *Bauernarbeit im Feudalismus: Landwirtschaftliche Arbeitsgeräte und-verfahren in Deutschland von des Mitte des ersten Jahrtausends u. Z. bis um 1800*, Berlin.

Beranova, M. 1984, "Types of Slavic agricultural production in the 6th–12th centuries AD", *Ethnologia Slavica* 16, 7–48.

Berglund, B. (ed.) 1990, *The cultural landscape during 6000 years in southern Sweden*, Lund.

Besteman, J. C. 1990, "North Holland AD 400–1200: turning tide or tide turned". In *Medieval archaeology in the Netherlands*, eds J. C. Besteman, J. M. Bos and H. A. Heidinga, Maastricht, 95–118.

Bois, G. 1992, *The transformation of the year one thousand: the village of Lournand from antiquity to feudalism*, Manchester.

Broberg, A. 1990, *Bönder och samhälle i statsbildningstid: en bebyggelsearkeologisk studie av agrarsamhället i Norra Roden 700–1350*, Uppsala.

Chapelot, J. and Fossier, R. 1985, *The village and house in the middle ages*, London.

Chernetsov, A. 1972, "On the origin and early development of the East European plough and the Russian sokha", *Tools and Tillage* 2:1, 34–50.

David, P. 1971, "The landscape and the machine: technical interrelatedness, land tenure and the mechanization of the corn harvest in Victorian Britain". In *Essays on a mature economy: Britain after 1840*, ed. D. McCloskey, Princeton, 145–205.

Duby, G. 1968, *Rural economy and country life in the medieval west*, London.

Dyer, C. 1980, *Lords and peasants in a changing society: the estates of the bishopric of Worcester 680–1540*, Cambridge.

Ek, S. B. 1963, *Väderkvarnar och vattenmöllor*, Stockholm.

Fenton, A. 1970, "The plough song: a Scottish source for medieval plough history", *Tools and Tillage* 1.3, 175–91.

Frandsen, K.-E. 1983, *Vang og taekt: Studier over dyrkningssystemer og agrarstrukturer i Danmarks landsbyer 1682–83*, Copenhagen.

Götlind, A. 1993, *Technology and religion in medieval Sweden*, Göteborg.

Granlund, J. 1961, "Harv", in *Kulturhistoriskt lexikon för nordisk medeltid*, vol. 6, Malmö, 239–40.

Hagar, H. 1970, "Seldon," in *Kulturhistoriskt lexikon för nordisk medeltid*, vol. 22, Malmö, 106–17.

Hansen, H.-O. 1969, "Experimental ploughing with a *døstrup* ard replica", *Tools and Tillage* 1.2, 67–92.

Hoff, A. 1990, "På sporet af vikingetidens landbrug?", in *Bol og by* 2, 7–49.

Hughes, T. 1987, "The evolution of large technological systems". In *The social construction of technological systems*, ed. W. E. Bieker, Cambridge Mass., 51–82.

Jirlow, R. 1970, *Die Geschichte des schwedischen Pfluges*, Stockholm.

Joerges, B. and Braun, I. 1994, "Grosse technische Systeme: erzählt, gedeutet, modelliert". In *Technik ohne Grenzen*, eds I. Braun and B. Joerges, Frankfurt am Main, 7–49.

Langdon, J. 1986, *Horses, oxen and technological innovation: the use of draught animals in English farming from 1066 to 1500*, Cambridge.

Lemonnier, P. 1992, *Elements for an anthropology of technology*, Michigan.

Lerche, G. 1994, *Ploughing implements and tillage practices in Denmark from the Viking period to about 1800 experimentally substantiated*, Copenhagen.

Leroi-Gourhan, A. 1971, *L'homme et la matière: évolution et techniques*, 2nd edn, Paris.

—— 1973, *Milieu et techniques: évolution et techniques*, 2nd edn, Paris.

Leser, P. 1931, *Entstehung und Verbreitung des Pfluges*, Münster.

Liestol A., Norberg, R. and Valonen, N. 1961, "Hestesko", in *Kulturhistoriskt lexikon för nordisk medeltid*, vol. 6, Malmö, 545–8.

Magnusson, G. (ed.) 1995a, *The importance of iron-making: technical innovation and social change*, Stockholm.

Magnusson, G. 1995b, "Järnmöllan i Tvååker—en teknisk innovation i Danmarks bergslag". In *Medeltida Danskt Järn*, ed. S-O. Olsson, Halmstad, 52–63.

Myrdal, J. 1985, *Medeltidens åkerbruk: agrar teknik i Sverige ca. 1000–1520*. Stockholm.

—— 1988a, "Agrarteknik och samhälle under två tusen år". In *Folkevandringstiden i Norden: en krisetid mellem ældre og yngre jernalder*, ed. U. Näsman, Aarhus, 187–226.

—— 1988b, "The plunge churn from Ireland to Tibet". In *Food and drink and travelling accessories*, eds A. Fenton and J. Myrdal, Edinburgh, 111–37.

—— 1989, "Jordbruk och jordägande: en aspekt av sambandet mellan agrarteknik och samhällsutveckling i äldre medeltid". In *Medeltidens födelse*, ed. A. Andrén, Lund, 35–49.

Myrdal, J. and Söderberg, J. 1991, *Kontinuitetens dynamik: agrar ekonomi i 1500-talets Sverige*, Stockholm.

Parain, C. 1966, "The evolution of agricultural technique". In *The Cambridge economic history of Europe 1: the agrarian life of the middle ages*, ed. M. M. Postan, 2nd edn, Cambridge, 125–79.

Persson, K. G. 1988, *Pre-industrial growth: social organization and technological progress in Europe*, Oxford.

Porsmose, E. 1987, *De fynske landbyers historie*, Odense.

Puhvel, J. 1964, "The Indo-European and Indo-Aryan plough: a linguistic study of technological diffusion", *Technology and Culture* 5.2, 175–91.

Randsborg, K. 1991, *The first millennium AD in Europe and the Mediterranean*, Cambridge.

Shiel, R. S. 1991, "Improving soil productivity in the pre-fertilizer era".

In *Land, labour and livestock*, eds B. M. S. Campbell and M. Overton, Manchester, 51–79.

Schovsbo, P. O. 1987, *Oldtidens vogne i Norden*, Bangsbo.

Søvberg, I. Ø. 1976, *Driftsmåter i vestnorsk jordbruk ca. 600–1350*, Oslo.

Smith, R. E. F. 1977, *Peasant farming in Muscovy*, Cambridge.

Steensberg, A. 1983, *Borup 700–1400: a deserted settlement and its fields in South Zealand, Denmark*, Copenhagen.

Szabó, M., Grenander-Nyberg, G., and Myrdal, J., 1985, *Die Holzfunde aus der frühgeschichtlichen Wurt Elisenhof*, Frankfurt am Main.

Thålin-Bergman, L. 1976, "Øks: Sverige och Skåne" in *Kulturhistoriskt lexikon för nordisk medeltid*, vol. 20, Malmö, 660–5.

Veibaek, O. 1974, *Ploven og dens betydning med saerlig henblik på landsbyorganisationen*, Skanderborg.

Wallerström, T. 1995, *Norrbotten, Sverige och medeltiden*, 1, Lund.

White, L. 1962, *Medieval technology and social change*, Oxford.

Windelhed, B. 1995, *Barknåre by: Markanvändning och bebyggelse i en uppländsk by under tusen år*, Stockholm.

8. FIELDS AND FIELD SYSTEMS IN SCANDINAVIA DURING THE MIDDLE AGES

Mats Widgren

The ideas and models for the development of medieval fields and field systems in Scandinavia between AD 500 and 1500 were for a long time mainly based on the retrogressive analysis of seventeenth- and eighteenth-century cadastral maps. The wealth of evidence provided by the Swedish, and to a lesser extent the Danish, large-scale village and farm maps (1:4000) stimulated very active research in historical geography. The regional coverage and the detail in these maps made possible both an evolutionary classification of contemporaneous forms in the agrarian landscape and in-depth studies of particular villages.[1] Since these studies made little use of medieval source material and had the post-medieval maps as their starting point, it was only possible in the main to model, rather than to reconstruct, the medieval landscape. In an attempt to solve this problem, David Hannerberg in the 1960s initiated a programme of surveying and mapping pre-seventeenth-century abandoned fields, using evidence on the ground, in order to test these models. Further studies of agrarian landscapes, based on survey evidence, and carried out by Hannerberg's pupils, followed in the later 1960s and 1970s. However, as it turned out, most of this information neither related to the medieval period nor gave an indication of how the landscape recorded in the cadastral maps had come into being, but rather gave evidence of the pre-medieval landscape. A number of geographical studies of Iron-Age landscapes were then launched, making use of a wide variety of techniques, including pollen analysis, phosphate analysis and archaeological excavation. These geographical investigations built upon a tradition of Iron-Age settlement archaeology (500 BC to AD 1050) which in all the Scandinavian countries had produced substantial results in the 1950s and 1960s.[2] The outcome is that in most Scandinavian countries agrarian landscapes of the first five centuries

---

[1] For an overview see Helmfrid 1972.

[2] Lindquist 1968 and 1974; Carlsson 1979; Widgren 1983. And see for example Hagen 1953; Stenberger 1955; Ambrosiani 1964.

AD can be reconstructed on a much fuller scale than those of the medieval period.

This overview of medieval fields and field systems, then, has to be seen in the light of the character and direction of this past research. Thus, any approach to the medieval agrarian landscape must be based on the much fuller pictures which can be painted for the preceding and subsequent periods.

Pre-Medieval Fields

Whole complexes of settlements and fields from the Roman Iron Age and the Migration Period (Birth of Christ to AD 600) are known in different regions of Scandinavia. Two main types can be distinguished. During the pre-Roman and early Roman Iron Age large complexes of "block-shaped" parcels existed in Denmark and on the Baltic island of Gotland. These fields correspond—in date as well as in function—with the similar Celtic fields of northern Germany and the Netherlands. Research into this field type was undertaken by the Danish geographer and archaeologist Gudmund Hatt in the 1930s and more recently by Viggo Nielsen.[3] Similar fields were discovered on Gotland in the late 1960s which led to extensive surveys and excavations by a team of historical geographers from Stockholm University.[4]

More recent research has shown that a similar type of farming also existed during the same period in large areas of the Swedish southern uplands and in eastern Norway. Today such fields are only seen as large (20 to 200 ha) areas of scattered clearance cairns. The individual fields are not easily identifiable, but seem to have been small irregular plots.[5] The exact nature of the cultivation techniques and agro-ecosystems of these prehistoric fields is not yet fully known but one can sketch the following development. The Celtic fields of Denmark and Gotland, as well as the clearance cairn fields, probably represent an extensive form of semi-permanent cultivation, where fields and settlements were moved within a large cultivable area. Different forms of nutrient restoration have been proposed: two-course

---

[3] Hatt 1949; Nielsen 1984.
[4] Lindquist 1974; Carlsson 1979.
[5] Pedersen 1990; Gren 1989; Holm 1995.

rotation followed by long fallow periods, bush-fallow, grass-fallow, or some kind of woodland/arable management (for example, coppicing). The fields in Denmark and Gotland were cultivated with an ard, while in the clearance cairn fields in southern Sweden and Norway hand implements, such as hoes or digging-sticks, may have played a more important role. Because of the lack of clearly distinguishable lynchets and field boundaries which are usually associated with ard or plough cultivation, the vernacular term for these abandoned fields is often *hackerör* (hoe cairns).

In some areas of Scandinavia the integration of a cattle and arable economy is indicated by complexes of walled enclosures, long stone-walled cattle ways, large hay meadows for the collection of winter fodder, and small cultivated patches. The settlements associated with these fields consisted of loosely grouped single farmsteads and hamlets, situated along the outer boundary or head dyke of a common field area. The landuse within this infield comprised both arable fields and hay meadows. Such settlement forms and fields are mainly found on Gotland, Öland, east-central Sweden (Uppland and Östergötland) and in the province of Jaeren in southwestern Norway.[6] The general conclusion about these fields, which date from the Late Roman Iron Age and the Migration Period (AD 200 to 600), is that they were manured, and then worked with an ard.

Early Modern Fields

During the middle ages these types of fields were, gradually or suddenly, transformed, resulting in what is recorded in the seventeenth- and eighteenth-century cadastral maps (mainly for Sweden and Denmark) and the written sources, which permit a detailed reconstruction of fields and field systems.

By the seventeenth/eighteenth centuries open-field farming, usually associated with three-field systems and village settlements, dominated on the Danish islands of Fyn and Zealand as well as in the adjoining parts of Scania in present-day Sweden. Ploughing was carried out with a wheel plough and this southern Scandinavian region thus formed the outer fringe of the general northwestern European area of open-field farming.

---

[6] Carlsson 1979; Fallgren 1993; Widgren 1983; Myhre 1972; Rønneseth 1974.

However, other regions of open-field farming can also be identified in different parts of Scandinavia during that period. But in these areas, neither the three-field arrangement nor the wheel plough was necessarily a part of the system. The type of field system also varied. Two- or three-field systems dominated, but there were also cases with one common field which was annually cropped. Such regions of open-field farming can be found in parts of Jutland (Denmark), central parts of Västergötland, Östergötland and the Mälar valley area in Sweden, as well as in eastern Norway. A similar variation holds for the tilling implements: as Myrdal shows in this volume, both lighter ploughs and ards were in use.

In the rest of Scandinavia infield systems with annually cropped fields alternated with different forms of regulated fallow systems (field-grass systems). A special region of interest is western Norway, where a type of intensively manured small infield was cultivated in a manner comparable to the plaggen cultivation (*Plaggenwirtschaft*) of the Dutch and German North Sea coast.[7]

Developments During the Middle Ages

In large parts of Scandinavia we can thus establish one cross-section in the mid-first millennium and another *c.* 1000 years later. For a long time ideas on the development of medieval field systems were more or less based on different methods of extrapolating from the pre-existing and successive landscapes. During the last decade work within archaeology, historical geography and palaeobotany has produced substantial results which have made possible a more precise chronology.

Dating the Two- and Three-Field Systems

The medieval development of fields and settlements in the areas where open fields came to dominate has been described by Müller-Wille as

---

[7] The German word "plaggen" refers to the use of turf as bedding for stalled cattle. The mixture of dung and turf was then used as manure, thus giving rise to a distinct anthropogenic soil type—plaggen soils. Long-term use of this process also usually produced raised fields with high lynchets. For an overview see Spek 1992.

the combined effect of four related phenomena: *Vergetreidung, Verzelgung, Vergewannung* and *Verdorfung* (the processes of increased grain cultivation, establishment of common fields, scattering of strips and the concentration of settlements into villages).[8] The scheme was based on the belief that changes in agrarian structures should be related to the intensification of cereal farming (*Vergetreidung*) which would lead to the establishment of three- or two-field-systems (*Verzelgung*). The creation of common-field systems also involved a subdivision of parcels into small strips which were distributed among the different fields (*Vergewannung*). The effect on settlement would be that former single farms and hamlets were concentrated into villages (*Verdorfung*). When the pre-medieval situation is compared with the seventeenth-century evidence such a sequence is, no doubt, what happened in most of the later open-field areas of Scandinavia. The problem is, however, when and why this development took place.

In the 1950s researchers such as Müller-Wille and Krenzlin proposed an interpretation of the development of settlement and fields which broke with the existing static and ethnic tradition in European landscape research and provided a coherent evolutionary framework for the analysis of the agrarian landscape. Krenzlin showed that the different forms in the pre-industrial agrarian landscape of Europe could be seen as steps in the increasingly intensive use of the land.[9] In Sweden, David Hannerberg and Staffan Helmfrid were instrumental in introducing these ideas and also carried out morphogenetic-based village investigations, where different steps in the development of the medieval village and its fields were reconstructed by the retrogressive approach.[10]

This idea of an evolution from the primitive, annually cropped, one-field system via a two-field system to the three-field system has been highly influential in this area of research. In Scandinavia this idea has deep roots and was first proposed by the Danish historian Kristian Erslev in 1898.[11]

However, as Karl-Erik Frandsen has pointed out, the actual evidence of a transformation of field systems has been scanty. From the early thirteenth century the regional Danish law codes seem to indicate a transition from individually farmed land to a common-field

---

[8] Müller-Wille quoted by Helmfrid 1985, 46.
[9] Krenzlin 1958, 250, 266.
[10] Helmfrid 1962; Hannerberg 1976.
[11] Frandsen 1988.

system. The thirteenth-century Swedish law codes mention the two-course rotation in the Östgötalag, while the earliest direct evidence of a two-field system is in a letter from 1338.[12] Frandsen draws from this the conclusion that not one case of a transition from one type of open-field system to another can be documented before the seventeenth century.[13]

While a strict reading of the sources makes this the only possible conclusion, others, for example Porsmose, have shown that the existing written evidence is at the least a very strong indication that the three-field system was replacing less regulated fallow systems during the eleventh to thirteenth centuries.[14] A critical point here is the vague terminology which is often used to describe field systems, and much earlier research confuses three different characteristics of the fully fledged open-field system: the two- or three-course rotation, the common field and the specific arrangement of fields in two, three or more separately enclosed fields.

When these three characteristics are treated separately it is possible to draw more detailed conclusions about the chronology of the development. The problem of the chronology of field systems has mainly been approached from the archaeological evidence of ancient field boundaries. In the 1970s Lindquist studied a deserted hamlet in Västergötland which, to judge from the abandoned field boundaries, was practising a three-field system in the fifteenth century. He dated the establishment of field systems through radiocarbon analyses of charcoal from the different clearance phases sealed underneath the field boundaries (figure 8.1). He concluded that the two-field system was introduced in the tenth and eleventh centuries, and that it was replaced by a threefield system in the thirteenth century.[15]

In east-central Sweden (Mälar valley area and Östergötland) the two-field system dominated at least from the time of the provincial laws in the early fourteenth century. This is also substantiated by the archaeological dating of field boundaries in two-field systems. Ulf Sporrong has published fourteen radiocarbon dates relating to different stages in the development of open fields in the Mälar valley (figure 8.2). The outer boundaries of the common fields were dated

---

[12] Compare the literature quoted by Frandsen 1988, 120, and Myrdal 1985, 73.
[13] Frandsen 1988, 121.
[14] Porsmose 1988, 289.
[15] Lindquist 1976.

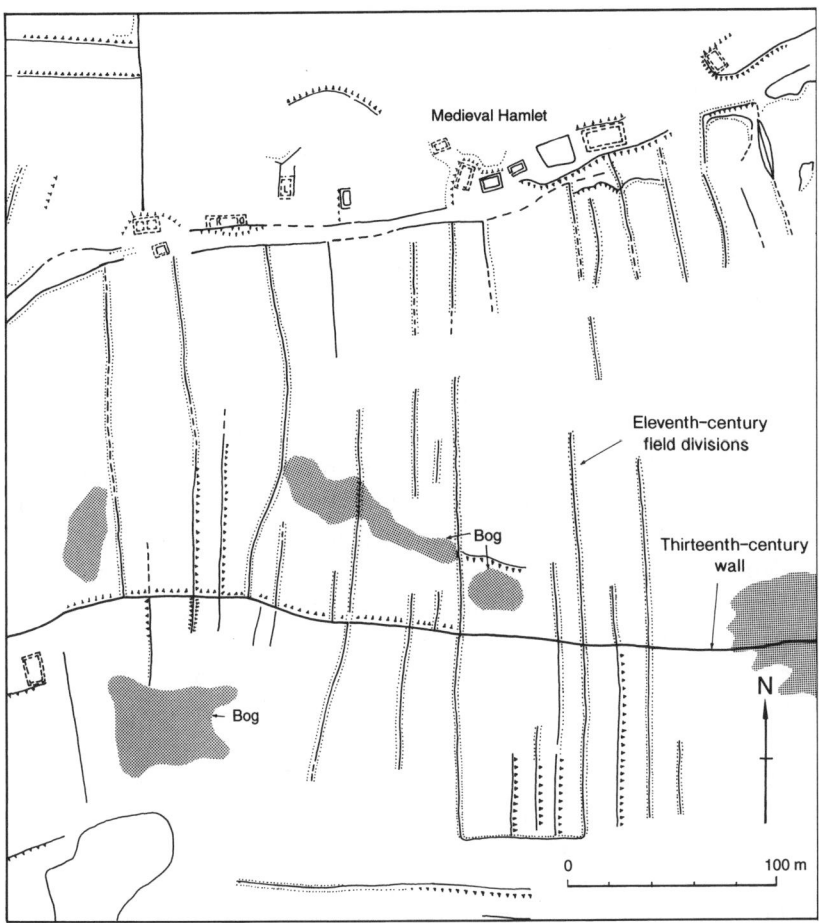

Figure 8.1. Brunsbo in Västergötland, Sweden. The strip fields running north-south were laid out in the tenth or eleven centuries, as part of a two-field system. A third field was created in the thirteenth century, and the strips overlain by a new, east-west, boundary (Lindquist 1976, Karta VII).

between AD 600 and 1200 (eight dates), while the field boundaries of strip fields within the common fields were dated between AD 1100 and 1600. Sporrong argues that the two-field system can be dated to the late Viking Age (tenth century) while the strip fields within the common fields were successively developed at a later stage.[16]

The most recent evidence on medieval crop rotations comes from investigations in the old Danish area in southern Scania. The inter-disciplinary work in Lund University's Ystad project confronted these problems with a new methodology. While no investigations of actual fields and field boundaries were carried out, the palaeobotanical evidence from settlements and the pollen analyses of lake sediments shed new light on medieval crop rotations. The survival in equal proportions of barley and rye among the charred seed remains indicates that in the Viking Age winter-sown rye alternated with spring-sown barley. Furthermore, the proportions of pollen from sheep's sorrel (*Rumex acetosa/acetosella*) during the Viking Age support the evidence that fallow was also part of the rotation. All the Ystad evidence has been used to argue for a three-course rotation, a conclusion in accordance with fieldwork results from northern Germany. The evidence also gives indirect indications of when a three-field system was introduced. In the pollen record Regnéll noted a significant change from the thirteenth century when the proportions of rye and barley and the occurrence of field weeds conform closely to what has been found in the period of the historically known three-field system, that is the seventeenth and eighteenth centuries.[17] The Ystad project thus shows that the three-course rotation was used in the Viking Age (ninth to tenth centuries AD), while the three-field system was only intro-duced in the thirteenth century.

We can thus see a development in some of the open-field regions of Scandinavia where regular crop rotations preceded the formal rearrangement of fields into a fully-fledged two- or three-field sys-tem. A more gradual view of development of the different com-ponents of the medieval village characterizes much of the present research into medieval fields and settlement. The medieval village was not born in its mature form in the tenth and eleventh centuries

---

[16] Sporrong 1985, 163. See also (in English) Sporrong 1986.
[17] The analysis of charred seeds was carried out by Roger Engelmark (1992, 373–4), while the pollen analysis was performed by Joachim Regnéll (1989). For the German evidence see Müller-Wille *et al.* 1988.

Boundaries of open fields

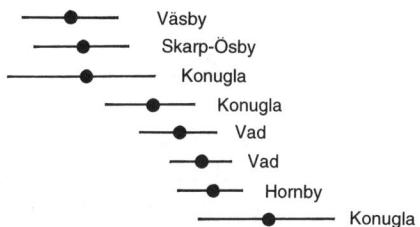

Boundaries between strips in open fields

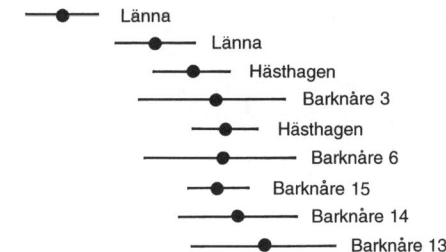

Small plots outside open fields

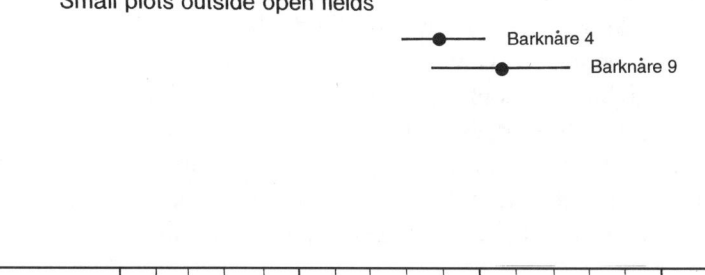

Figure 8.2. Radiocarbon datings from the Mälar valley area, Sweden. Outer boundaries of the common fields were dated between AD 600 and 1000; field boundaries between strips in the open fields were dated between AD 1100 and 1600. Later small individual plots were added outside the open fields. (Sporrong 1985 and 1986, 81).

as was previously argued, but was gradually developed during the middle ages.[18]

The Origin of Subdivided Fields

Müller-Wille and Krenzlin argued that strip parcelling was closely connected with the establishment of field systems. Over much of Europe strip fields have for a long time been regarded as intimately connected with the development of medieval field systems. Although the wheel plough is no longer considered to have determined the character of medieval field systems, there is still a tendency to associate the use of the plough with the need for strips. Both assumptions can be questioned on the basis of recent Scandinavian research.

In what follows I will attempt an overview of the occurrence of the three main types of strip fields and consider the significance of these data for understanding the origin of subdivided fields. Broadly defined, three main types of strip fields can be found in the open-field areas of Scandinavia: narrow ridges, ridge and furrow, and flat strips with banks.

Narrow Ridges

Different types of narrow and rather flat ridges exist in Scandinavia and they are generally associated either with recent spade cultivation or with the introduction of a light horse-drawn plough in the late eighteenth century (compare the English term horse-rigs). However, in two cases narrow ridges have been found in a medieval context. The best known case is that of Lindholm Höje (Denmark), where sixty-four ridges, varying in width between 0.75 and 1.25 m, and 0.25 m high, were found in association with a Viking-Age village.[19] The height of the ridges suggests that they were only formed by one season of ploughing and, though evidently cultivated with a wheel plough, they have therefore been interpreted as beds for intensive cultivation rather than the typical open-field type of ridge and fur-

---

[18] This view characterizes two recent monographs about village development in Uppland and Scania respectively: Windelhed 1995; Riddersporre 1995.
[19] Lindquist 1976.

row.[20] At Kungsmarken outside Lund similar narrow ridges can be seen, 3 to 5 metres in width and less than 0.2 m high. While these ridges cannot be given an absolute date, Christophersen convincingly argues on the basis of the stratification that they date from the eleventh to twelfth centuries.[21]

Ridge and Furrow

Ridge and furrow (over 6 m wide and 0.3 m high) is found in large parts of Denmark, eastern Norway and western Sweden. Its distribution coincides with those areas of Scandinavia where different forms of plough, rather than ards, were used from the late thirteenth century. The oldest dated ridge and furrow comes from an excavation in western Jutland where a twelfth-century house had been built over a couple of 11–14 m wide ridges.[22] Ridge and furrow has only been dated in a few localities. In Närke in central Sweden an area of seventy ridges has been investigated, each 15–20 m wide and 0.4–1 m high. Radiocarbon dates of a secondary clearance horizon on top of the ridges show that the fields were cultivated in that form before the fifteenth century.[23]

In south-eastern Norway ridge and furrow can be found on a number of deserted medieval farms. Three of these have been surveyed in Bohuslän (present-day Sweden) where the widths vary from 4 to 12 m and the heights from 300 to 700 mm. The written and cartographic evidence shows that these fields were abandoned in the late fourteenth or the early fifteenth century.[24]

Eastern Norway is usually considered to have been dominated by single farmsteads with fields held in severalty during the medieval period. The evidence from deserted single farms shows that strip fields in the form of ridge and furrow were a common phenomenon in Bohuslän. Taken together with the fourteenth-century evidence of farm subdivision within settlements and the information from the seventeenth-century maps, there are very strong indications that a certain number of concentrated settlements with subdivided fields with

---

[20] Compare the overview of Møller 1990, 102, and 1995.
[21] Christophersen 1981.
[22] Excavated by Vejbaek; and compare the overview of Møller 1990.
[23] Sporrong 1978.
[24] Widgren forthcoming.

ridge and furrow existed already in the thirteenth and fourteenth centuries.

Bohuslän, which came under Swedish rule in 1658, is unusual compared with the rest of Norway because cadastral maps exist from the late seventeenth century onwards and these permit a detailed comparison with the medieval records and later maps. It is thus possible that nucleated settlements and strip fields were a common, though not necessarily a dominant, phenomenon in large parts of eastern Norway.

In all respects these new investigations suggest that eastern Norway followed a general European trend towards the concentration of settlement and the subdivision of fields. But it should be remembered that this development occurred within the framework of annually cropped fields and a single common field rather than a two- or three-field system.[25]

Flat Strips with Banks

A large number of medieval fields in Scandinavia are flat or slightly concave rather than ridged, which reflects the widespread use of a lighter plough or ard. The fields were defined by low earthen banks, rows of stones or lynchets.

This type of strip field was investigated by Axel Steensberg at Borup, where they were dated between the late Viking Age and the twelfth century.[26] Similar fields can also be found in parts of Sweden, where in one locality they have been dated to the tenth to eleventh centuries.[27] These fields are often concave in profile, indicating that they were cultivated with an ard in an adaptation of the criss-cross cultivation known from the Iron Age. Since the strips were long and narrow a 90 degrees criss-cross ploughing was not feasible, so instead the second cultivation was done obliquely, approximately at 45 degrees to the first.[28]

More recent evidence has, however, shown that there does not necessarily have to be a connection between strip fields and a drawn implement—be it ard or plough. In Sweden strip fields dating to the Early Iron Age (500 BC to AD 500) have been found, probably as-

---

[25] Widgren forthcoming.
[26] Steensberg 1983.
[27] Lindquist 1976, 155.
[28] Compare Myrdal 1985, 95ff.

sociated with a similar extensive kind of cultivation as that of the clearance cairn fields.[29] This type of extensive cultivation might be associated with a delayed integration of arable and livestock farming in some woodland districts.

During the first millennium the very widespread cultivation indicated by the clearance cairn fields was successively concentrated into more regular field systems based on what were later to become the historically-known hamlets and villages. These field systems covered areas much larger than the historically-known arable. The conclusion has therefore been drawn that cultivation must still have been based on some kind of long-term fallow system. The best cultivable land was divided into several strips varying in width between 7 and 45 metres. Some of the strips are not fully cultivated and in many cases there is evidence of small rounded plots, probably hoe-cultivated, inside the strips.

The only viable interpretation of these fields is that they reflect the subdivision of holdings within a farming community. Strip-parcelling makes its appearance around the Birth of Christ, but recent research has also shown that in peripheral areas the laying-out of such fields goes well into the Viking Age. These strips do not represent any adjustment to accommodate the plough or any other drawn implement: they are thus land-holding parcels rather than working parcels. The relation between this prehistoric type of strip-parcelling and the medieval open-fields remains to be investigated, but the evidence supports the idea that subdivision of fields and the creation of strip fields have a logic of their own which need be related neither to the establishment of regular field systems (*Verzelgung*) nor to the introduction of ox-drawn implements, be it ard or plough.[30]

Intensification in Southwestern Norway

While extensive types of cultivation thus prevailed in upland parts of southern Scandinavia, the coastal areas of western Norway followed a contrasting line of development. In the nineteenth century we can find there a type of plaggen cultivation comparable to what was

[29] Holm 1995, 142, has shown that clearance cairn fields in Norway were in use also in the late medieval period.
[30] Widgren 1990.

practised in other areas along the North Sea coast and the Atlantic islands. The arable consisted of a small infield, while the major part of the enclosed field was used for hay-making. The small arable field was manured with a mixture of cow dung, soil and turf from the surrounding heath land. Through the application of this organic matter the fields were raised and a lynchet of considerable height was formed on the outer fringe of this *Gamleågeren* (the old field; figure 8.3). The field was generally cultivated with a spade.[31]

The age of this form of cultivation has been much debated. The numerous deserted Migration-Period farms in the same area do not usually have traces of such farming practices: the infields indicate some kind of arable husbandry with fields alternating with grasslands. Furthermore, in pre-medieval agriculture there are strong indications that the ard was the main implement used for the cultivation.

The archaeological and palynological evidence of the later, intensive, type of cultivation now seems to indicate that plaggen soils started to accumulate on certain farms in the eighth century AD and it has also been proved that the intensive type of cultivation was the rule on the numerous farms deserted in the fourteenth and fifteenth centuries. Without drawing too much on the earliest evidence, one can at least say that this farming system was expanding during the eleventh to fourteenth centuries.[32]

In these parts of Scandinavia we can thus trace a type of increased land productivity which was based on the shrinking of arable land, a move from fallow systems to annual cropping, increased manuring, and a shift from ard cultivation to spade cultivation. The logic behind a similar farming system in Scotland has recently been discussed by Dodgshon.[33]

Fields and Technological Development: Some Conclusions

In the material presented here one can easily see lines of development similar to the generally accepted image of medieval northern European agrarian development: common fields were developed,

---

[31] Rønneseth 1974, 87ff.

[32] The age of this type of cultivation is documented in Kvamme 1988, 93, and in Myhre 1985.

[33] Dodgshon 1992.

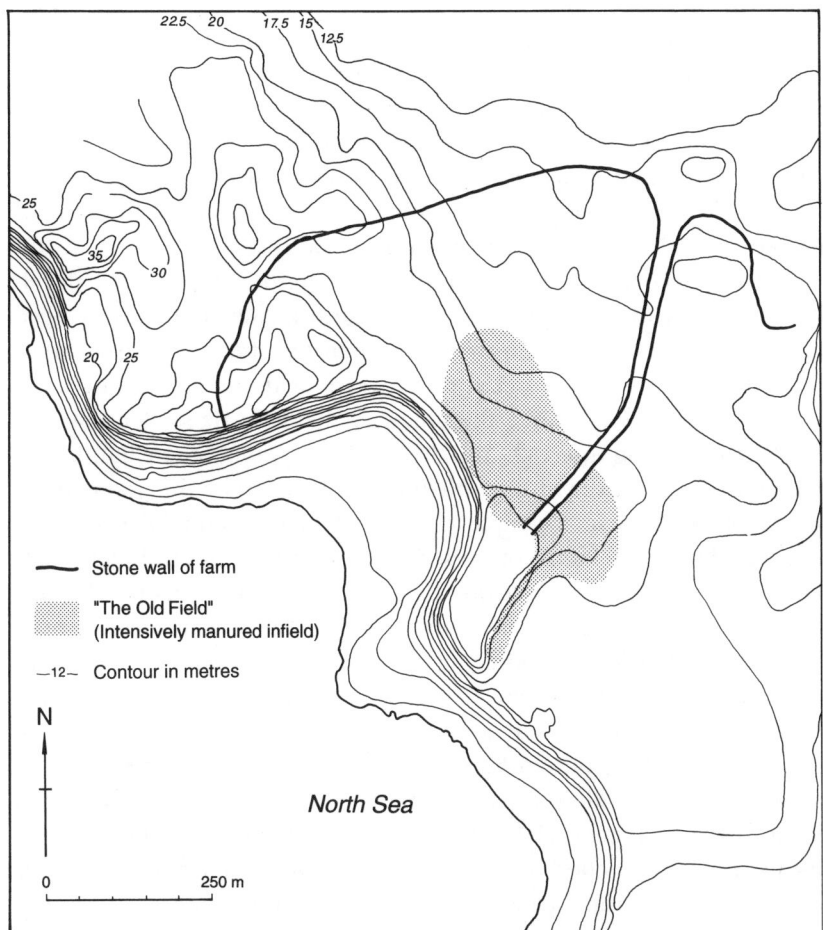

Figure 8.3. Obrestad in Jæren, Norway. The shaded area shows the intensively used infield in the nineteenth century. In the coastal areas of Norway the arable consisted of a small, manured, infield while the major part of the enclosed field was used for hay-making. This type of 'intensive' *plaggen* cultivation was probably introduced in the Viking Age (Rønneseth 1974, 105).

mainly as two- or three-field systems. The ploughed fields seem to
have been developed from narrow ridges to the broader ridge and
furrow, a development which has also been suggested for England.[34]
The suggested sequence of intensified cereal cultivation, introduction
of regular field systems, nucleation of settlement and the subdivision
of fields into strips at least gives a fair description of the post-Viking
Age development. The introduction of the plough in southern and
western Scandinavia, as well as the improvements of the ard used in
eastern Sweden as suggested by Myrdal, are processes closely connected with this hypothetical intensification.[35]

But the account above has also shown contrasting lines of development. This may be partly because agrarian technology has been
approached from the viewpoint of the fields. One has to bear in
mind the very intricate relation between fields and agrarian technology. On the one hand the shape of fields is dependent on the
implements being used to till them. But on the other hand, this
relation is never a direct one. Neither the profile nor the geometrical
shape of fields have a simple chronological or technological deter-
minant. Different kinds of raised and ridged beds and fields are
known in Scandinavia, cultivated with hoes, spades, ards or ploughs.
Furthermore, strip-fields were cultivated with a light ard in east-central
Sweden during the late middle ages and even up to the late eighteenth
century. And, as the evidence from Västergötland shows, subdivided
strip fields have their own rationale, regardless of any connection to
open-field farming or ox-drawn implements. One has therefore no
reason to expect a close relation between the development of agrarian
technology at a general level and the development of field forms.
Fields also serve other purposes of which the social one is perhaps
the more important. Fields have social and cultural imprints as well
as technological ones.

A more general tendency towards increased area productivity seems
to have worked through a number of different factors, which did not
necessarily lead to the same development on the ground. It seems as
if knowledge, implements, and field shapes can be assembled in an
infinite variety of agro-ecosystems to serve different purposes in different ecological and social contexts.

Instead of concentrating on the nature of the evidence for fields,

[34] Astill 1988, 62–85.
[35] Myrdal 1988, 207ff.

let me finish by discussing the significance of the evidence itself. We started off by trying to monitor changes between AD 500 and AD 1500. Most of the evidence of changes in the structure of individual parcels or of field systems which has been recounted above relates to the latter part of that period. The earlier part of that period, in Scandinavian chronology the Late Iron Age, has produced less evidence in the form of fossil fields and settlement. From the little of what we know about demographic changes during the first millennium AD, it seems that a certain decline in population occurred in the fifth and sixth centuries AD. In the areas of extensive remains of Roman Iron Age settlements and fields (Jutland, Öland, Gotland, Östergötland, Jaeren) there is little evidence that any large changes in the type of farming system occurred in the subsequent periods. From a landscape point of view the period can thus be described as one of contraction in cultivated land.[36]

These events in the landscape relate very closely to the conclusions drawn by Myrdal.[37] He launched a hypothesis that changes in agrarian technology which led to increased land productivity have alternated with changes designed to increased labour productivity. According to him the Late Iron Age (AD 500 to 1050) was characterized by a group of innovations in ploughing and harvesting which increased labour productivity rather than land productivity. Changes in agrarian technology were characterized by the increased use of iron in agrarian implements. The iron ard share was introduced and the longer scythe made possible more efficient hay harvesting. For Myrdal the dominant characteristic of these changes was that they increased labour productivity. This may also explain why so few changes are seen in the landscape. After a possible fall in population during the sixth century, the increasing demands could thus have been met within the same farming system and the same spatial structure.

The later part of the middle ages (AD 1050 to 1500) presents another picture. Great landscape changes can be documented at all levels: from the single parcel to field systems and to the settlement expansion in woodlands. These changes can also be related to the innovations in agrarian technology: the introduction of the plough in southern and western Scandinavia, the introduction of an ard better

---

[36] Widgren 1983, 124
[37] Myrdal 1988, 217ff., and in this volume.

suited for breaking the fallow in eastern Scandinavia and the development of spades with an iron share.[38] All these innovations seem to
be of a character that not only increased the productivity of land,
but also induced changes in field layout, facilitated the introduction
of regular fallow and of the cultivation of virgin land.

Bibliography

Ambrosiani, B. 1964, *Fornlämningar och bebyggelse*, Uppsala.
Astill, G. 1988, "Fields". In *The countryside of medieval England*, eds G. Astill
    and A. Grant, Oxford, 62–85.
Carlsson, D. 1979, *Kulturlandskapets utveckling på Gotland*, Visby.
Christophersen, A. 1981, "Några reflexioner kring de högryggade åkrarna i
    Kungsmarken", *Ale* 3, 1–18.
Dodgshon, R. 1992, "Farming practice in the western highlands and islands
    before crofting: a study in cultural inertia or opportunity costs?", *Rural
    History* 3, 173–89.
Engelmark, R. 1992, "A review of the farming economy in south Scania
    based on the botanical evidence". In *The archaeology of the cultural landscape*,
    ed. L. Larsson, Lund, 369–76.
Engelmark, R. and Viklund, K. 1991, "Makrofossilanalys av växtrester—
    kunskap om odlandets karaktär och historia", *Bebyggelsehistorisk tidskrift* 19,
    33–42.
Fallgren, H. 1993, "The concept of village", *Current Swedish Archaeology* 1,
    59–86.
Frandsen, K.-E. 1988, "The field systems of southern Scandinavia in the
    seventeenth century: a comparative analysis", *Geografiska Annaler Series B*,
    70B, 117–22.
Gren, L. 1989, "Det småländska höglandets röjningsröseområden", *Arkeologi
    i Sverige 1986*, Riksantikvarieämbetet, Stockholm, 73–95.
Hagen, A. 1953, *Studier i jernalderens gårdssamfunn*, Oslo.
Hannerberg, D. 1976, "Models of medieval and pre-medieval territorial
    organization, with a select bibliography", *Journal of Historical Geography* 2,
    3–34.
Hatt, G. 1949, *Oldtidsagre*, Copenhagen.
Helmfrid, S. 1962, *Östergötland "Västanstång". Studien über die ältere Agrarlandschaft
    und ihre Genese*, Stockholm.
—— 1972, "Historical geography in Scandinavia". In *Progress in historical
    geography*, ed. A. Baker, London, 63–89, 225–8.
—— 1985, *Europeiska agrarlandskap*, (mimeo) Stockholm.

---

[38] Myrdal 1985, 156ff., and 1988, 207ff.

Holm, I. 1995, "Trekk av Vardals agrare historie", *Varia* 31, Universitets Oldsaksamling, Oslo.

Krenzlin, A.-L. 1958, "Blockfluhr, Langstreifenfluhr und Gewannfluhr als Funktion agrarischer Nutzungssysteme in Deutschland", *Berichte zur deutsche Landeskunde* 20 (2), 250–66.

Kvamme, M. 1988, "Lokale pollendiagram og bosetningshistoria". In *Folkvandringstiden i Norden*, eds U. Näsman and J. Lund, Aarhus, 75–114.

Lindquist, S.-O. 1968, *Det förhistoriska kulturlandskapet i östra Östergötland*, Stockholm.

—— 1974, "The development of the agrarian landscape on Gotland", *Norwegian Archaeological Review* 7, 6–32.

—— 1976, "Fossilt kulturlandskap som agrarhistorisk källa", *Västergötlands fornminnesförenings tidskrift* 1975–6, 119–64.

Myhre, B. 1972, *Funn, fornminnen og Ødegårder*, Stavanger.

—— 1985, "Arable fields and farms structure". In *Honorem Evert Baudou. Archaeology and Environment 4*, ed. M. Backe, Umeå, 69–82.

Myrdal, J. 1985, *Medeltidens åkerbruk. Agrarteknik i Sverige ca. 1000 till 1520*, Stockholm.

—— 1988, "Agrarteknik och samhälle under tvåtusen år". In *Folkevandringstiden i Norden*, eds U. Näsman and J. Lund, Arhus, 187–226.

Møller, P. G. 1990, "Højryggede agre—forskning og bevaring", *Bol og by* 1990:1, 90–118.

—— 1995, "Højryggede agre i fymske skove", *Fortid og Nutid* 54, 295–312.

Müller-Wille, M., Dörfler, W., Kroll, H. and Meier, D. 1988, "The transformation of rural society, economy and landscape during the first millennium AD", *Geografiska Annaler, Series B* 70B, 53–68.

Nielsen, V. 1984, "Prehistoric field boundaries in eastern Denmark", *Journal of Danish Archaeology* 3, 135–63.

Pedersen, E. A. 1990, "Rydningsrøysfelter og gravminner spor av eldre bosetningsstruktur på Østlandet", *Viking* 53, 50–66.

Porsmose, E. 1988, "Middelalder". In *Det danske landbrugs historie 4000 f. Kr.–1536*, Odense.

Regnéll, J. 1989, *Vegetation and land use during 6000 years. Palaeoecology of the cultural landscape at two lake sites in southern Skåne, Sweden*, Lund.

Riddersporre, M. 1995, *Bymarker i backspegel*, Lund.

Rønneseth, O. 1974, "'Gard' und Einfriedigung", *Geografiska Annaler Series B*, special issue 2.

Spek, T. 1992, "The age of plaggen soils. Tijdschhrift van de Belgische Vereniging voor Aardrijksskundige Studies", *Bulletin de la Société Belge d'Etudes Géographiques* 61, 72–91.

Sporrong, U. 1978, "Ryggade åkrar", *RIG* 61, 81–90.

—— 1985, *Mälarbygd. Agrar bebyggelse och odling ur ett historisk-geografiskt perspektiv*, Stockholm.

—— 1986, "Agrarian settlement and cultivation from an historical geographical perspective", *Geografiska Annaler* 68, 69–94.

Steensberg, A. 1983, *Borup AD 700–1400*, Copenhagen.

Stenberger. 1955, *Vallhagar: a migration period settlement on Gotland, Sweden*, Copenhagen.

Widgren, M. 1983, *Settlement and farming systems in the Early Iron Age*, Stockholm.

—— 1990, "Strip fields in an Iron Age context", *Landscape History* 12, 5–24.

—— forthcoming, *Bysamfällighet och tegskifte i Bohuslän*.

Windelhed, B. 1995, *Barknåre by*, Stockholm.

9. AN ARCHAEOLOGICAL APPROACH TO THE DEVELOPMENT OF AGRICULTURAL TECHNOLOGIES IN MEDIEVAL ENGLAND

Grenville Astill

British archaeologists have in general been reluctant to use their data to understand rural technologies, which is surprising considering that the reconstruction of technological processes has traditionally been regarded as one of the strengths of archaeological evidence. This reluctance can in part be explained by the way medieval archaeologists in Britain have worked. Whereas in northern and central Europe considerable progress has been made in the study of agricultural techniques through extended work on field remains, and on implements (see, for example, Widgren and Myrdal in this volume), in Britain these are areas of research which have attracted few archaeologists, perhaps because there was no established tradition of working in these areas. The extensive work on the development of open fields in Britain, for example, was until recently largely conducted within a documentary framework which did not really need information from the physical remains.[1] A dearth of evidence could explain the lack of work on agricultural implements, but the curious absence of a strong research tradition in material culture may explain the reluctance to use the plentiful iconographic data.[2] In the recent past aspects of agricultural technologies have only been briefly considered in reviews of medieval farming which summarize the results of archaeological and environmental research.[3]

Agricultural technologies have, therefore, to be approached obliquely from the available evidence. The numerous reports from the excavations of particular building complexes within medieval settlements can be used, but many of these did not include a consideration of the associated field arrangements, and this is in contrast to the research done on the remains of prehistoric and Roman fields.

---

[1] See, for example, Rowley 1981; Hall 1995; and Taylor 1975.
[2] This is in contrast to work done in Scotland and Ireland: Fenton 1963, 1994; Brady 1994.
[3] For example Fowler 1981; Addyman 1976.

The task has been made easier by the archaeological field surveys and especially the multi-disciplinary landscape studies which have become so characteristic of medieval rural research in Britain over the last twenty years. These mark an attempt to understand how particular blocks of land were managed during the middle ages and before.[4]

We are therefore in a position to identify some of the major periods of change, and in particular periods of intensification, which occurred in the English countryside. Often the evidence is sufficient to allow us to interpret the character, and to suggest the context, of such changes, and in so doing consider the implications for technological development. Archaeological interpretations can, however, be conditioned by longer established documentary research. So, for example, archaeological data could be viewed in the context of the macro-economic and demographic trends constructed by historians: between the eleventh and late thirteenth centuries most technological changes could be seen as essentially "land-saving", whereas those changes from the fourteenth century could be regarded as "labour-saving". Differences in approach between the two disciplines inevitably generate a tension which is seldom discussed usefully, let alone reconciled. The tension is particularly apparent in the case of medieval England, where it is now possible to juxtapose the vast body of documentary evidence with a significant, and growing, body of archaeological data.[5]

At the most basic level, it is possible to compare the information derived from contemporary sources, as has been done, for example, by considering that archaeological evidence which can illustrate an eleventh-century agricultural text, *The Sagacious Reeve*. This approach tends to suppress rather than acknowledge any potential tensions.[6] Many inter-disciplinary discrepancies, of course, only illustrate the very real differences in the nature of the data base, or the differing stages of enquiry which the two disciplines have reached. At present, archaeological evidence is best at charting long-term, and major, changes. It points, for example, to significant differences in general animal and crop management, such as the cultivation of different

[4] See the surveys which developed as part of the Wharram Percy Project: Beresford and Hurst 1990; the work at Raunds, and Hanbury: Dyer 1991; and at Shapwick: Aston and Costen 1993; and on Bodmin Moor: Johnson and Rose 1994.

[5] For a recent discussion of the relationship, see Austin 1990.

[6] Addyman 1976.

types of cereal or a change in the size of animals. Documentary research, on the other hand, tends to emphasize the smaller-scale, but nevertheless significant, nature of medieval technological improvements, particularly in the later middle ages. So the frequency of ploughing and weeding, the sowing rates and the folding of animals all reflect the nature of technological change rather than the introduction of new species; but, it should be remembered that this information is only available from the thirteenth century and even then patchily (see Campbell and Mate in this volume).

Perhaps the most pressing area of dispute, and evident in this volume, is the social context of technological change. There is a notable archaeological bias which emphasizes the role of the aristocracy in exploiting innovations; this is largely because "high-status" sites have produced the most chronologically refined, and most distinctive and prolific, material assemblages. The archaeological emphasis on a seigneurial stimulus to technological change reflects the interpretations of an earlier generation of historians, particularly that of Marc Bloch.[7] More recently historians, who also labour under similar social biases in their information, have attempted to use their essentially seigneurially inspired data to discover the role of the peasantry in technological change, with the result that the latter class is now credited with a more active role in the process (see Campbell and Langdon, this volume).

What follows is—of necessity—a cursory, chronological, survey of agricultural development in England which highlights particular periods of change—from the eighth to the tenth centuries, the later twelfth century and the later fourteenth and fifteenth centuries. This is to simplify and to overstate the case. These particular periods *are* distinctive because it seems that certain innovations were introduced at these particular times, perhaps as part of a technological complex. But it is also clear that the adoption or diffusion of specific innovations was by no means assured, because this was often dependent on the receptivity of particular regions or communities: there was thus potentially a long interval between the appearance of an innovation and its widespread adoption.

---

[7] Bloch 1967, 151–60.

The Eighth to the Tenth Centuries

In common with other areas of northern Europe, it is now generally assumed that there was no profound dislocation of farming practices during the fifth and sixth centuries in Britain (see Raepsaet, this volume). The evidence is, however, slight and is largely environmental. While a small number of pollen analyses indicate a regeneration of woodland, the majority show a continuation of the Roman landuse, whether arable or pasture, although in some areas there was a change from arable to pastoral farming.[8] However, most data come from locations which are regarded as peripheral to the main arable areas, about which we know little. Even the majority of excavated migration-period settlements, although sited within the enclosures of Roman field systems, were located in "marginal" areas.[9]

It is, however, clear that agricultural practices mostly had not changed. Spelt was still the principal type of wheat grown and there was no significant variation in size of the three most common animals—cattle, sheep and pig. Because many of the early Anglo-Saxon settlements were located in close proximity to Roman settlements, or within their lands, it is assumed that some of the associated fields continued to be cultivated using an ard. At Sutton Hoo a "silhouette" burial was excavated where the person appeared to be pushing an ard (figure 9.1). The abandonment of Roman agricultural buildings and some fields indicates a reduced scale of activities, but we have no independent information about early medieval fields.[10] Early Anglo-Saxon animal management did, however, differ from the Roman in a way which was to become increasingly important. Sheep were looked after for longer, until they had yielded several clips of wool, before they were killed—the new emphasis was therefore on wool production rather than just a source of protein.[11]

If fifth- and sixth-century agriculture then is to be characterized as a scaled-down version of what happened in the late Roman period, it is perhaps surprising that by the mid-seventh century there were clear signs of expansion and intensification which appeared to exceed Roman endeavours. Field survey on the fenlands around the Wash has located large numbers of middle Saxon settlements on the

[8] Williamson 1993, 58–9;.Rackham 1994, 8; Murphy 1994, 25–6.
[9] For a review of the evidence see Welch 1992, 14–42.
[10] Carver 1986, 41–6; Murphy 1994; Crabtree 1994.
[11] Crabtree 1994.

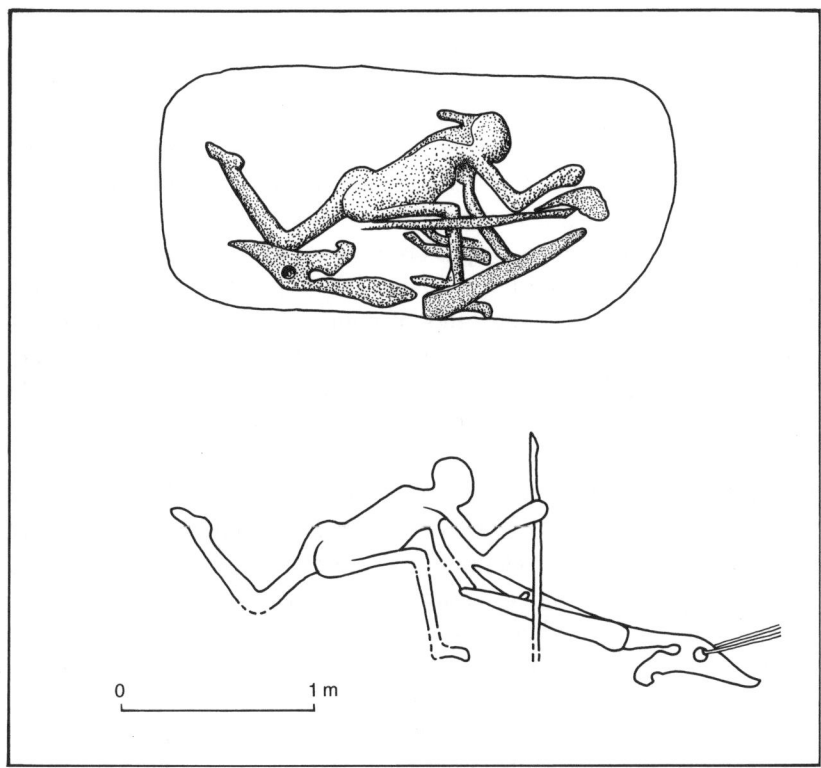

Figure 9.1. The Sutton Hoo silhouette burial (Carver 1986).

marine silts and in areas which had been abandoned in the second century AD.[12] In river valleys such as Upper Teesdale (Durham) which are at a much higher altitude than the current limit of cultivation, eighth-century settlements have been excavated and demonstrate a combination of cereal cultivation and iron smelting.[13]

Evidence also exists for intensification within well-settled areas. At Hockham Mere (Norfolk), for example, pollen analysis demonstrated a substantial increase in cereal (rye) cultivation between 650 and 850.[14] This is matched by exploitation and management of water resources: for the first time we have evidence of fish weirs and sluices in the rivers Trent and Kennet.[15]

---

[12] Hayes 1988.
[13] Coggins *et al.* 1983.
[14] Bennett 1983.
[15] Losco-Bradley and Salisbury 1988, 329–38; Butterworth and Lobb 1992, 175–7.

One possible social context for this increased productivity was the establishment of aristocratic, both secular and ecclesiastical, control over lands and the exaction of renders.[16] Renders were obtained from large, "multiple" estates on which it is sometimes suggested there was a certain amount of specialized production of a particular crop or stock most suited to a particular environment. The environmental evidence from Wessex, however, shows little evidence of specialization (beyond, for example, a preference for barley as the main cereal on calcareous soils); most settlements were practising a subsistence style of production, although this of course did not stop lords specifying particular crops or animals as acceptable renders.[17] The late seventh and eighth centuries is the time when evidence of large-scale processing of agricultural surpluses starts to appear in the archaeological record. The most significant is the late seventh-century watermill from Old Windsor, a *villa regalis*. This corn mill had three vertical waterwheels and a leet over one kilometre in length.[18] This is the earliest medieval watermill in the country. Of course it is possible that water-powered milling continued uninterrupted after the Roman period, but it is more likely that the particular circumstances of the eighth century, when there was a new need to process surpluses, provided the stimulus for mill building. By European standards this is not a particularly early example of the use of a watermill. In Ireland, for example, thirteen mills have been dated between the later sixth and the tenth centuries and, while the majority of these were of the horizontal-wheeled type, at least two of the earliest had vertical wheels; this suggests that the two types were in use at the same time, but in different topographical, and perhaps tenurial, situations.[19]

The ninth-century horizontally-wheeled mill from Tamworth has also been interpreted as an adjunct to a defended royal estate centre.[20] Further evidence for the processing of renders comes from the large-scale corn-driers from eighth-century Hereford, already an episcopal and royal centre; and the iron smelting works at Ramsbury (Wiltshire), Gillingham and perhaps Wimborne (Dorset); all these sites had either a royal or ecclesiastical character.[21]

---

[16] Hinton 1990, 42–64.
[17] Hinton 1994, 38.
[18] Wilson and Hurst 1958, 183–5; Schove 1979.
[19] Rynne 1989, 21–4.
[20] Rahtz and Meeson 1992, 1–5, 158.
[21] Shoesmith 1982, 28–31; Haslam 1980; Hinton 1994, 35–8.

The most substantial evidence for an increase in agricultural productivity occurs on the coastal trading sites such as Hamwic and Ipswich. Such sites used to be interpreted primarily as international trading centres specializing in the exchange of exotic products across the Channel and the North Sea, reinforcing a North Sea cultural zone (see Langdon, this volume).[22] More recently analysis of the faunal and ceramic assemblages has demonstrated that these emporia also had a strong link with the interior and drew large quantities of bulk agricultural products from the countryside, perhaps for export or for consumption within the port, in particular sheep and cereals—the latter were brought in already cleaned and only needed grinding into flour.[23]

This phase of intensification appears to have been achieved by extending existing techniques. The use of machinery to process the surplus is, however, probably new and related to the increasing power of the aristocratic landowners. It also took place within the context of growing royal power and beginnings of the English state. The emporia are usually interpreted as royal foundations and probably represent a stage in the increasing royal control of trade and taxation. The performance of army-, borough- and bridge-work were obligations attached to land grants at this time, and indicates a perception of the need to develop the country's infrastructure—indicated by defended royal centres, but also now by a bridge over a major river (the Trent).[24]

Indeed, there is some evidence to suggest that the eighth and early ninth centuries were a period of pronounced change in the countryside. Crops were changed; spelt appears to have been abandoned in favour of the naked wheats such as rivet and bread wheat, which were generally more adaptable to variable weather conditions, more resistant to rust fungus, easier to thresh and produced better quality flour. Flax is well represented at many sites, and at West Cotton there is some suggestion of a three-course rotation which included the cultivation of vetches and horse beans. Here the strip fields were also divided by baulks which were grassland, and the extensive weed infestation suggests that ploughing may have only been carried out twice a year.[25]

---

[22] Astill 1991, 101–3.
[23] Bourdillon 1988; Brisbane 1988; Wade 1988; Crabtree 1994, 46.
[24] Brooks 1971; Salisbury 1995.
[25] Campbell 1994, 77–81.

It was also a period of settlement reorganization: most of the excavated early Saxon settlements were abandoned, and this is thought to be associated with a reallotment of lands as a result of the dissolution of multiple estates into smaller units.[26] In some parts the successor settlements have yet to be located, but in others, such as Berkshire and Norfolk, there was a period of settlement shift, which in some cases did not cease until the eleventh century. Settlements were surrounded by large paddock-like enclosures which were unlike the later village crofts.[27]

Early medieval fields have yet to be adequately defined because of lack of data and dating evidence. Hall argues that long strip fields, arranged in two or three sectors as an open-field system, were laid out from the beginning, that is from when the associated (nucleated) settlement was established in the ninth or even eighth century, but his information, deriving mainly from Northamptonshire, could be dated anywhere between the eighth and eleventh centuries.[28] Others argue that such extreme rearrangement would have only been necessary after an earlier type of field system had finally proved to be unworkable as a result of a combination of factors, such as having reached the limits of assarting, extreme subdivision of plots and a shortage of pasture.[29] The evidence for the form of such an earlier type of field system is elusive, but the indications suggest either enclosures held in severalty, or fields which were divided into broad strips (which would often have had different tenurial histories) with the waste beyond, as in an infield/outfield arrangement.[30] The decision to make a fundamental reallotment of resources in order to produce what we would recognize as an open-field system would only have come from what may have been a long process of intensification. We have the potential evidence for such reorganizations in northern England in the eleventh century, but it is appears that some areas of good-quality agricultural land, such as the downland valleys in Hampshire, did not start to be exploited until the eleventh century.[31] The process of field creation (preceded by, or associated with, settlement nucleation in some regions) may have lasted over three

---

[26] Welch 1985; Hamerow 1991.
[27] Astill and Lobb 1989, 82–5; Williamson 1993, 110–11.
[28] Hall 1995, 125–39, xi.
[29] Thirsk 1964; Campbell 1981.
[30] Astill 1988b, 68–85.
[31] Harrison and Roberts 1989, 86; but see Palliser 1993, 6–7; Cunliffe 1972.

centuries, and may be similar to the long sequences suggested for Scandinavia (Myrdal, Widgren and Poulsen in this volume).

The sparse evidence for ploughs and the configuration of fields would support a long sequence. Strip fields are traditionally thought to have been worked using a mouldboard plough because this implement can create the ridging which was necessary for strip demarcation and drainage. At least four examples of eleventh-century or earlier ridge and furrow are known, which would confirm the late Saxon origins of this type of ploughing, as do the references to headlands and baulks in late tenth-century charters.[32] Seven (English) manuscript illustrations of ploughing dating to the late tenth and eleventh centuries exist; all appear to have been well observed. They show a variety of implements: both mouldboard ploughs and ards are illustrated.[33] The archaeological evidence for ploughs is very scarce and entirely consists of iron shares and coulters. In Ireland there appear to have been two sizes of share: one that is up to 170 mm long, which is interpreted as coming from an ard; the other is thicker and up to 260 mm long and is thought to come from a plough, especially as these shares are sometimes associated with a coulter; it is suggested that the longer share was used from the tenth century.[34] In England five shares and one coulter have been found in tenth- or eleventh-century contexts (figure 9.2). On the basis of the Irish evidence, all would be interpreted as shares for ards.[35] The inference from this small amount of evidence may well be that the need for strip fields, let alone an open-field system, was by no means universal and was more dependent on the internal dynamics of each community.[36] Such a particular, essentially local, explanation may account for the increased use of uplands for agricultural purposes, as for example in the

---

[32] Astill 1988b, 73–4; Fox 1981, 83–8.

[33] Fowler 1981, 269.

[34] Brady 1994, 135–8.

[35] Two were found in iron hoards, at Westley Waterless and Nazeing: Morris 1983, 32–7; two were found in the urban centre of Thetford: Rogerson and Dallas 1984, 81–2; Andrews 1995, 97; and one was from a rural site: Addyman 1973, 93–4. These contexts, then, may suggest that this collection was not typical of the implements used in the countryside. None of these reports records the weights of the pieces, as is customary in Scandinavia (see Myrdal, this volume). The documentary and archaeological evidence for late Saxon open fields and ridge and furrow is so widespread that it is difficult to support the idea that the mouldboard plough was a Scandinavian introduction; see Myrdal in this volume for a discussion of this.

[36] As Raepsaet reminds us in this volume, the practised use of an ard could produce something akin to furrows.

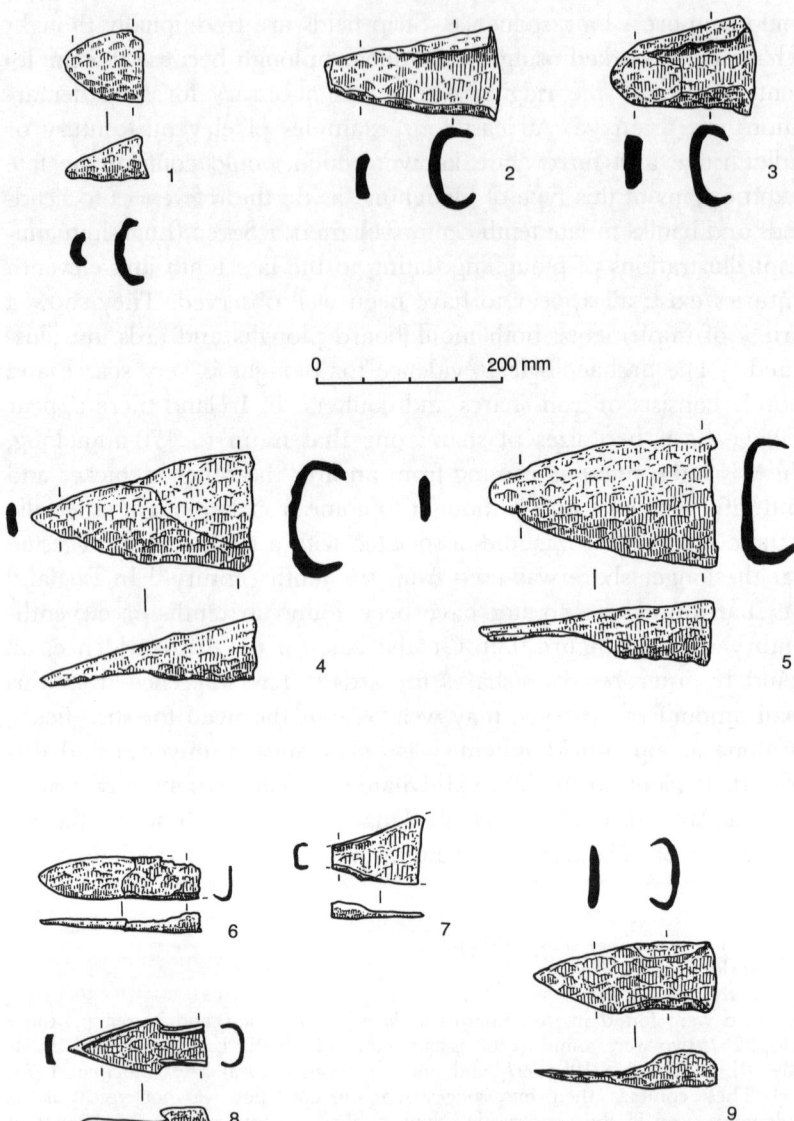

Figure 9.2. Excavated ploughshares from Ireland and England.
1. Whitechurch, County Waterford (ninth-century). *2* and *3.* Fishamble Street, Dublin (tenth-century). *4.* Dundrum, County Down (end of first millennium AD). *5.* Massereene, County Antrim (sixteenth-century). All after Brady 1994, 134. *6.* Nazeing (tenth-century). *7* and *8.* Thetford (tenth-century). *9.* St Neots (tenth-century).

Pennines, and on Dartmoor where settlements were associated with an infield which had been divided into three and then subdivided into strips.[37]

The agrarian rearrangements suggested from the eighth century were also accompanied by other developments: it is noticeable, for example, that building construction techniques changed to produce houses which departed from the type established in the fifth century. Buildings were increasingly constructed using post-in-trench or plank-in-trench techniques, which seems to have enabled the creation of larger structures.[38] The trend of aristocratic involvement in the processing of agricultural surplus and in industrial production continued and emphasizes the importance of such estate centres as Cheddar and Netherton with their iron and fine metalworking and Goltho with its large weaving sheds.[39]

This interpretation differs from that which emphasizes the ninth and early tenth centuries as the period when the English state was formed on the back of an "urban explosion" and an "industrial revolution", which in turn was underpinned by a massive increase in agricultural productivity.[40] The relationship between agricultural innovation, urban development and the market is of course critical, but it is fundamental to discover when towns were sufficiently well established to exercise a profound influence on the countryside in terms of providing marketing facilities and drawing in agricultural surplus. While many "centres of authority" of a royal or ecclesiastical nature, which no doubt had some exchange function, existed in the ninth century, the lack of evidence for an extensive population or a diverse economic base or a widespread currency would suggest these were centres of production and consumption for the aristocracy alone, and often were little different in terms of function from *villae regales* such as Cheddar or indeed thegnly residences like Goltho. It is only in the later tenth or even eleventh centuries that there are clear signs of extensive urban development over most of the country, and this coincides with a more widespread use of currency and the revival of long-distance trade.[41]

---

[37] King 1978; Fleming and Ralph 1982, although there is a possibility that the Dartmoor evidence may be later, to judge from the Bodmin Moor data.

[38] Rahtz 1976, 84–6; Marshall and Marshall 1991, 42.

[39] Rahtz 1979, 252–3; Fairbrother 1990, 244–72; Beresford 1987, 68.

[40] Hodges 1989, 186–202.

[41] Discussed in Astill 1991.

The later tenth century then should be seen as the time when there is some integration between the town and the countryside and when there was a clear need to increase productivity. That this was achieved is suggested by the faunal assemblages of towns like Lincoln, where the steady consumption of meat was apparently met by careful husbandry without distorting the breeding pattern.[42]

Further indirect indications of the need to increase productivity can be seen in the evidence for the reorganization of landscapes, often in association with reclamation. Large areas of the Fens, for example, were permanently settled and cultivated, and the same case can be made for parts of Essex.[43] The plentiful documentary and archaeological evidence for low-lying settlements by the eleventh century would argue for a prior period of reclamation which would have been accomplished by the construction of sea walls: environmental evidence suggests that cereal cultivation in saline conditions took place soon after reclamation. Similar evidence for an extensive drainage scheme, also probably of the late Saxon period, has been proposed for north Somerset. As with most drainage schemes, it is usual to credit monastic communities with this enterprise, in this case Glastonbury Abbey, but lay landlords were also known to be busy colonizing the Levels.[44] It is, however, the case that some monasteries inaugurated a widespread reorganization of their estates during this period. In Glastonbury's case it seems to have involved not only the creation of villages, but also improvements in communications by the cutting of canals, something in which Abingdon Abbey was also involved, in order to improve the navigation of the Thames.[45]

The Twelfth and Thirteenth Centuries

There are some grounds for regarding the late Saxon period, from the tenth century, as marking the high watermark of agricultural achievement when one looks at the evidence for the twelfth century and later. At this time there are a series of reverses in trends which had been initiated in the later Saxon period. Most remarkable is the

---

[42] O'Connor 1982, 47; and see York: O'Connor 1994.
[43] Silvester 1988, 56–60; Rippon 1991.
[44] Murphy 1993; Rippon 1994, 246–52.
[45] Rahtz 1993, 112–8; Costen 1992, 118–19; Bond 1979, 69–71.

decline in the size of sheep and cattle between the twelfth and fifteenth centuries; the animals did not start to regain the stature of their Roman and Saxon predecessors until the sixteenth century. In one respect this is not surprising, because of the shortage of pasture in some parts of the country; the cases of periodontal disease found in thirteenth-century sheep, for example, confirm that over-grazing took place. But one of the most puzzling aspects is the lack of unambiguous evidence for the increase in animal size in the later fourteenth and fifteenth centuries when there was little shortage of pasture, and when there was a clear change in animal husbandry because more meat from younger animals was consumed. The documentary evidence also makes clear how demesne animals were exchanged in order to avoid too much in-breeding. While management technique was proficient in some aspects, it may have fallen short in others, such as the failure to give supplementary feed to young animals or to concentrate on the fattening of mature animals.[46]

A similar reversal, at least for the eleventh and most of the twelfth centuries, is in building techniques, for structures within agricultural settlements appear to become more ephemeral; eleventh- and twelfth-century phases of settlement from village sites, for example, are usually revealed by collections of ceramics rather than extensive structural remains, even when there is no superimposition of later buildings.[47] One explanation of the poor quality of building materials and techniques could be the pressure on such resources, the majority of which may have been diverted into the construction of ecclesiastical and military buildings. In addition the market may not have been sufficiently developed to allow materials or expertise to have been brought in from a distance.

The character of seigneurial sites, secular ones in particular, also changed during the eleventh century. Palaces and castles were no longer the sites where industrial working or the processing of agricultural surpluses took place. The productive functions had disappeared on these sites, which were now mainly concerned with consumption: production seems to have been concentrated within the towns, which also, of course, supplied the aristocracy's luxury needs.[48]

---

[46] For this paragraph see Grant 1988, 158, 176–8.

[47] Hinton 1990, 106–7.

[48] Compare, for example, the tenth- and eleventh-century phases with those of the twelfth century and later at Cheddar and Goltho: Rahtz 1979, 374–8; Beresford 1987, 8–14.

During the twelfth century there were clear signs of the expansion of arable cultivation into upland areas, areas which were previously used for rough grazing. The expansion is perhaps most obvious archaeologically in the southwest, where surveys of Bodmin Moor and Dartmoor have recorded hamlets with associated infields, some of which have subdivisions.[49] These results complement similar evidence for reclamation and colonization from the Fens and parts of East Anglia, as well as the assorting which took place in woodland-pasture regions in the midlands.[50]

It is usually accepted that those aspects of agricultural technology which are susceptible to immediate archaeological enquiry, such as agricultural tools, appear to have changed little during this period: the range of tools was small, with many tools being multi-purpose, and there was a remarkable consistency over most of the country.[51] There were, however, variations in implements such as ploughs, whose character was essentially determined by soil type and topography. In regions where it was important to control the depth of ploughing, for example in areas of well-drained soils such as the chalk downlands, the wheeled plough was preferred. In areas of heavier soils where there was still a need to regulate ploughing depth, foot ploughs were used; where the ploughing could be controlled entirely by the farmer, the swing plough was used.[52] Manorial accounts show that ploughs needed regular maintenance, and that some estates had annual contracts with smiths who were to supply and fashion the iron into shares and coulters; this was not an inconsiderable amount—16 lb (7.3 kg) for a share and between 8.75 and 10 lb (4–4.5 kg) for a coulter. Even with such attention, it was rare for a demesne plough to last for more than one year.[53] Wear could be reduced by using different ploughs: it was for instance usual to use the heaviest plough for breaking up the fallow. Pebbles were also inserted into the sole and axle of a plough to reduce the wear on the wooden parts. Worn and striated pebbles are common finds in Scandinavia, and in Ireland, where they are mainly dated to the thirteenth century, but they have also been found in Scotland and in England north of the Humber;

---

[49] Johnson and Rose 1994, 77–98; Fleming and Ralph 1982.
[50] Silvester 1988, 60–2; Dyer 1991, 27–32.
[51] Langdon 1988, 95–107.
[52] This information only concerns demesne ploughs: Langdon 1986, 129–41.
[53] Britnell 1991, 204–5.

none are recorded to the south, and this may again indicate a regional variation in plough types.[54]

The mouldboard plough was, however, not necessarily the universal cultivation tool; cross-ploughing, presumably using an ard, was still practised in the twelfth century, and it is important to remember that spades were used to cultivate large areas.[55] We are most aware of spade cultivation in the development of "plaggen" soils in infields such as those discussed by Widgren in this volume, but it was clearly also a common practice in the islands of Scotland. In his discussion about the potentially post-medieval introduction of open-field cultivation on the islands, Dodgshon argues that the medieval fields were more akin to infields. He also reminds us of the superiority of spade cultivation over ploughing, and raises the possibility of spading as a real alternative method of cultivation at times of maximum population pressure.[56] Iron-shod spades are relatively common archaeological finds from the tenth century, and wooden shovels with detachable, unshod, blades have been found, mainly dated from the twelfth century onwards.[57]

Further indications of pressure on existing resources have been observed in the earthwork evidence of fields, but this is difficult to date, although some field arrangements are known to have reached their final form by the thirteenth century. The creation of new furlongs, which can be seen as earthworks overlying the previous arrangements, is the most common example. More dramatic are those cases where steep slopes which show signs of soil slippage have been ploughed over.[58]

Given our limited knowledge about medieval tools, it might be beneficial to approach the subject by considering if there was any potential for improving the quality of the available implements, and in the process discover more about the control of technological processes. The easiest way to improve the effectiveness and durability of wooden tools was to sheath the working parts with iron, and I now intend to consider the problem of the availability and supply of iron.

Smithies are generally regarded as one of the most common features of medieval villages, and this view is largely based on the records

[54] Lerche 1970; Clarke 1972; Brady 1988.
[55] Webster and Cherry 1972, 205.
[56] Dodgshon 1994.
[57] Goodall 1981, 55–6; Addyman 1976, 319–20; Morris 1980.
[58] Hall 1995, 135; RCHM 1975, 84, 123–5.

of iron-working services owed by tenants which occur in twelfth- and thirteenth-century estate surveys.[59] Archaeologists have drawn similar conclusions, and therefore it is surprising to discover how few have been excavated.[60] Of the eleven excavated smithies in England, only four were located in villages and all are dated to the later fourteenth or fifteenth centuries: the low number and the date are both significant because the excavated archaeological sequences of most villages end before the later fourteenth century. Village smithies, therefore, have come from the least well represented periods in our stock of excavations; this would suggest that the late medieval smithies reflect a real chronological distribution rather than fortuitous discoveries. The archaeological work on villages has also demonstrated the absence of specialized buildings until, that is, the appearance of smithies on the few which survived into the later middle ages.[61] Does this indicate that the demand for iron within a village was insufficient to warrant a smith, and that perhaps it was satisfied by itinerant ironworkers or through other agencies?

It is of course difficult, if not impossible, to assess how much iron was used in the countryside. Ideally it would be useful to record the amount of iron which was used in the most common implements, such as plough shares and coulters, as has been done in Sweden (Myrdal, this volume), but very few survive for England after the eleventh century. An alternative may be to study the occurrence and weight of horseshoes. Items of horse furniture are some of the most frequent artefacts found on village sites, but the lack of stratigraphy often makes it difficult to date them. This difficulty does not exist for urban assemblages however, and the most recent typology shows that shoes between the eleventh and early fourteenth centuries weighed between 70 and 130 grams and were fixed by six nails. From about 1300 the weight of the shoe increases to, and then exceeds, 250 grams and was fixed by eight nails; oxen appear to have been shoed more frequently during the same time.[62] Increased weight and fixings clearly indicate an increase in the use of iron during the later middle ages, and that this might be a general trend is suggested by the increase in iron and steel imports during the fifteenth century.[63]

---

[59] Harvey 1990.
[60] Hurst 1988, 926–7.
[61] For the data on which this is based, see Astill 1995.
[62] Clark 1986.
[63] Childs 1981.

Where was the iron obtained if there were no local smiths? The open market was the obvious choice, but how much iron was actually available? To judge from the vast numbers of high-status buildings that were constructed, iron was not in short supply; most of this, however, appears to have been specially commissioned and wrought on site rather than bought on the market. Manorial accounts also show that iron was forged to order, and indeed that smiths were sometimes responsible for finding their own raw materials. But it is also true that the same accounts show that significant quantities of iron, especially nails, were bought in, presumably on the open market. The crown's right to levy goods from the countryside, especially to support military campaigns ("purveyance") includes iron, and the large amounts demanded in such a short time could be used to demonstrate the ready availability of the material.[64]

This is another case of different messages coming from the documentary and archaeological material. Archaeological evidence could be used, for example, to suggest that there was a general dearth of metalwork, and especially iron, in the eleventh and twelfth centuries. The few eleventh-century hoards of ironwork consist entirely of scrap-metal, some of which dates from the Roman period, indicating the need to recycle stocks and a shortage of new iron. It is also clear that the towns where large quantities of iron were smelted in the late Saxon period, such as Stamford and Norwich, had a much reduced industry in the eleventh and twelfth centuries.[65] It has also been estimated that even the exceptionally large demands for iron to equip a military campaign could have been met by a short and intense period of smelting and forging; if such heavy and irregular demands could be satisfied in this way, there was no real incentive to develop the iron-producing industry.[66] The other factor in the equation would be of course the iron produced in towns, but even now, after a long period of urban excavation, it is still "difficult to establish the scale of the enterprise and the actual processes being carried out". Even in a comparatively well-explored town like Winchester, ironworking was confined to smithing, was small-scale and served "both visitors and permanent residents from scattered workshops"; Biddle also raises the possibility that a proportion of smithies

---

[64] See, for example, Langdon 1992, 69–71; Maddicott 1975, 15–24.
[65] Morris 1983; Morris 1988; Hinton 1990, 106–64.
[66] Hinton 1990, 155.

were located within the private royal, and especially ecclesiastical, urban households.[67]

To return to our sample of eleven excavated smithies, the remaining seven were on seigneurial sites, the majority (five) on monastic properties, and date from the twelfth to the fourteenth centuries. It is possible, therefore, that there was not a great deal of iron available on the open market for smiths to work and that the bulk had to be obtained from seigneurial sources which may have only catered for the needs of their estates.

One of the excavated smithies was part of a metalworking watermill at the Cistercian abbey of Bordesley, and dates from the later twelfth to later fourteenth centuries. The preponderance of documentary references to corn mills implies that the investment in the construction of mills was normally only recovered if they were used for cereals because the demand for cloth or metalwork was insufficient to justify industrial watermills. Potentially the archaeological evidence may be at odds with such an interpretation. A mere five medieval watermills have been excavated, of which only two have produced definite evidence for cereal processing; the other two are assumed to be corn mills in view of the general documentary work on mills. It is critical to gain positive evidence of function because we now realize that the actual metalworking sites are often notable for the absence of the residues (such as slags) we would logically expect to be present: in fact such materials were invariably recycled. There is thus a potential danger that the archaeological evidence for industrial mills could be overlooked because of the character of the documentary record, especially as regards the indirect evidence for fulling mills (see below). This is, of course, not to deny that non-corn mills must always have been in the minority, but we should be aware that even a comparatively tiny number could disproportionately affect the use of iron in particular areas. It is probable, for example, that the metalworking mill at Bordesley Abbey had a capacity to produce metalwork far in excess of what was required by that monastic community, and that the excess was sold (figure 9.3). The mill also provided facilities to recycle broken implements. It is even conceivable, given the paucity

---

[67] Schofield and Vince 1994, 104; Biddle 1990, 138. The archaeological evidence is especially scarce for the small medieval market towns which could have been a source for much peasant ironwork.

Figure 9.3. A reconstruction of the twelfth-century metalworking watermill from Bordesley Abbey, Worcestershire, England (Astill 1993).

of ironwork, that metalworking remained a seigneurial privilege, as indeed was corn milling, during the period of direct farming.[68]

Even if this speculation is not accepted, the preponderance of seigneurial smithies excavated, and a potential shortage of iron, would argue that the supply of metalwork was more closely influenced by the aristocracy than the market, and this could have resulted in most of the population having less efficient tools during the thirteenth century. The high valuation placed on iron-shod tools in contemporary manorial accounts and court rolls suggests the costliness of the material.[69]

---

[68] For the previous paragraph, see Astill 1993, 299–304.
[69] Dyer 1989, 170–2.

At the beginning of the "long thirteenth century", however, there are clear indications of a technological breakthrough which is comparable to the situation in the eighth century. During the last quarter of the twelfth century there is a significant change in woodworking technology. The techniques for full timber framing—such as pegged mortices—appear in association with the first evidence for the use of saws and chisels (as opposed to the previous use of adzes). The development of full timber framing meant that two-stage, and more, buildings could be constructed. At precisely the same time there is additional evidence for a more substantial building tradition—one that abandons the earthfast construction technique in favour of framed buildings resting on dwarf stone walls or padstones. Ceramic roof tiles were also introduced at the same time, and it is remarkable that a much more sophisticated tradition of pottery production, involving decorated and painted forms, also appeared. In earlier periods such a break in the material culture would have been sufficient to posit an invasion. The evidence for such a change is limited at the moment to London and some monastic sites, but, as far as the regions are concerned, it could indicate that these innovations were socially specific. For example, the use of ceramic tiles or padstones for buildings is to be seen in a region's monasteries some fifty years before they were adopted in the major towns, let alone in villages.[70]

Indeed the late twelfth century should be seen as a time of immense innovation, the evidence for which is largely confined to seigneurial properties. The most impressive examples are probably the water engineering feats which were necessary in order to create suitable locations for, and to supply, monastic precincts, but we should also remember the canals which were cut at the same time as part and parcel of town plantation schemes.[71] A combination of documentary research and archaeological fieldwork has also identified settlements which had a specialized function. The most obvious examples are the Cistercian granges, but other sites with a very particular arrangement have been identified as vaccaries and bercaries.[72] This specialization is also evident in industrial contexts. In the northwest, for example, complex water diversions have been interpreted

---

[70] Brigham 1992, 93; Astill 1993, 295–8.

[71] Bond 1989; for example Rhuddlan and Alresford: Beresford 1967, 37, 109, 177.

[72] Moorhouse 1989, 44–8.

as fulling mill sites, associated with tenter banks and potash pits, which can be dated to the twelfth and thirteenth centuries and which were in the possession of not only monasteries but also local lay lords. There was also a firm geographical association between these fulling sites and flax-retting ponds, and so it is possible that a linen industry was connected with a demesne wool enterprise, and that the linen production may have been combined with the cloth in the estate accounts, giving the impression that "industrial crops" were not exploited (figure 9.4).[73]

The new form of timber jointing, the pegged mortice, which allowed the development of a fully framed timber building may also help explain the genesis of the main invention of the middle ages, the windmill. The pegged mortice was first used in the 1180s, and allowed the construction of an extremely rigid frame which was used not only for buildings but also for bridges.[74] It is also the kind of joint that would have been necessary in order to construct the base frame on which a windmill rotated, that is the crosstrees and the diagonal braces or quarter bars. The first explicit documentary references to windmills in England occur in the decade between 1180 and 1190, and their invention and rapid adoption at this time may have been due to the development of this new form of jointing.[75]

The archaeological evidence, then, would suggest that there are strong indications of a seigneurial initiative for rural specialization in the late twelfth and thirteenth centuries, that is during the period of high farming. Recent surveys of medieval settlements and fields in parts of Lincolnshire and East Anglia have produced a considerable amount of evidence for the rearrangement of settlements and their associated fields which, it is suggested, occurred as late as the later twelfth century. Although the initiative for such alterations will continue to be debated, they should perhaps be considered in the context of the other contemporary changes discussed above.[76]

---

[73] Higham 1989.

[74] Brigham 1992; Rigold 1975, 88.

[75] For an excavated site of a windmill and reconstruction: Mynard and Zeepvat 1992, 104–7; Holt 1988, 20, 171–5.

[76] Everson *et al.* 1991, 36–7.

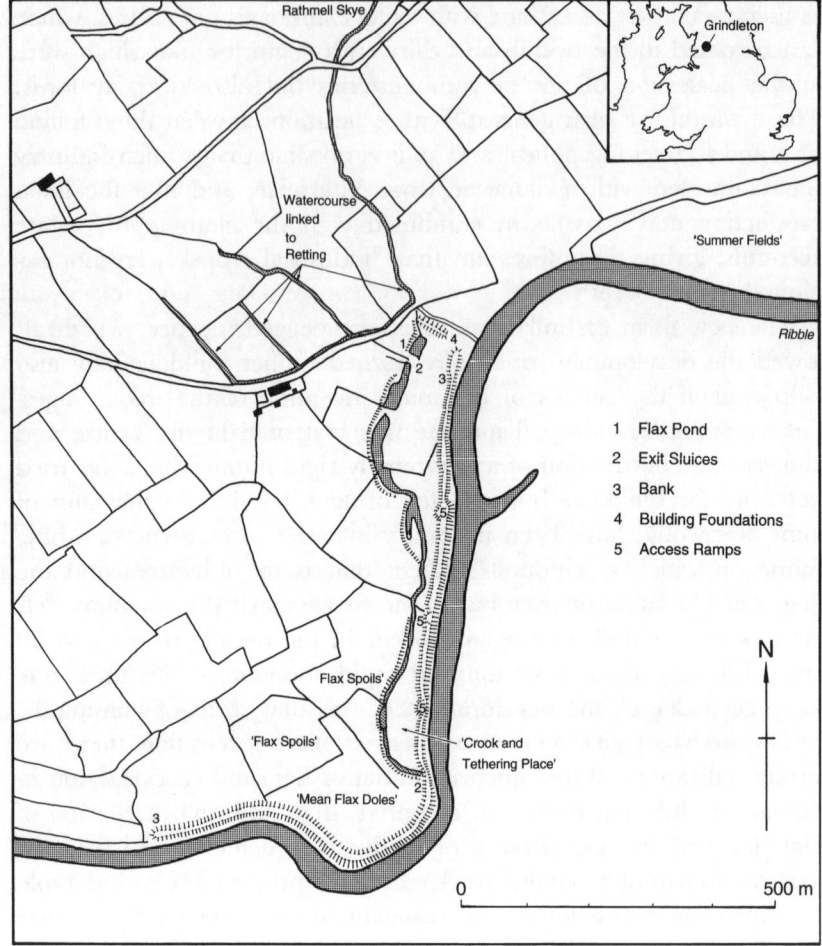

Figure 9.4. Evidence for medieval flaxworking in Lancashire, England (Higham 1989).

The Later Fourteenth to Sixteenth Centuries

The reduction in population and the dramatic shift from arable to pastoral farming in many parts of the country from the mid-fourteenth century provide the context for significant changes in agricultural practice which are to be seen archaeologically for the first time at the non-seigneurial level. It is only in the later fourteenth and fifteenth centuries that there is evidence of specialization in villages, with all its implications of capital investment. In areas where animal husbandry dominated, crew yards for the over-wintering of cattle to conserve pasture, as at Barton Blount, Goltho and Wawne, became a particular characteristic of rural settlements. The trend towards specialization was also to be seen in areas which remained primarily cereal producing regions; in this case large barns were constructed, some with grain-drying facilities, to ensure long-term storage of cereals, as at Grenstein and Caldecote (figure 9.5). The plentiful documentary references to barns in the later middle ages, particularly in the fifteenth century, represent an investment at the tenant level which the aristocracy were making during the thirteenth and fourteenth centuries. These, and of course the occurrence of smithies, point to a period of investment which was designed to improve the efficiency of agriculture.[77]

It is this kind of specialization at the peasant level, coupled with the improvement of the communications and marketing infrastructure, that allowed areas of England to specialize in the production of surpluses which were most suited to that particular region. In other words, we are starting to see the development of distinctive farming regions which were released from the demands of subsistence production.[78]

The development of the water-powered blast furnace enabled more efficient iron production and helped to confine such activities to the areas most suited to the current methods.[79] Such innovations depended on an efficient marketing infrastructure, and this also probably facilitated the increased exchange of breeding stock which may explain the remarkable change in the size of sheep which anticipates the known developments of the eighteenth century.[80]

The foregoing discussion has emphasized the role of the aristocracy in the process of agricultural innovation, particularly in the eighth to ninth centuries, and during the "long thirteenth century". This is in potential conflict with the current thrust of documentary research, some of which is presented in this volume, which stresses the innovatory role of the peasantry. The nature of the archaeological record may of course be partly responsible for the discrepancy: it is possible that technological innovations may only be visible archaeologically in those situations where there was considerable capital investment. Some innovations, for example, may have been practised on a smaller

[77] Astill 1988a, 54–7.
[78] Astill and Grant 1988, 224–9.
[79] Crossley 1990, 153–62.
[80] Dobney et al. nd, 59–61; Albarella and Davis 1994, 42–58.

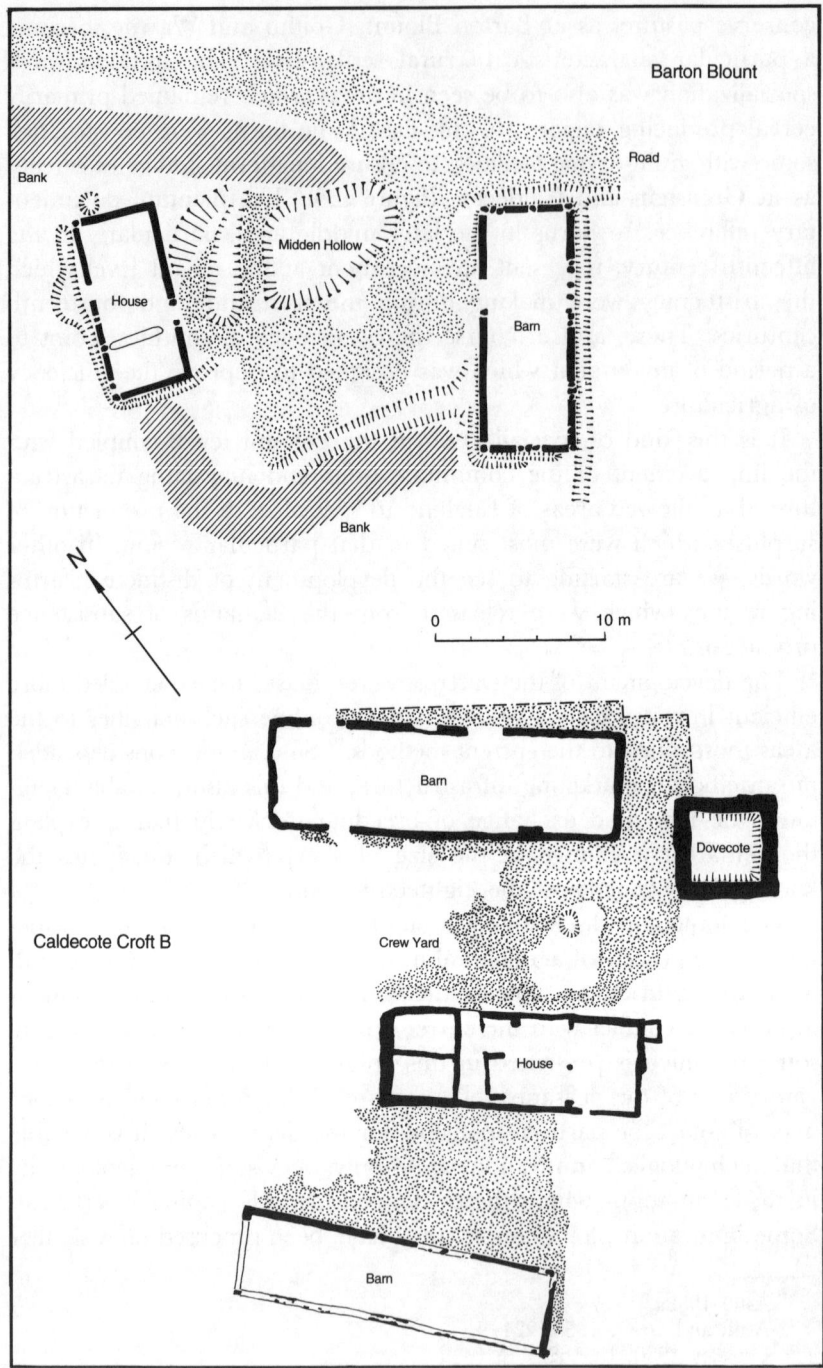

Figure 9.5. Examples of specialization in late medieval agriculture. A crew-yard from Barton Blount (Derbyshire) and barns from Caldecote (Hertfordshire), England (Astill 1988a).

scale at the peasant level and not left any archaeological trace. However, it must be said that most of our archaeological information about medieval peasant society comes from work done on nucleated settlements. What is so remarkable about the excavated village sites is the lack of evidence (until the later fourteenth century) for any social or economic differentiation. This had clear advantages: the uniformity of the buildings allowed flexibility, so that subsidiary buildings could be converted into additional residential buildings for older family members, or for premises to dry grain. It is just this flexibility which has been commented upon in discussions about the increased peasant use of the horse for ploughing and hauling at this time.[81] The evidence for some aristocratic investment in more specialized operations is in contrast to the, admittedly more limited, peasant information which stresses the flexibility which can be obtained from non-specialized investment.

A further consideration must be the extent to which the village community exercised a brake on individual enterprise. The community, if only through its regulation of the farming year, clearly exercised a strong influence on peasant behaviour; the well-defined tofts might be interpreted as an individual family's statement of privacy or independence from the community. But the remarkable homogeneity of peasant buildings is in stark contrast to the considerable documentary evidence for an economically differentiated peasantry; this might suggest that there were communally agreed ways of expressing wealth and individuality, and, that if these norms were transgressed, the offending families were punished, as has been documented in contemporary South American peasant societies.[82] It is remarkable that when there is a documented weakening of the village community from the later fourteenth century, that is also the time when there is evidence in the archaeological record of peasant innovation. This is another reason for investigating the archaeology of dispersed settlements, because if this hypothesis about the influence of the village community is correct, there should be more signs of peasant innovation in the non-nucleated regions of England.

The aristocracy's role in technological development may well have been more managerial than innovatory (see Dyer, this volume), but it is important to remember that some diffusion of technological ideas

---

[81] Astill 1988a, 58–9; Langdon 1986, 283–4.
[82] Astill 1988a, 53; Wilk 1990, 38–42.

probably took place at an aristocratic level, whether it was via the elevated dissemination of an international court culture or through the regulation of monastic orders.

This survey also suggests a more ambivalent role for the market and urbanization in the process of technological innovation. It is possible for example that increased agricultural productivity, as implied in the rearrangement of settlements and fields, was a long process, with pronounced regional variations because there was little external stimulus, until the rapid growth of towns which may not have occurred until the late tenth or early eleventh century. And the consideration of the availability of iron suggests that, even when there is plentiful evidence for marketing which involved all sections of the population, some commodities could not have been purchased because the supply was limited and/or was strictly controlled or was being diverted to non-agricultural projects.

Bibliography

Addyman, P. V. 1973, "Late Saxon settlements in the St. Neots area, part 3", *Proceedings of the Cambridge Antiquarian Society* 64, 45–99.

—— 1976, "Archaeology and Anglo-Saxon society". In *Problems in economic and social archaeology*, eds G. Sieveking, I. Longworth and K. Wilson, London, 309–22.

Albarella, U. and Davis, S. J. M. 1994, "Mammals and birds from Launceston Castle, Cornwall: decline in status and the rise of agriculture", *Circaea* 12, 1–156.

Andrews, P. 1995, "Excavations at Redcastle Furze, Thetford, 1988–9", *East Anglian Archaeology* 72.

Astill, G. G. 1988a, "Rural settlement: the toft and the croft". In *The countryside of medieval England*, eds G. G. Astill and A. Grant, Oxford, 36–61.

—— 1988b, "Fields". In *The countryside of medieval England*, eds G. G. Astill and A. Grant, Oxford, 62–85.

—— 1991, "Towns and town hierarchies in Saxon England", *Oxford Journal of Archaeology* 10, 95–117.

—— 1993, *A medieval industrial complex and its landscape: the metalworking watermills and workshops of Bordesley Abbey*, York.

—— 1995, "Iron smithing in medieval England: a review". In *The importance of ironmaking: technological innovation and social change*, ed. G. Magnusson, Stockholm, 183–93.

Astill, G. G. and Grant, A. 1988, "The medieval countryside: efficiency, progress and change". In *The countryside of medieval England*, eds G. G. Astill and A. Grant, Oxford, 213–34.

Astill, G. G. and Lobb, S. J. 1989, "Excavation of prehistoric, Roman and Saxon deposits at Wraysbury, Berkshire", *Archaeological Journal* 146, 68–134.

Aston, M. A. and Costen, M. D. (eds) 1993, *The Shapwick project. A topographical and historical study. Fourth report*, Bristol.

Austin, D. 1990, "The 'proper study' of medieval archaeology". In *From the Baltic to the Black Sea; studies in medieval archaeology*, eds L. Alcock and D. Austin, London, 9–42.

Bennett, K. 1983, "Devensian late-glacial and Flandrian vegetational history at Hockham Mere, Norfolk, England", *New Phytologist* 95, 457–87.

Beresford, G. 1987, *Goltho: the development of an early medieval manor*, London.

Beresford, M. W. 1967, *New towns of the middle ages*, London.

Beresford, M. W. and Hurst, J. G. 1990, *Wharram Percy deserted medieval village*, London.

Biddle, M. 1990, *Winchester studies 7. Artefacts from medieval Winchester. Part 2. Object and economy in medieval Winchester*, Oxford.

Bloch, M. 1967, *Land and work in medieval Europe*, London.

Bond, C. J. 1979, "The reconstruction of the medieval landscape: the estates of Abingdon Abbey", *Landscape History* 1, 59–75.

—— 1989, "Water management in the rural monastery". In *The archaeology of rural monasteries*, eds R. Gilchrist and H. Mytum, Oxford, 83–111.

Bourdillon, J. 1988, "Countryside and town: the animal resources of Saxon Southampton". In *Anglo-Saxon settlements*, ed. D. Hooke, Oxford, 176–96.

Brady, N. 1988, "The plough pebbles of Ireland", *Tools and Tillage* 6, 47–60.

—— 1994, "Labor and agriculture in early medieval Ireland: evidence from the sources". In *The work of work: servitude, slavery and labor in medieval England*, eds A. J. Frantzen and D. Moffat, Glasgow, 125–45.

Brigham, T. 1992, "Reused house timbers from the Billingsgate site, 1982–3". In G. Milne, *Timber building techniques in London c. 900–1400*, London.

Brisbane, M. 1988, "Hamwic (Saxon Southampton): an eighth-century port and production centre". In *The rebirth of towns in the west AD 700–1050*, eds R. Hodges and B. Hobley, London, 101–8.

Britnell, R. 1991, "Farm practices and techniques: eastern England". In *The agrarian history of England and Wales, volume 3, 1348–1500*, ed. E. Miller, Cambridge, 194–210.

Brooks, N. 1971, "The development of military obligations in eighth- and ninth-century England". In *England before the Conquest. Studies in primary sources presented to Dorothy Whitelock*, eds P. Clemoes and K. Hughes, Cambridge, 69–84.

Butterworth, C. A. and Lobb, S. J. 1992, *Excavations in the Burghfield area, Berkshire. Developments in the Bronze Age and Saxon landscapes*, Salisbury.

Campbell, B. M. S. 1981, "Commonfield origins—the regional dimension". In *The origins of open-field agriculture*, ed. T. Rowley, London, 112–30.

Campbell, G. 1994, "The preliminary archaeobotanical results from Anglo-Saxon West Cotton and Raunds". In *Environment and economy in Anglo-Saxon England*, ed. J. Rackham, York, 65–82.

Carver, M. O. H. (ed.) 1986, *Bulletin of the Sutton Hoo Research Committee 4*.

Childs, W. R. 1981, "England's iron trade in the fifteenth century", *Economic History Review* 34, 25–47.

Clark, J. 1986, "Medieval horseshoes", *Datasheet 4*, Finds Research Group.

Clarke, D. V. 1972, "A plough pebble from Coulston, Scotland", *Tools and Tillage* 2, 50–1.

Coggins, D., Fairless, K. and Batey, C. 1983, "Simy Folds: an early medieval settlement site in Upper Teesdale", *Medieval Archaeology* 27, 1–27.

Costen, M. 1992, *The origins of Somerset*, Manchester.

Crabtree, P. 1994, "Animal exploitation in East Anglian villages". In *Environment and economy in Anglo-Saxon England*, ed. J. Rackham, York, 40–54.

Crossley, D. 1990, *Post-medieval archaeology in Britain*, Leicester.

Cunliffe, B. W. 1972, "Saxon and medieval settlement pattern in the region of Chalton, Hampshire", *Medieval Archaeology* 16, 1–12.

Dobney, K. M., Jaques, S. D., and Irving, B. G. nd, *Of butchers and breeds. Report on vertebrate remains from various sites in the city of Lincoln*, Lincoln.

Dodgshon, R. A. 1994, "Rethinking highland field systems". In *The history of soils and field systems*, eds S. Foster and T. C. Smout, Aberdeen, 53–65.

Dyer, C. C. 1989, *Standards of living in the later middle ages*, Cambridge.

—— 1991, *Hanbury: settlement and society in a woodland landscape*, Leicester.

Everson, P. L., Taylor, C. C., and Dunn, C. J. 1991, *Change and continuity. Rural settlement in north-west Lincolnshire*, London.

Fairbrother, J. R. 1990, *Faccombe Netherton. Excavations of a Saxon and medieval manorial complex*, London.

Fenton, A. 1963, "Early and traditional cultivating implements in Scotland", *Proceedings of the Society of Antiquaries of Scotland* 96, 264–317.

—— 1994, "Field systems and cultivating implements". In *The history of soils and field systems*, eds S. Foster and T. C. Smout, Aberdeen, 75–82.

Fleming, A. and Ralph, N. 1982, "Medieval settlement and land use on Holne Moor, Dartmoor: the landscape evidence", *Medieval Archaeology* 26, 101–37.

Fowler, P. J. 1981, "Farming in the Anglo-Saxon landscape: an archaeologist's view", *Anglo-Saxon England* 9, 263–80.

Fox, H. S. A. 1981, "Approaches to the adoption of the midland system". In *The origins of open-field agriculture*, ed. T. Rowley, London, 64–111.

Goodall, I. H. 1981, "The medieval blacksmith and his products". In *Medieval industry*, ed. D. W. Crossley, London, 51–62.

Grant, A. 1988, "Animal resources". In *The countryside of medieval England*, eds G. G. Astill and A. Grant, Oxford, 149–87.

Hall, D. 1995, *The openfields of Northamptonshire*, Northamptonshire Record Society 38, for 1990–2, Northampton.

Hamerow, H. F. 1991, "Settlement mobility and the 'middle Saxon shift': rural settlements and settlement patterns in Anglo-Saxon England", *Anglo-Saxon England* 20, 1–17.

Harrison, B. J. D. and Roberts, B. K. 1989, "The medieval landscape". In *The North York Moors: landscape heritage*, eds D. A. Spratt and B. J. D. Harrison, Newton Abbot.

Harvey, P. D. A. 1990, "Non-agrarian activities in twelfth-century English estate surveys". In *England in the twelfth century*, ed. D. Williams, Woodbridge, 101–11.

Haslam, J. 1980, "A middle Saxon smelting site at Ramsbury, Wiltshire", *Medieval Archaeology* 24, 13–68.

Hayes, P. 1988, "Roman to Saxon in the south Lincolnshire Fens", *Antiquity* 62, 321–6.

Higham, M. C. 1989, "Some evidence for twelfth- and thirteenth-century linen and woollen textile processing", *Medieval Archaeology* 33, 38–52.

Hinton, D. A. 1990, *Archaeology, economics and society. England from the fifth to the fifteenth century*, London.

—— 1994, "The archaeology of eighth- to eleventh-century Wessex". In *The medieval landscape of Wessex*, eds M. A. Aston and C. Lewis, Oxford, 33–46.

Hodges, R. A. 1989, *The Anglo-Saxon achievement*, London.

Holt, R. A. 1988, *The mills of medieval England*, Oxford.

Hurst, J. G. 1988, "Rural building in England". In *The agrarian history of England and Wales, volume 2*, ed. H. E. Hallam, Cambridge, 854–930.

Johnson, N. and Rose, P. 1994, *Bodmin Moor: an archaeological survey. Volume 1: the human landscape to c. 1800*, London.

King, A. 1978, "Gauber High Pasture, Ribblehead—an interim report". In *Viking age York and the north*, ed. R. Hall, London, 21–5.

Langdon, J. 1986, *Horses, oxen and technological innovation*, Cambridge.

—— 1988, "Agricultural equipment". In *The countryside of medieval England*, eds G. G. Astill and A. Grant, Oxford, 86–107.

Langdon, J. 1992, "The birth and demise of a medieval windmill", *History of Technology* 14, 54–76.

Lerche, G. 1970, "Pebbles from wheelploughs", *Tools and Tillage* 1, 150.

Losco-Bradley, P. M. and Salisbury, C. R. 1988, "A Saxon and Norman fish weir at Colwick, Nottinghamshire". In *Medieval fish, fisheries and fishponds in England*, ed. M. Aston, Oxford, 329–51.

Maddicott, J. R. 1975, *The English peasantry and the demands of the crown, 1294–1341*, Oxford.

Marshall, A. and Marshall, G. 1991, "A survey and analysis of the buildings of early and middle Anglo-Saxon England", *Medieval Archaeology* 35, 29–43.

Moorhouse, S. 1989, "Monastic estates: their composition and development". In *The archaeology of rural monasteries*, eds R. Gilchrist and H. Mytum, Oxford, 29–81.

Morris, C. A. 1980, "A group of early medieval spades", *Medieval Archaeology* 24, 205–10.

—— 1983, "A late Anglo-Saxon hoard of iron and copper-alloy artefacts from Nazeing, Essex", *Medieval Archaeology* 27, 27–39.

—— 1988, "Note on iron objects 331–42". In T. Darvill, "Excavations on the site of the early Norman castle at Gloucester, 1983–4", *Medieval Archaeology* 32, 32–9.

Murphy, P. 1993, "Anglo-Saxon arable farming on the silt fens—preliminary results", *Fenland Research*, 75–9.

—— 1994, "The Anglo-Saxon landscape and rural economy: some results from sites in East Anglia and Essex". In *Environment and economy in Anglo-Saxon England*, ed. J. Rackham, York, 23–39.

Mynard, D. C. and Zeepvat, R. J. 1992, *Excavations at Great Linford, 1974–80*, Aylesbury.

O'Connor, T. P. 1982, *Animal bone from Flaxengate, Lincoln c. 870–1500*, London.

—— 1994, "8th–11th century economy and environment in York". In *Environment and economy in Anglo-Saxon England*, ed. J. Rackham, York, 136–47.

Palliser, D. M. 1993, "Domesday Book and the 'Harrying of the North'", *Northern History* 29, 1–23.

Rackham, O. 1994, "Trees and woodland in Anglo-Saxon England: the documentary evidence". In *Environment and economy in Anglo-Saxon England*, ed. J. Rackham, York, 1–6.

Rahtz, P. A. R. 1976, "Buildings and rural settlement". In *The archaeology of Anglo-Saxon England*, ed. D. M. Wilson, London, 49–98.

—— 1979, *The Saxon and medieval palaces at Cheddar*, Oxford.

—— 1993, *Glastonbury*, London.

Rahtz, P. A. R. and Meeson, R. 1992, *An Anglo-Saxon watermill at Tamworth*, London.

RCHM 1975, Royal Commission on Historic Monuments, *North-east Northamptonshire*, London.

Rigold, S. E. 1975, "Structural aspects of timber bridges", *Medieval Archaeology* 19, 48–91.

Rippon, S. 1991, "Early planned landscapes in south-east Essex", *Essex History and Archaeology* 22, 46–60.

—— 1994, "Medieval wetland reclamation". In *The medieval landscape of Wessex*, eds M. A. Aston and C. Lewis, Oxford, 239–53.

Rogerson, A. and Dallas, C. 1984, "Excavations in Thetford, 1948–59 and 1973–80", *East Anglian Archaeology* 22.

Rynne, C. 1989, "The introduction of the vertical watermill into Ireland: some recent archaeological evidence", *Medieval Archaeology* 33, 21–31.

Salisbury, C. 1995, "An eighth-century Mercian bridge over the Trent at Cromwell, Nottinghamshire, England", *Antiquity* 69, 1015–18.

Schofield, J. and Vince, A. 1994, *Medieval towns*, Leicester.

Schove, D. J. 1979, "Dark age tree-ring dates", *Medieval Archaeology* 23, 219–23.

Shoesmith, R. 1982, *Hereford city excavations 2: excavations on and close to the defences*, London.

Silvester, R. J. 1988, "The Fenland project number 3: Norfolk survey, Marshland and Nar valley", *East Anglian Archaeology* 45.

Taylor, C. C. 1975, *Fields in the English landscape*, London.

Thirsk, J. 1964, "The common fields", *Past and Present* 29, 3–29.

Wade, K. 1988, "Ipswich". In *The rebirth of towns in the west AD 700–1050*, eds R. Hodges and B. Hobley, London, 93–100.

Webster, L. and Cherry, J. 1972, "Medieval Britain in 1971", *Medieval Archaeology* 16, 147–212.

Welch, M. 1985, "Rural settlement patterns in the early and middle Anglo-Saxon periods", *Landscape History* 7, 13–25.

—— 1992, *Anglo-Saxon England*, London.

Wilk, R. R. 1990, "The built environment and consumer decisions". In *Domestic architecture and the use of space; an interdisciplinary, cross-cultural study*, ed. S. Kent, Cambridge, 34–42.

Williamson, T. 1993, *The origins of Norfolk*, Manchester.

Wilson, D. M. and Hurst, J. G. 1958, "Medieval Britain in 1957", *Medieval Archaeology* 2, 183–213.

10. ECONOMIC RENT AND THE INTENSIFICATION OF ENGLISH AGRICULTURE, 1086–1350

Bruce M. S. Campbell

The eleventh, twelfth, and thirteenth centuries witnessed a European-wide expansion in population and economic activity.[1] Although the trend is unmistakable, in England alone is it possible to quantify the scale on which landuse was transformed. Thus, on the evidence of the number of ploughteams recorded by Domesday Book, the arable area may be estimated at approximately 8.5 million acres (3.4 million hectares) in 1086.[2] By 1300, after two centuries of active reclamation and colonization, it is unlikely to have exceeded the 11.5 million acres (4.7 million ha) attained at the height of the ploughing-up campaign of the Napoleonic Wars.[3] By dint of much hard effort—assarting, draining and reclaiming, and making good the devastation wrought by William I in the north—the cultivated area was thus extended by approximately a third.[4] Yet over the same period the population grew from an estimated 1.5–2.5 million in 1086 to an estimated 3.8–7.2 million *c.* 1300.[5] Although there is much debate over which are the most acceptable of these various global estimates, few would dispute that the population at least doubled and may even have trebled.[6] As a result the amount of arable land *per caput* roughly halved over this 200-year period. The decline in the *per caput* supply of grassland and woodland is likely to have been even greater.

Such a substantial reduction in the ratio of land to people lends substance to those who have argued that the twelfth and thirteenth centuries exemplify the classic Malthusian scenario of population outstripping available food supplies. Indeed, M. M. Postan and J. Z. Titow

---

[1] Postan 1966a, 291–659.

[2] The estimate of ploughteams and total arable area is based on that given by Lennard 1959, 393, correcting for the fact that his estimate omits eleven of the English counties.

[3] For the arable area *c.* 1300 (and the population which it was capable of supporting) see Campbell *et al.* 1993, 44–5.

[4] For a survey of colonization and reclamation in this period see Donkin 1973, 98–106; also Darby *et al.* 1979, 249–56.

[5] Smith 1988, 189–91; Smith 1991, 47–50.

[6] For a dissentient view see Bridbury 1992b, 121–5.

both maintain that once reserves of colonizable land were depleted continued population growth was largely sustained at the price of a serious erosion of peasant living standards. Nor does Postan concede that the adoption of more intensive methods of production, especially among the peasantry who made up the mass of cultivators, did much to compensate for the declining supply of land.[7] Yet some intensification there must have been; after all, it was inherent to the very process of land-reclamation. The progressive upgrading of land from marsh, to pasture, and eventually meadow, which H. S. A. Fox has described at Podimore in Somerset, for instance, was only achieved through the expenditure of much labour.[8] Its reward was a significant addition to that township's limited stock of pastoral resources. The same kind of thing was going on in countless other townships, repeating in miniature what was being undertaken on a large scale in the Somerset levels, Romney Marsh, the Fens of East Anglia, and, most spectacularly, across the North Sea in the polder lands of the Low Countries. In every case, labour, capital, and enterprise (in the form of organization) were being lavished upon land in order to make it more productive.[9] Nor was this a one-off investment; converting land from pastoral to arable use may have delivered a higher food yield per unit area but it also incurred a permanent increase in unit labour costs.[10] It was not, however, intensification through the expansion of agricultural/arable land that Postan doubted; rather, he questioned whether a given unit of arable land itself could be made more productive. If not, as Titow points out, "the quantity of food produced per head of population must have been declining".[11]

Raising productivity per unit of arable land necessarily entailed both the improvement of existing techniques (involution) and some measure of technological change (innovation). In the absence of innovation some degree of involution is almost always possible. Ricardo's landuse model implies that the intensity of production rises with the demand for land, as higher levels of economic rent justify the expenditure of more labour.[12] This chimes with Boserup's empirical obser-

---

[7] Titow 1969, 64; Postan 1972, 44.

[8] Fox 1986, 544–5.

[9] Compare Thoen and Hoppenbrouwers this volume. On the often quite complex organization and collaboration involved in the reclamation of the Fens see Hallam 1965, 16–22, 218–20.

[10] Clark 1991, 230–1.

[11] Titow 1969, 72

[12] On Ricardo, see Blaug 1978, 91–112.

vation that at a given level of technology there is a positive correlation between population density, the intensity of agriculture, and agricultural output per unit area.[13] Yet Postan and Titow deny that any such technological advance took place over this period.[14] Whether this was because of a genuine paucity of new technical ideas, the communal organization of agriculture, or the manorial regime's stultifying effect upon investment at all levels, Postan is undecided.[15] Robert Brenner, however, is less diffident:

> the inability of the serf-based agrarian economy to innovate in agriculture ... is understandable in view of the interrelated facts, first, of heavy surplus extraction by the lord from the peasant and, second, the barriers to mobility of men and land which were themselves part and parcel of the unfree surplus-extraction relationship. ... At the same time, given his unfree peasants, the lord's most obvious mode of increasing income from his lands was not through capital investment and the introduction of new techniques, but through squeezing the peasants, by increasing either money rents or labour services.[16]

Such a verdict, of course, presumes that medieval farmers did indeed fail to match at least in part rising population with rising output per unit area. There are grounds now for doubting whether this was in fact the case.

In an exclusively organic age, higher land productivity was contingent upon the evolution of increasingly intensive, self-sustaining, mixed-farming systems. In the seventeenth and eighteenth centuries the development of such systems is associated with a degree of technological novelty: new crops, new breeds, novel implements and methods, and the physical and tenurial reorganization of farms and fields.[17] Medieval farmers enjoyed fewer such options. The establishment of rabbit warrens, erection of windmills, and substitution of horses for oxen were the main technological novelties of the age.[18] More productive husbandry systems therefore had mostly to be fashioned from the existing range of crops and animals. Much could nevertheless be achieved. Rather than "a few technological leaps" the process of

---

[13] Boserup 1981, 15–28.
[14] "... the inertia of medieval agricultural technology is unmistakable": Postan 1972, 44. See also Titow 1969, 72.
[15] Postan 1976b, 42–4.
[16] Brenner 1987, 31.
[17] For the post-medieval development of English agriculture see Thirsk 1967–89.
[18] Sheail 1971; Bailey 1988; Holt 1988, 20–1; Langdon 1986; also Langdon this volume.

agricultural change during the twelfth and thirteenth centuries consisted of "a long chain of small improvements".[19] It assumed seven
main forms:

(1) A refocusing of production upon those agricultural food chains
    which were most productive of food and energy per unit area,
    namely (a) crops rather than animals and animal products, (b) pas-
    toral regimes based upon a combination of grazing and fodder
    cropping rather than grazing alone, (c) coppiced timber rather
    than natural woodland. This was reflected in the expansion of
    arable at the expense of grassland, coupled with a greater em-
    phasis upon fodder cropping, hay meadows, and coppiced wood-
    land, all of which required higher factor inputs per unit area.[20]

(2) The substitution of crops and animals of higher financial, food,
    and/or energy yield for those of lower: industrial and horticul-
    tural crops for grain crops, food grains for drink grains, pottage
    grains for bread grains,[21] legumes for bare fallows, draught horses
    for draught oxen, dairy animals for meat animals, cattle for sheep,
    and sty-fed for pannage-fed pigs. In this way the aggregate value
    of agricultural output was raised per unit area.[22]

(3) The closer integration of arable and pastoral husbandry to create
    mixed-farming systems in which the two sectors were complemen-
    tary rather than competitive in their respective landuse require-
    ments. This was crucial if the resultant arable-based mixed-farming
    systems were to prove ecologically sustainable in more than the
    short-term. It was achieved via better control of fallow grazing
    and folding, the development of convertible-farming systems in
    which land alternated between arable and temporary pasture, a
    greater reliance upon fodder crops (principally legumes and oats)
    in conjunction with the stall and sty feeding of animals, especially

---

[19] Persson 1988, 28.
[20] On agricultural food chains see Grigg 1982, 68–80; on woodland management
see Rackham 1980.
[21] Consuming grains, like oats, as pottage, rather than, as in the case of wheat,
grinding them into flour and processing the flour into bread, maximizes the kilo-
calorie extraction rate: that is, provided that wheat and oats yield equally well,
more people per unit area can be supported on a diet of porridge than upon a diet
of bread.
[22] Campbell *et al.* 1993, 41–2; Simmons 1974, 20–2, 170–2.

in winter, and a more systematic recycling of nitrogen via the collection and application of animal wastes to the arable fields.[23]

(4) The diversification of rotations (and, where necessary, modification of field systems through the creation of extra field divisions), reduction of fallows, and increase in the frequency of cropping. As a corollary it became necessary to pay greater attention to (a) the preparation of the seed-bed via repeated ploughings, (b) the quality and quantity of seed sown, (c) the maintenance and improvement of soil structure and fertility through the cultivation of nitrogen-fixing legumes, systematic folding of sheep, and applications of farmyard manure, marl, lime, night-soil and the like, and (d) weed control, via systematic weeding, heavier seeding rates, and multiple summer ploughings of bare fallows.[24]

(5) The intensification and rationalization of labour processes to achieve higher standards of arable and pastoral management. On seigneurial demesnes this tended to comprise increased labour inputs, more careful management and supervision of workers, greater specialization of labour, and the partial or complete substitution of waged for servile labour to improve work motivation. It was facilitated by the widespread adoption of a system of annual accounting.[25]

(6) Improvements to tools and implements and fuller investment in farm-buildings intended for the storage of harvests and protection of stock and livestock, including barns, stables, byres, sties, and sheepcotes. Also, investment in equipment and machinery intended for the processing of agricultural products.[26]

(7) Exploitation of the opportunities afforded by market expansion to specialize according to comparative advantage, thereby securing the benefits of a greater spatial division of labour.[27]

---

[23] Bailey 1990; Searle 1974, 272–91; Campbell and Overton 1993, 61–2; Campbell 1988; Biddick 1989, 116–25; Campbell 1983a; Smith 1943, 128–65; Hallam 1988, 272–496.

[24] For example, Campbell 1983a; Mate 1985; Brandon 1972; Postles 1989.

[25] For example, Thornton 1991, 201–7; for the spread of annual accounting see Harvey 1976.

[26] Hurst 1988, 859, 867–8, 888–98; Holt 1988; Campbell 1992, 112.

[27] Overton and Campbell 1991, 19–22; Persson 1988, 10–12, 31.

All of these methods and strategies can be documented as having been employed to some extent in some part of England by the beginning of the fourteenth century. Their selective adoption is reflected in a greater differentiation of mixed-farming types, as exemplified by the eight basic systems which have been identified as in operation on seigneurial demesnes at this time (table 10.1). The range of farming systems on peasant holdings was undoubtedly wider. With higher labour to land ratios than most demesnes peasant holdings were potentially far more intensively cultivated. Peasants, for instance, outpaced landlords in the replacement of the ox with the horse and, on the evidence of the *Nonarum inquisitiones* of 1342 (which for some counties detail the value of the small tithes on flax, hemp, and cider), were significant producers of industrial crops.[28]

Although the peasant sector is the larger and more crucial it is the demesne sector that is the better documented thanks to the survival of manorial accounts in large numbers. Table 10.1 is based upon a national sample of such accounts from 388 demesnes.[29] On its evidence the more intensive seigneurial mixed-farming systems—"intensive mixed-farming" and "light-land intensive"—remained outnumbered roughly two to one by such relatively extensive systems as "sheep-corn husbandry", "extensive mixed-farming", "extensive arable husbandry" and "oats and cattle". Features diagnostic of intensification are most conspicuous on the pastoral side. For instance, intensive demesnes tended to make greater use of the faster, stronger, but more expensive horse than the slower, cheaper ox. The dividend was the far higher proportion of non-working animals which this enabled them to stock, particularly dairy cattle, but also sheep (especially on "light-land intensive demesnes") and some swine. The livestock profiles of intensive demesnes were therefore particularly well developed, to the extent that this often translated into an above average ratio of livestock to crops. This had obvious benefits for the recycling of nitrogen and maintenance of soil fertility. Although these were the farming systems which made greatest use of the horse they were also the systems which devoted the smallest proportions of their cropped acreage to oats (thus refuting the widely canvassed notion that greater use of the horse promoted an increase in the oats acreage).[30] Enough

---

[28] Langdon 1986, 172–253; Vanderzee 1807; Sutton 1989; Evans 1985, 41–6.
[29] Omitted were solely pastoral regimes: for example, Atkin 1994.
[30] For example, Parain 1966, 162; Persson 1988, 30.

Table 10.1. The principal English seigneurial mixed-farming systems 1250–1349 (mean characteristics).

	Farming type								All
	1	2	3	4	5	6	7	8	
% total sown acres:									
wheat	22	8	34	33	36	33	43	1	31
rye	4	19	7	3	2	2	1	8	5
winter mixtures	1	2	6	1	5	1	<1	0	2
barley	35	43	4	16	12	13	9	10	16
oats	13	15	42	28	26	43	38	78	34
spring mixtures	5	2	2	6	9	2	3	0	4
legumes	20	10	4	14	10	5	5	3	8
Total sown acres	180	180	230	182	177	216	144	107	185
% total livestock units*									
horses	17	19	18	33	12	8	15	14	16
oxen	16	9	36	39	39	32	84	58	43
mature cattle	29	29	22	5	1	19	<1	18	14
immature cattle	18	12	15	0	0	12	0	7	8
sheep	14	28	6	5	48	27	0	2	17
swine	5	2	3	17	1	3	<1	0	3
Total livestock units*	84	51	57	28	51	110	24	40	59
Stocking density⁺	46	32	27	19	32	61	18	40	35

Notes:
1 intensive mixed-farming (66 demesnes)
2 light-land intensive (34 demesnes)
3 mixed-farming with cattle (72 demesnes)
4 arable husbandry with swine (19 demesnes)
5 sheep-corn husbandry (45 demesnes)
6 extensive mixed-farming (65 demesnes)
7 extensive arable husbandry (80 demesnes)
8 oats and cattle (7 demesnes)
* (horses × 1.0) + ([oxen + mature cattle] × 1.2) + (immature cattle × 0.8) + ([sheep + swine] × 0.1).
⁺ total livestock units per 100 sown acres.

Source and note: Power and Campbell 1992, 234.

oats were grown to satisfy the requirements of traction and haulage and no more. Oats had the lowest cash value of any of the grains hence pride of place within the spring-sown schedule was allocated to barley, which commanded a substantially higher relative price and was preferred for brewing.[31] The emphasis upon barley was part-and-parcel of a general bias towards spring-sown crops, which often occupied well over two-thirds of the cropped acreage. This bias provides a clue to the generally intensive and flexible character of rotations, especially in the case of "intensive mixed-farming". Demesnes practising this most exacting form of husbandry devoted the lion's share of the winter course to wheat, the most demanding crop of all, and partly to replenish soil nitrogen, partly for fodder, and partly for food, grew legumes on a larger scale than in any other farming system. This is consistent with virtually continuous cropping of the arable and the near elimination of fallows which independent investigation reveals to have been the case on some of the most intensively cultivated of this group of demesnes.[32] In contrast, "light-land intensive" demesnes favoured rye rather than wheat and grew smaller acreages of legumes. Their soils would not support such demanding rotations hence they practised a variety of irregular rotations, including the periodic alternation of land between crops and temporary pasture.

Typically, the more extensive the farming system the less closely integrated the arable and pastoral sectors, which were sometimes conducted as virtually separate enterprises.[33] Even when stocking densities were relatively high, as in the case of "extensive mixed farming", this usually owed more to an abundance of permanent grassland than fodder cropping or the operation of some kind of convertible regime. Again, the composition of the pastoral sector is revealing. Usually grass-fed oxen outnumbered the grass- and fodder-fed horse by over four to one. Working animals made up practically 40 per cent of total livestock units, and a significant proportion of the remainder were devoted to rearing replacement draught animals (hence the more balanced ratio between mature and immature cattle than that maintained on more intensively-managed demesnes where the prime function of cattle herds was milk production). Among the non-working

---

[31] Nationally, the relative price per bushel of the principal grains *c.* 1300 was: wheat 1.00, rye 0.75, barley 0.70, and oats 0.41 (Campbell 1991, 169); see also Comet this volume.

[32] Campbell 1983a, 28–36; Campbell 1983b, 390–4.

[33] Biddick with Bijleveld 1991, 115.

animals, sheep—the most grassland dependent animals of all—were particularly prominent, especially on "sheep-corn" demesnes, with ewes often supplanting cows as the principal dairy animal.[34] "Extensive arable" demesnes, in contrast, effectively stocked working animals alone (again primarily oxen). For replacement draught animals they either relied upon transfers from other manors or purchase. Irrespective of the size of their pastoral sectors, most of these demesnes often operated some version of two- or three-course cropping, with wheat the predominant winter crop and oats the predominant spring. Some legumes were grown, but on too small a scale to make much contribution to available supplies of soil nitrogen. Regular fallowing must therefore have played a key role in the maintenance of soil fertility, with cropping restricted to a maximum of two consecutive courses.[35]

A wide productivity gulf separated the most from the least intensive of these farming systems, as exemplified by arable rental values of 12–36 pence an acre on the most intensive mixed-farming demesnes c. 1300, compared with valuations of 4 pence, 3 pence, or even as little as 2 pence an acre on demesnes operating extensive mixed-farming systems.[36] In Norfolk at this time "intensive mixed-farming" demesnes commonly obtained mean gross yields per acre for all the principal grain crops in the range 15–25 bushels per acre (1,350–2,250 litres per hectare), rising to 20–30 bushels per acre (1,800–2,700 lit/ha) on the most productive of all. The latter constitute the highest recorded English medieval yields, comparable with yield levels more usually associated with the era of the agricultural revolution in the late eighteenth and early nineteenth centuries and almost on a par with the exceptionally high yields documented by J. Derville in parts of northern France at the close of the thirteenth century.[37] Yields on more extensively farmed demesnes were generally much lower, the unimpressive and, sometimes, dismal yields obtained on many of the midland demesnes of Westminster Abbey and southern demesnes of the bishopric of Winchester, St. Swithin's Priory, Winchester, and Glastonbury Abbey, being largely responsible

---

[34] On ewe-dairying see Hallam 1981, 129–30, 248; Biddick with Bijleveld 1991, 115–18.

[35] Shiel 1991, 70–3.

[36] These figures are derived from an unpublished analysis of the extents attached to *inquisitiones post mortem*, held at the Public Record Office, London.

[37] As reported in Thoen this volume. Corresponding yield *ratios* were nevertheless well below those obtained on these French farms.

for medieval agriculture's reputation for low productivity.[38] For instance, aggregate output per arable acre on the bishop of Winchester's
demesne of Rimpton in Somerset was only a third that prevailing on
the prior of Norwich's intensively farmed demesne of Martham in
eastern Norfolk.[39] Moreover, whereas it has been claimed that yields
tended to fall on many of the Winchester demesnes as pressure upon
the land mounted during the second half of the thirteenth century,
mean yields actually rose in intensively-cropped Norfolk over the same
period.[40] In other words, there is a strong positive association between the intensity and the sustainability of husbandry.[41]

Why was this productivity gap so wide? Certainly, by the end of
the eighteenth century it had closed dramatically, and nowhere did
mean yields per county deviate far from the national mean.[42] And
why, if food was in such short supply, had so few demesnes intensified, innovated, and raised their productivity to the maximum sustainable given available technology? Patently, medieval cultivators did
not lack the technological means to create sustainable systems which
produced more from the land. As Postan mused, "the real problem
of medieval technology is not why new technological knowledge was
not forthcoming, but why the methods, or even the implements, known
to medieval men were not employed, or not employed earlier or
more widely than they in fact were".[43] In east Norfolk and northeast
Kent medieval husbandmen had solved the central dilemma of how
to reconcile the conflicting landuse requirements of the arable and
pastoral sectors without jeopardizing the fragile ecological basis of
reproduction. They did this by integrating crop and livestock production into a mutually reinforcing mixed-farming regime in which
fodder cropping, and especially the cultivation of nitrogen-fixing legumes, played a crucial role. Half a millennium later rapid and widespread diffusion of an improved version of this mixed-farming system
was to be one of the cornerstones of the so-called agricultural revolution.[44] The late thirteenth and early fourteenth centuries, however,
experienced no such agricultural revolution, and seigneurial agricul-

---

[38] Lennard 1922, 12–27; Postan 1966b, 556–9; Titow 1972; Farmer 1983; Hybel
1989.
[39] Thornton 1991, 191–3.
[40] Titow 1972, 12–33; Campbell 1991, 159–74.
[41] Campbell 1991, 144–6.
[42] See the yield figures summarized in Allen and Ó Gráda 1986, 42–4.
[43] Postan 1972, 42.
[44] Campbell and Overton 1993, 88–95.

ture at least remained more extensive than intensive, with low rather than high productivity the norm. The explanations most commonly advanced to account for this are either Whiggish or Marxist, and stress either the conservatism of cultivators and their resistance to change or the disincentives to investment and innovation provided by the feudal system and the licence it gave lords to raise revenues instead by raising feudal exactions.[45] Unfortunately, such explanations do not sit easily with the clear empirical evidence that medieval cultivators—both peasants and lords—could and did innovate and adopt new practices when it suited them. Rather than a failure of supply, the problem may have lain more with demand and a lack of sufficient incentives to invest, innovate, and intensify.

Demand is translated into landuse and farming systems via the medium of economic rent. Economic rent is the return due for the use of the land alone as a factor of production. It represents that part of a farmer's revenue above production costs, but excluding remuneration derived from the three other main factors of production, namely labour, capital, and enterprise. So long as self-sufficiency was the predominant objective of most cultivators, economic rent—as Ricardo recognized—was largely a function of land quality and the demand for land (that is population density mediated via institutional controls upon rent levels and access to land). Where competition for land for subsistence was strongest and that land was inherently most productive, rents would be highest. But once markets developed and agriculture became more commercialized so, as J. H. von Thünen demonstrated, economic rent was increasingly determined by the cost of transporting goods to market, rents rising with proximity to the market.[46] In both cases, the higher the economic rent the greater the incentive and justification for raising inputs of labour, capital, and enterprise. Where, close to major cities, von Thünen rents generally exceeded Ricardian rents it was the market which determined the character and intensity of production, whether for consumption or exchange. At a greater distance, however, the rent for subsistence production (with no effective transport costs to bear) may have exceeded that for commercial production, resulting in different types and intensities of production between the subsistent and commercial sectors. In a medieval context this meant that in areas of low von

---

[45] For a summary see Persson 1988, 3–7, 63–4; also Langdon this volume.
[46] Chisholm 1962, 20–32; Grigg 1982, 135–40; Bailey 1989.

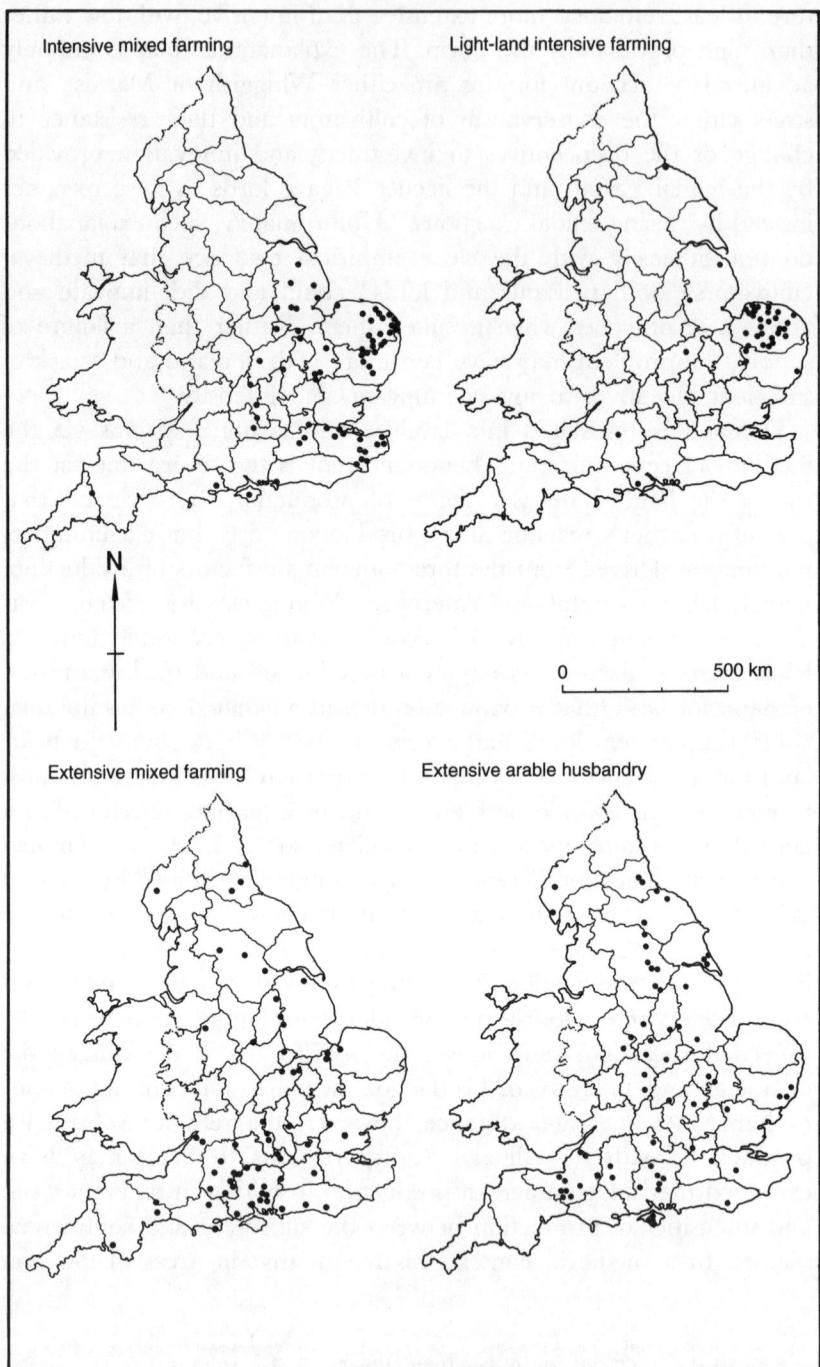

Figure 10.1. Intensive and extensive seigneurial mixed-farming systems in England, 1250-1349.

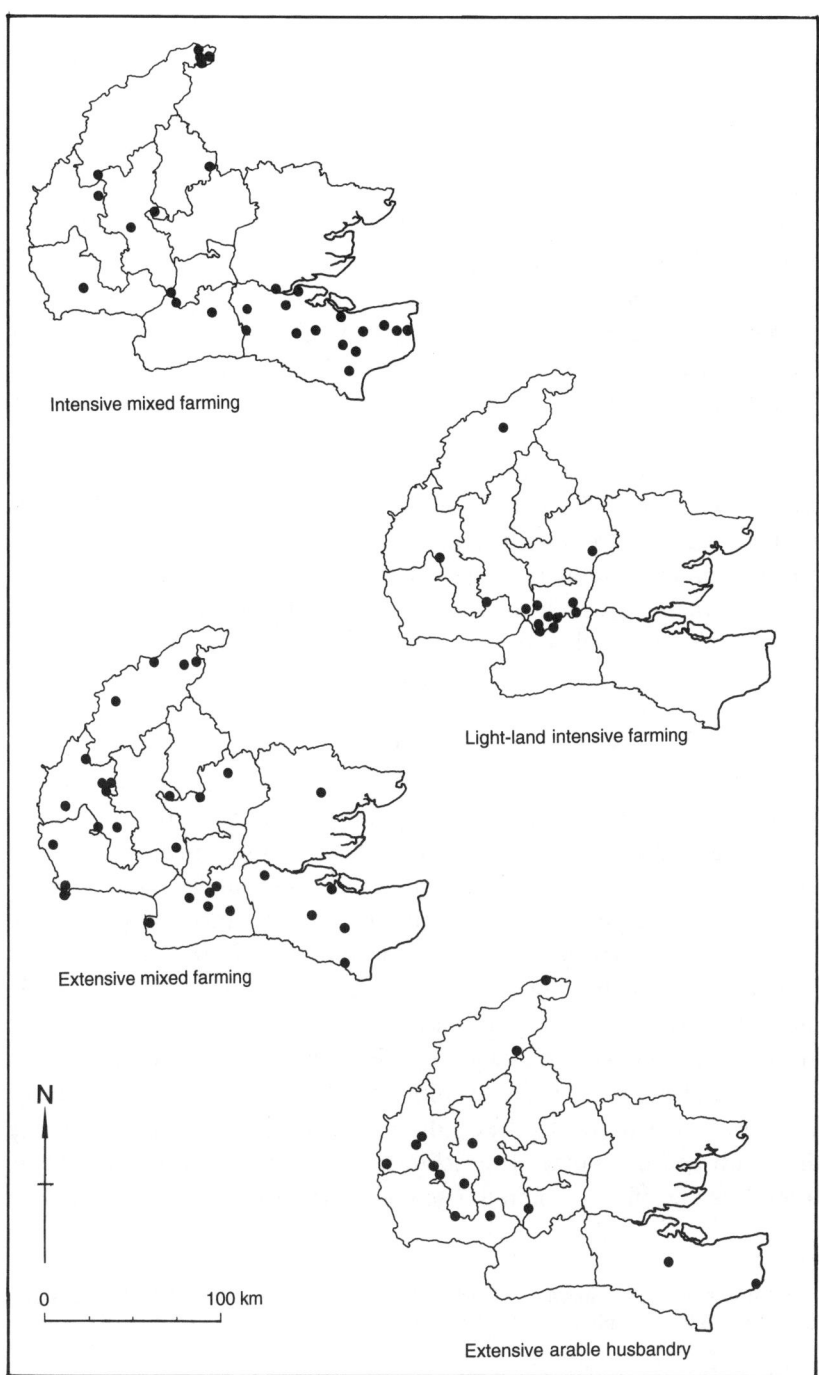

Figure 10.2. Intensive and extensive seigneurial mixed-farming systems in the hinterland of London, 1288-1315.

Thünen economic rent peasants producing for consumption may have
been more intensive in their methods than demesnes producing for
exchange. Areas of high economic rent—be it because of good soils,
strong demand for land, and/or favourable access to markets—should
therefore have been characterized by more intensive and productive
husbandry systems than areas of low economic rent. A test of whether
this was in fact the case is provided by the geographical distribution
of the principal demesne-farming systems.

As will be seen from figures 10.1 and 10.2, the two most intensive
farming systems—"intensive mixed-farming" and "light-land intensive farming"—were both comparatively limited in distribution. Notable concentrations of "intensive mixed-farming" demesnes occurred
in eastern Norfolk and eastern Kent and, to a lesser extent, in the
Soke of Peterborough and on the better soils of the lower Thames
valley just upsteam from London. Independent studies of agriculture
in these localities confirm the intensive and integrated nature of the
mixed-farming systems employed and the high returns per unit area
thereby obtained, which were also reflected in exceptionally high per
acre valuations.[47] All four localities were characterized by naturally
fertile and readily cultivated loam soils, access to substantial local
markets, and wider access, via navigable rivers and coastal trading
ports, to major external markets.[48] East Norfolk and northeast Kent
were also weakly manorialized and exceptionally densely populated.
In short, they were characterized by an extreme coincidence of high
Ricardian and von Thünen economic rent with the result that it was
here that English medieval agriculture attained its greatest peak of
intensity and productivity (the superior intensity and productivity of
husbandry in parts of Flanders and northern France at this date
implying even higher levels of economic rent, as was consistent with
their more specialized and urbanized economies).[49] Outside of these
localities "intensive mixed-farming" demesnes were to be found at a
scatter of locations in the east midlands (Lincolnshire, Leicestershire,
Rutland, and Cambridgeshire), the upper Thames valley, and along
the south coast. Studies of agriculture in several of these areas have
shown that it often exhibited a tendency towards the kinds of inten-

---

[47] Campbell 1983a; Smith 1943, 128–65; Mate 1985; Biddick 1989, 50–77; Brandon 1988, 320; Campbell *et al.* 1993, 128–44.
[48] Langdon 1993.
[49] Compare Thoen this volume.

sive methods which found their fullest and most productive expression in eastern Norfolk and eastern Kent.[50] "Intensive mixed-farming" demesnes were, however, conspicuously absent from the rest of the country.

"Light-land intensive" demesnes were, if anything, even more specific in distribution. First and foremost, this was a husbandry system associated with Norfolk, where it was especially characteristic of the county's lighter soils (a distribution which spills over into the Breckland of north-west, and the Sandlings of east Suffolk). The concentration of "light-land intensive" demesnes in the immediate vicinity of Norwich—the second city after London with a population of perhaps 25,000—is especially notable and is probably to be explained by the coexistence of light soils and strong urban demand for the rye and barley which were the principal crops grown (neither of which was as capable as wheat of bearing the cost of carriage from a distance).[51] The need to produce relatively bulky, low-value crops close to consuming centres—as von Thünen's landuse model predicts—no doubt helps to explain the corresponding cluster of "light-land intensive" demesnes which occurs in the immediate vicinity of London (figure 10.2), where rye and barley were also grown in quantity and there was a premium to be gained from maximizing the ratio of non-working to working animals.[52] Holywell, an outlying "light-land intensive" demesne on the outskirts of Oxford, shares the same close association with a major urban centre. The few other isolated outliers—in Lincolnshire, Northamptonshire, and Wiltshire—are mostly explicable in environmental and/or institutional terms.

Both the most intensive farming systems therefore display a close association with areas of high economic rent, especially where this was borne of access/proximity to major urban markets at home and overseas. Significantly, the very areas where these two systems were most developed were those from which "extensive mixed-farming" and "extensive arable husbandry" were most conspicuously absent. These more extensive farming systems are symptomatic of lower levels of economic rent and, since they are comparatively widespread

---

[50] Ravensdale 1974, 116–20; Brandon 1972; 1988, 318–24.

[51] Rutledge 1988.

[52] Campbell *et al.* 1993, 111–25. London's demand for oats and fat cattle may also account for the presence of "oats and cattle" demesnes (no. 8 in the farming type classifications: see table 10.1) at Harrow, Middlesex, and Esher and Lambeth, Surrey (a farming type otherwise more characteristic of moist, upland areas).

(figure 10.1), there is a clear implication that moderate to low economic rent was the norm throughout the greater part of the country. Examples of "extensive mixed farming" are to be found from the far north to the extreme south of the country, wherever demesnes were well endowed with both arable and grassland. More numerous and as widespread are demesnes practising "extensive arable husbandry", which display a particular bias towards Somerset, the midlands, and the north of England. These areas were remote from major concentrations of demand and characterized, for the most part, by below average population densities, circumstances which evidently encouraged neither the fuller exploitation of pastoral resources nor the closer integration of crop and livestock production.

On this analysis, extensive farming systems of one sort or another were very much the norm throughout much of the country, and intensive farming systems the exception, confined to a few favoured areas. By implication, therefore, low to moderate levels of economic rent were far more typical than high. This is consistent with what is known about the number, size, and location of the leading urban centres and the extent of their provisioning zones at the climax of medieval demographic, urban, and commercial expansion *c.* 1300. By that date there were probably at least sixteen English cities with 10,000 or more inhabitants.[53] With a population of perhaps 80,000–100,000, London was by far the largest English city and second only to Paris among the cities north of the Alps.[54] It owed its impressive size to its precocious economic and political primacy and growing centrality.[55] Already it was the focus of the country's road network and was well served by river and coastal communications. These rendered it the country's busiest port and enabled it to exercise a magnetic pull upon a wide area for the supply of foodstuffs and raw materials.

In years of normal harvest London regularly drew on an area of 4,000 square miles in extent for its grain supplies. This area was irregular in shape, reflecting the availability of water transport, and included at its furthest extent several ports on the south and east coasts which were 100 miles or more distant from the city.[56] Livestock,

---

[53] Campbell *et al.* 1993, 9–11.
[54] Keene 1984; Keene 1985a. But see also Nightingale 1996, 95–6, who argues that the city's medieval population never exceeded 60,000.
[55] Keene 1989; again, for a contrary view, see Nightingale 1996, 95–6.
[56] Campbell *et al.* 1993, 46–77.

which were capable of walking to market, and the higher-valued livestock products were probably drawn in from further afield. The city's supply lines were well developed and it was served by a sophisticated and well-articulated marketing network. In years of abnormal harvest this extended outwards to tap a wider area. In 1317, for instance, at the height of the worst medieval harvest failure on record, the king ordered his sheriffs to procure essential provisions for the royal household at Westminster. Hay, one of the bulkiest of commodities, was to be obtained from the counties closest to London: Middlesex, Essex, Hertfordshire, Surrey, and Sussex. Grain, better able to withstand the costs of carriage, was to come from a much wider geographical area, comprising Kent, Surrey, Sussex, Hertfordshire, Essex, Suffolk, Norfolk, Cambridgeshire, and Huntingdonshire (in the last two cases presumably shipped to London via the major grain entrepot of Kings Lynn). Finally, livestock were to be procured from a wide scatter of inland counties from as far afield as Gloucestershire and Somerset to the west and Cambridgeshire and Huntingdonshire to the north, and thence driven overland to Westminster.[57]

This example illustrates how London drew upon different areas at different distances for different commodities in much the way that von Thünen's model predicts. It also demonstrates that even in one of the worst years on record London's provisioning zone remained confined to certain very specific counties. In normal years its hinterland was even more circumscribed, embracing—for all commodities— perhaps a fifth of the country's total land area. Of this, less than half was engaged in the regular supply of grain to the city. Much of the country, therefore, lay beyond the stimulus of the capital's influence. Nor did the needs of other urban centres provide adequate compensation since they were few and far between and their provisioning hinterlands were even smaller. Winchester, for example, lacking a navigable river and situated in a region of below average productivity, was supplied with grain from within a radius of about 12 miles.[58] Exeter, smaller than Winchester but situated within an even less productive hinterland, regularly drew its supplies from up to 20 miles or more away.[59] In a world where, even on the most generous estimates, a maximum of 417,200 people and 10.4 per cent of the

[57] *Calendar of Close Rolls, 1313–18*, 513–14. I am grateful to Dr. Derek Keene for directing me to this reference.
[58] Keene 1985b, 251–5.
[59] Kowaleski 1995, 28–31.

country's population lived in towns of at least 10,000 inhabitants, it was inevitable that the agricultural impact of urban markets remained restricted and selective.[60] Under these circumstances, as David Farmer has emphasized, "the local markets and the communities around them were the more important outlets for the produce of the country-side".[61] Such markets were, nevertheless, incapable of stimulating economic rent and agricultural intensification to the same extent as major urban concentrations of demand, with the result that across much of the country extensive methods of production remained the most rational form of landuse, especially on the extensive demesne holdings of lords.

Nor, with certain notable exceptions such as the Cornish stannaries, does it appear that areas beyond the reach of major markets had yet hit upon the solution of producing manufactured goods and marketing these at a distance (thereby turning their low land values and cheap food to advantage).[62] This was not to occur until the close of the middle ages when it was to transform the fortunes of many hitherto under-developed areas in the midlands, north and west of the country. Indeed, John Langton and Göran Hoppe have stressed the mutually beneficial tripartite relationship which subsequently developed during the early modern period between expanding metropolitan demand, the evolution of capitalist agriculture, and the growth of proto-industrialization.[63] But in the middle ages, the corresponding relationship remained dual rather than tripartite, since proto-industrialization existed only in embryo, and the capacity of urban demand alone to stimulate agrarian development was further qualified by the lesser scale of the metropolis and other urban centres. The rural-urban nexus was consequently a less powerful agent of change in the Middle Ages than it was to be in later centuries.[64] It was in Flanders—northern Europe's most urbanized region—rather than England that its impact was greatest in this period.[65]

Moreover, even within provisioning range of cities high economic rent was the exception rather than the rule. According to the von

---

[60] Campbell *et al.* 1993, 10–11; Britnell 1995, 9–12.

[61] Farmer 1991, 329.

[62] Hatcher 1970.

[63] Langton and Hoppe 1983.

[64] Wrigley 1967 and 1985; Overton and Campbell 1991, 35–44; Campbell and Overton 1993, 88–105.

[65] Persson 1988, 73–6.

Thünen model intensive landuse systems of one sort or another occupied only the inner 25 per cent of a city's hinterland, systems of lesser intensity occupying the remainder.[66] Thus, in the case of London c. 1300 only 25 per cent of documented demesnes within provisioning range of the capital practised either "intensive mixed-farming" or "light-land intensive" husbandry; all other demesnes operated more extensive systems. Most cities therefore relied more on the extensive than the intensive production of grain and other provisions. Beyond their provisioning range, wool—in many respects the most extensive agricultural commodity of all—was practically the only product capable of being marketed at a distance and was therefore a commercial lifeline for many rural producers otherwise largely dependent upon the dispersed demand of local markets. A. R. Bridbury has estimated that by the opening of the fourteenth century the wool of approximately 8 million sheep was being exported in one form or another, the volume of this trade testifying to the low economic rent that prevailed across so much of the country.[67]

On this evidence, it required cities a great deal larger than the largest medieval cities to raise economic rent over a sufficiently wide area to encourage agricultural intensification by more than a minority of farmers in a few favoured localities. When appropriate incentives existed medieval cultivators were not backward in adopting more intensive and productive methods, but across much of the country the impulse to change was weak with the result that limited technological development and low productivity remained the order of the day on the majority of demesnes. Had London and the leading provincial cities been larger there can be little doubt that their correspondingly enlarged hinterlands would have experienced little difficulty in producing and supplying the additional foodstuffs and raw materials demanded, for where market signals were strong specialization and intensification almost invariably resulted and marketing links and commercial institutions were forthcoming.[68] At the culmination of medieval economic expansion c. 1300 this was most conspicuously the case in the east and southeast of England, and it can be no coincidence that it is here that adoption of the various new technological

---

[66] Campbell *et al.* 1993, 5–6.
[67] Bridbury 1992a, 185–6.
[68] Compare the situation in relatively highly urbanized Flanders at the time: Thoen this volume.

innovations of the age—windmills, rabbits, horse traction and haulage, vetches and legumes—made greatest progress. It was also here
that medieval farmers achieved their greatest technological breakthrough of all by evolving an integrated mixed-farming system capable
of a sustained high level of production. Intriguingly, however, it was
not in the immediate vicinity of London or even of Norwich that
this breakthrough was made, but in eastern Norfolk and eastern Kent,
localities accessible to, but at some remove from both those cities.

What these two localities offered over the more immediate en-
virons of these two major cities were good soils, a relatively free and
enterprising rural population, and riverine and maritime access to a
variety of different markets at home and overseas. They serve as a
salutary reminder that environmental, institutional, and cultural factors also exercised an important influence upon the course of agricultural development. But these were also localities dominated by a
numerous small-holding peasantry and it is possible that here, as in
the Low Countries,[69] it was they rather than the lords who pioneered
adoption of more intensive methods. Peasants generally had a more
favourable ratio of labour to land than the larger demesne holdings
(whose distinguishing feature is more likely to have been a superior
productivity of labour than of land) and fewer problems of work
motivation. As pressure mounted to raise agricultural output in line
with population they were therefore in a stronger position to substi-
tute labour for land. Moreover, it was upon their holdings that they
largely had to rely for their income, be that in the form of food
produced for direct consumption or cash received from marketed
goods (except where this could be augmented by the sale of either
surplus labour for wages or craft goods produced with that labour).
In this sense most peasants were primarily subsistence rather than
commercial producers. This distinction is important for there were
undoubtedly many parts of the country where the economic rent for
subsistence production of a commodity was superior to that for its
commercial production, thus inducing and sustaining more intensive
methods of production.[70] The impression of medieval agriculture
formed from the evidence of demesnes may therefore be unduly
weighted towards extensive methods of production and low per unit
area levels of productivity.

---

[69] Verhulst 1985 and 1990, 25.
[70] Bailey 1989, 4.

It may have been peasants rather than lords who were most energetic and successful at reconciling the widening gap between the expanding population to be fed and the shrinking area *per caput* from which to feed it. At late thirteenth-century demesne yield levels, and with the same basic product mix, there would have been little difficulty in feeding a population in 1086 of 2.5 million from the estimated arable area of approximately 8.5 million acres.[71] By *c.* 1300, however, the same yields and production mix would have had difficulty in feeding a population of more than 4.0 million from an arable area which is unlikely to have exceeded 11.5 million acres and may well have been less.[72] Yet most recent estimates of the population *c.* 1300 favour a figure in the region of 6.0 million.[73] Clearly, something in the equation does not fit; the estimates of either the total population, the total arable area, or the output per unit area are wrong. Of course, none of the estimates is robust, but the most plausible explanation of the inconsistency may well be the demesne sector's unrepresentativeness of the productivity of agriculture as a whole. If, for instance, it is assumed that intensification and specialization during the twelfth and thirteenth centuries not only raised demesne productivity but raised peasant productivity further, to a level significantly above that of the demesnes, then the estimates of total population could more easily be reconciled with those of the arable area available to support it. Only further research can resolve this crucial enigma.

Bibliography

Allen, R. C. and Gráda, C. Ó. 1986, "On the road again with Arthur Young: English, Irish, and French agriculture during the Industrial Revolution" (University of British Columbia, Department of Economics discussion paper no. 86–38).

Atkin, M. A. 1994, "Land use and management in the upland demesne of the de Lacy estate of Blackburnshire", *Agricultural History Review* 42, 1–19.

Bailey, M. 1988, "The rabbit and the medieval East Anglian economy", *Agricultural History Review* 36, 1–20.

---

[71] Bridbury (1992b, 121–5) has, in fact, suggested that the Domesday population may have been significantly higher.

[72] Campbell *et al.* 1993, 44–5.

[73] For example, Smith 1991, 48–9.

—— 1989, "The concept of the margin in the medieval English economy", *Economic History Review* 42, 1–17.

—— 1990, "Sand into gold: the evolution of the foldcourse system in west Suffolk, 1200–1600", *Agricultural History Review* 38, 40–57.

Biddick, K. 1989, *The other economy: pastoral husbandry on a medieval estate*, London.

Biddick, K. with Bijleveld, C. J. H. 1991, "Agrarian productivity on the estates of the bishopric of Winchester in the early thirteenth century: a managerial perspective". In *Land, labour and livestock: historical studies in European agricultural productivity*, eds B. M. S. Campbell and M. Overton, Manchester, 95–123.

Blaug, M. 1978, *Economic theory in retrospect*, 3rd edn, Cambridge.

Boserup, E. 1981, *Population and technology*, Oxford.

Brandon, P. F. 1972, "Cereal yields on the Sussex estates of Battle Abbey during the later Middle Ages", *Economic History Review* 25, 403–20.

—— 1988, "Farming techniques: south-eastern England". In *The agrarian history of England and Wales*, vol. II, *1042–1350*, ed. H. E. Hallam, Cambridge, 312–25.

Brenner, R. 1987, "Agrarian class structure and economic development in pre-industrial Europe". In *The Brenner debate: agrarian class structure and economic development in pre-industrial Europe*, eds T. H. Aston and C. H. E. Philpin, Cambridge, 10–63.

Bridbury, A. R. 1992a, "Before the Black Death". In his *The English economy: from Bede to the Reformation*, Woodbridge, 180–99.

—— 1992b, "The Domesday valuation of manorial income". In his *The English economy: from Bede to the Reformation*, Woodbridge, 111–32.

Britnell, R. H. 1995, "Commercialisation and economic development in England, 1000–1300". In *A commercialising economy: England 1086 to c. 1300*, eds R. H. Britnell and B. M. S. Campbell, Manchester, 7–26.

Calendar of Close Rolls, 1313–18 (HMSO, London, 1893).

Campbell, B. M. S. 1983a, "Agricultural progress in medieval England: some evidence from eastern Norfolk", *Economic History Review* 36, 26–46.

—— 1983b, "Arable productivity in medieval England: some evidence from Norfolk", *Journal of Economic History* 43, 379–404.

—— 1988, "The diffusion of vetches in medieval England", *Economic History Review* 41, 193–208.

—— 1991, "Land, labour, livestock and productivity trends in English seignorial agriculture, 1208–1450". In *Land, labour and livestock: historical studies in European agricultural productivity*, eds B. M. S. Campbell and M. Overton, Manchester, 144–82.

—— 1992, "Commercial dairy production on medieval English demesnes: the case of Norfolk", *Anthropozoologica* 16, 107–18.

Campbell, B. M. S., Galloway, J. A., Keene, D. and Murphy, M. 1993, *A medieval capital and its grain supply: agrarian production and distribution in the London region c. 1300*, Historical Geography Research Series 30.

Campbell, B. M. S. and Overton, M. 1993, "A new perspective on medieval and early modern agriculture: six centuries of Norfolk farming, c. 1250–c. 1850", *Past and Present* 141, 38–105.

Chisholm, M. 1962, *Rural settlement and land-use: an essay on location*, London.

Clark, G. 1991, "Labour productivity in English agriculture, 1300–1860". In *Land, labour and livestock: historical studies in European agricultural productivity*, ed. B. M. S. Campbell and M. Overton, Manchester, 211–35.

Darby, H. C., Glasscock, R. E., Sheail, J. and Versey, G. R. 1979, "The changing geographical distribution of wealth in England 1086–1334–1524", *Journal of Historical Geography* 5, 247–62.

Donkin, R. A. 1973, "Changes in the early Middle Ages". In *A new historical geography of England*, ed. H. C. Darby, Cambridge, 75–135.

Evans, N. 1985, *The East Anglian linen industry: rural industry and local economy 1500–1850*, Pasold studies in textile history 5, Aldershot.

Farmer, D. L. 1983, "Grain yields on Westminster Abbey Manors, 1271–1410", *Canadian Journal of History* 18, 331–47.

—— 1991, "Marketing the produce of the countryside, 1200–1500". In *The agrarian history of England and Wales*, vol. III, *1348–1500*, ed. E. Miller, Cambridge, 324–430.

Fox, H. S. A. 1986, "The alleged transformation from two-field to three-field systems in medieval England", *Economic History Review* 39, 526–48.

Grigg, D. 1982, *The dynamics of agricultural change: the historical experience*, London.

Hallam, H. E. 1965, *Settlement and society: a study of the early agrarian history of south Lincolnshire*, Cambridge.

—— 1981, *Rural England 1066–1348*, London.

—— (ed.) 1988, *The agrarian history of England and Wales*, vol. II, *1042–1350*, Cambridge.

Harvey, P. D. A. (ed.) 1976, *Manorial records of Cuxham, Oxfordshire* (Oxfordshire Record Society 50, and Royal Commission on Historical Manuscripts 23, joint publication), London.

Hatcher, J. 1970, *Rural economy and society in the Duchy of Cornwall 1300–1500*, Cambridge.

Holt, R. 1988, *The mills of medieval England*, Oxford.

Hurst, J. G. 1988, "Rural building in England and Wales: England". In *The agrarian history of England and Wales*, vol. II, *1042–1350*, ed. H. E. Hallam, Cambridge, 854–965.

Hybel, N. 1989, *Crisis or change. The concept of crisis in the light of agrarian structural reorganization in late medieval England*, trans. J. Manley, Aarhus.

Keene, D. 1984, "A new study of London before the Great Fire", *Urban History Yearbook, 1984*, 11–21.

—— 1985a, *Cheapside before the Great Fire*, London.

—— 1985b, *Survey of medieval Winchester*, Winchester Studies 2, Oxford.

—— 1989, "Medieval London and its region", *London Journal* 14, 99–111.

Kowaleski, M. 1995, "The grain trade in fourteenth-century Exeter". In *The salt of common life: individuality and choice in the medieval town, countryside, and church: essays presented to J. Ambrose Raftis*, ed. E. B. DeWindt, Kalamazoo, 1–53.

Langdon, J. 1986, *Horses, oxen and technological innovation: the use of draught animals in English farming from 1066–1500*, Cambridge.

—— 1993, "Inland water transport in medieval England", *Journal of Historical Geography* 19, 1–11.

Langton, J. and Hoppe, G. 1983, *Town and country in the development of early modern western Europe*, Historical Geography Research Series 11, Norwich.

Lennard, R. V. 1959, *Rural England 1086–1135*, Oxford.

—— 1922, "The alleged exhaustion of the soil in medieval England", *Economic Journal* 32, 12–27.

Mate, M. 1985, "Medieval agrarian practices: the determining factors?", *Agricultural History Review* 33, 22–31.

Nightingale, P. 1996, "The growth of London in the medieval English economy". In *Progress and problems in medieval England: essays in honour of Edward Miller*, eds R. Britnell and J. Hatcher, Cambridge, 89–106.

Overton, M. and Campbell, B. M. S. 1991, "Productivity change in European agricultural development". In *Land, labour and livestock: historical studies in European agricultural productivity*, eds B. M. S. Campbell and M. Overton, Manchester, 1–50.

Parain, C. 1966, "The evolution of agricultural technique". In *The Cambridge economic history of Europe*, ed. M. M. Postan, vol. I, 2nd edn, Cambridge, 125–79.

Persson, K. G. 1988, *Pre-industrial economic growth: social organization and technological progress in Europe*, Oxford.

Postan, M. M. (ed.) 1966a, *The Cambridge economic history of Europe*, vol. I, 2nd edn, Cambridge.

—— 1966b, "Medieval agrarian society in its prime: England". In *The Cambridge economic history of Europe*, ed. M. M. Postan, vol. I, 2nd edn, Cambridge, 548–632.

—— 1972, *The medieval economy and society*, London.

Postles, D. 1989, "Cleaning the medieval arable", *Agricultural History Review* 37, 130–43.

Power, J. P. and Campbell, B. M. S. 1992, "Cluster analysis and the classification of medieval demesne-farming systems", *Transactions of the Institute of British Geographers* 17, 227–45.

Rackham, O. 1980, *Ancient woodland: its history, vegetation and uses in England*, London.

Ravensdale, J. R. 1974, *Liable to floods: village landscape on the edge of the Fens AD 450–1850*, Cambridge.

Rutledge, E. 1988 "Immigration and population growth in early fourteenth-century Norwich: evidence from the tithing roll", *Urban History Yearbook, 1988*, 15–30.

Searle, E. 1974, *Lordship and community: Battle Abbey and its banlieu, 1066–1538*, Toronto.

Sheail, J. 1971, *Rabbits and their history*, Newton Abbot.

Shiel, R. S. 1991, "Improving soil fertility in the pre-fertilizer era". In *Land, labour and livestock: historical studies in European agricultural productivity*, eds B. M. S. Campbell and M. Overton, Manchester, 51–79.

Simmons, I. G. 1974, *The ecology of natural resources*, London.

Smith, R. A. L. 1943, *Canterbury Cathedral Priory: a study in monastic administration*, Cambridge.

Smith, R. M. 1988, "Human resources". In *The countryside of medieval England*, eds G. Astill and A. Grant, Oxford, 188–212.

—— 1991, "Demographic developments in rural England 1300–1348: a survey". In *Before the Black Death: studies in the "crisis" of the early fourteenth century*, ed. B. M. S. Campbell, Manchester, 25–77.

Sutton, A. 1989, "The early linen and worsted industry of Norfolk and the evolution of the London Mercers' Company", *Norfolk Archaeology* 40, 201–25.

Thirsk, J. (gen. ed.) 1967–89, *The agrarian history of England and Wales*, vols. IV–VI, Cambridge.

Thornton, C. 1991, "The determinants of land productivity on the bishop of Winchester's demesne of Rimpton, 1208–1403". In *Land, labour and livestock: historical studies in European agricultural productivity*, eds B. M. S. Campbell and M. Overton, Manchester, 183–210.

Titow, J. Z. 1969, *English rural society, 1200–1350*, London.

—— 1972, *Winchester yields: a study in medieval agricultural productivity*, Cambridge.

Vanderzee, G. (ed.) 1807, *Nonarum inquisitiones in Curia Scaccarii tempore regis Edwardi III*, Record Commission, London.

Verhulst, A. 1985, "L'intensification et la commercialisation de l'agriculture dans les Pays-Bas méridionaux au XIIIᵉ siècle". In *La Belgique rurale du Moyen Âge à nos jours. Mélanges offerts à Jean-Jacques Hoebanx*, Brussels, 89–100.

—— 1990, "The 'Agricultural Revolution' of the middle ages reconsidered". In *Law, custom and the social fabric in medieval Europe. Essays in honor of Bryce Lyon*, eds B. S. Bachrach and D. Nicholas, Kalamazoo, 17–28.

Wrigley, E. A. 1967, "A simple model of London's importance in changing English society and economy, 1650–1750", *Past and Present* 37, 44–70.

—— 1985, "Urban growth and agricultural change: England and the continent in the early modern period", *Journal of Interdisciplinary History* 15, 683–728.

11. AGRICULTURAL TECHNOLOGY IN SOUTHEAST ENGLAND, 1348–1530

Mavis Mate*

Historians know a great deal more about agricultural techniques than they do about technological mentalities. Why, for example, did some farmers start using horses as plough beasts, whereas others continued to use oxen until the end of the middle ages? Why, in some areas, was land farmed very intensively, with seed sown at a high rate, and legumes used as an alternative to fallow, whereas in other areas no changes were made and the land was exploited much less intensively? The techniques that could make farming more efficient and more productive were well-known, but were none the less not universally adopted. Landlords with scattered estates did not follow common policies for all their manors, but in matters such as seeding rates or use of legumes followed the practices that were generally used in that area.[1] One possible answer to these questions has been provided by Bruce Campbell in this volume. In discussing the incentives and disincentives to investment, innovation and intensification, he comes to the conclusion that intensive and productive husbandry systems were most likely to occur in areas of high economic rent.

In the twelfth and thirteenth centuries, when both the population and the market was expanding, it was easier for land to achieve a high economic rent. Labour costs were low and prices were generally rising. In the late middle ages, however, after the onslaught of the Black Death and subsequent outbreaks of plague, producers were faced with a shortage of labour and a consequent rise in wages. Yet in the late fourteenth century the economy as a whole remained buoyant and a sharp burst of inflation in the 1360s and 1370s allowed some lords to continue to receive high profits. It was not until the mid-fifteenth century that the economy as a whole hit rock bottom and the country faced a severe recession. High returns from

* I am grateful to the staffs of the Public Record Office, East Sussex Record Office, and Arundel Castle Library. I would also like to thank Bruce Campbell, and Christopher Dyer for helpful comments on earlier drafts of the text.
    [1] Campbell 1983b; Mate 1985b, 27–31.

agricultural activities, and hence high economic rents, were much more difficult to accomplish. During the two centuries after the Black Death lords and peasants continually had to decide whether to focus on producing food for their household or whether to produce for the market.[2] In making these decisions factors such as the cost of transportation, the prevailing price level, distance from the market, and the general buoyancy or otherwise of trade all played their part.

Not everyone responded in the same way. Moreover any decision, once made, could be reassessed and changed in the light of new economic circumstances. A close look at how both demesne and peasant farmers responded to the exogenous forces affecting the English economy in this period can shed some light on agricultural mentalities. Southeast England—the counties of Kent, Sussex and Surrey—is a particularly good area to study in depth (figure 11.1), since it includes within it six of the eight different farming types identified by Campbell (this volume). Not enough evidence has survived to allow a complete analysis of changes taking place in the late middle ages within each farming type, but enough accounts are available to show the variety of responses that occurred and the importance of technological adjustments generally in the development of late medieval agriculture. To this end, changes in agricultural practice from the middle of the fourteenth century to the beginning of the sixteenth will be examined firstly on demesnes, dividing them according to the six farming types evident in the region, and then secondly, for the region as a whole, on peasant farms.

Demesne Agriculture

Intensive Mixed Farming

As Campbell points out, in the period before the Black Death northeast Kent contained a high concentration of "intensive mixed farming" demesnes. It was an area of fertile and easily cultivated loam soils, with a high population density and ready access to a market—both in the immediate neighbourhood with the city of Canterbury and further afield, via the port of Faversham, to London and the

---

[2] For a good discussion of the issue of production for consumption or the market, see Biddick 1989 and 1991.

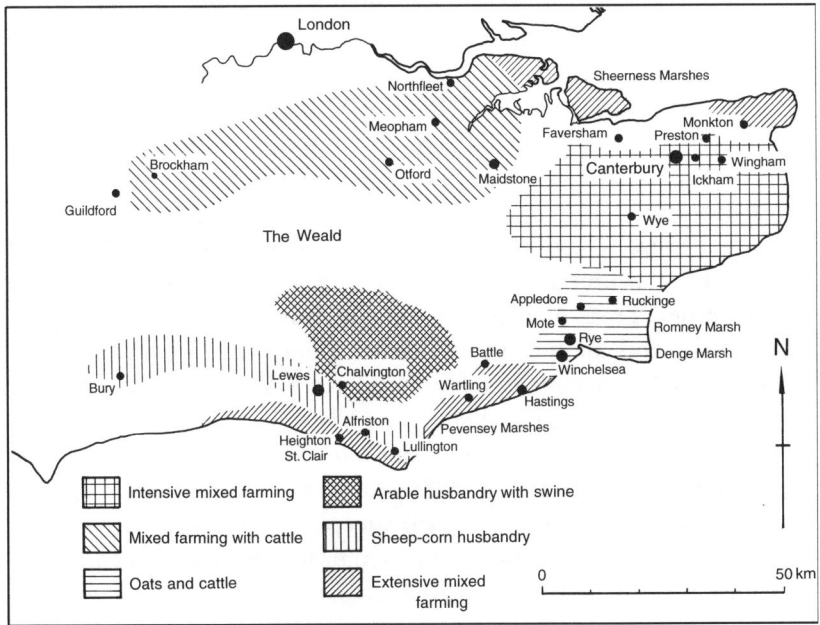

Figure 11.1. Map of southeast England showing manors and farming regions mentioned in the text.

continental markets. After 1348 demesnes continued to sow at a high rate of seed—wheat at 4 bushels an acre (about 360 litres per hectare), oats at 6 bushels an acre (540 lit/ha) and barley at 5–6 bushels an acre (450–540 lit/ha)—and to plant 20–30 per cent of their acreage with legumes. These leguminous crops were clearly used as an alternative to fallow. Much of the vetch was left in the ground so that it could be consumed by sheep or cattle that were folded on the fields by night. In this way their manure became especially valuable since it would contain a high nitrogen content.

In the 1360s and 1370s agriculture became in some respects even more intensive as demesnes placed greater reliance on horse power. Even before the Black Death both ecclesiastical and lay lords were using horses for hauling and for ploughing, although in the latter case, it was more common to use them in mixed teams of horses and oxen.[3] After 1348 some estates abandoned mixed teams and

---

[3] The use of horses and oxen within the region took place within the broad pattern outlined in Langdon 1986, but close analysis of individual demesnes reveals subtle variations.

switched over totally to horses as their main draught animals. When the monks of Christ Church, Canterbury, for example, took back their manor of Monkton in 1383, after the expiration of a lease, it contained six stots (workhorses) and eleven oxen. They immediately sold all the oxen and replaced them with twelve horses.[4] Lay lords such as Simon de Burley on his estate at Preston by Wingham kept two carthorses, and ten stots, but no oxen.[5]

Since a horse ploughs faster than an ox, fewer horses were needed. Horses also consumed grain rather than grass. This had two benefits for demesnes. First, there was room for more non-working animals. When wool and animal prices rose in the 1360s and 1370s, so did the size of some sheep-flocks. On the Christ Church manor of Ickham there had been around 300 sheep in 1349–51. The flock slowly increased until it peaked at 499 animals in 1370 before being cut back by outbreaks of murrain. Similarly on the Battle Abbey manor of Wye the flock expanded from 660 sheep and lambs in 1350 to 964 animals in 1371–2.[6] Second, fewer plough teams were needed. Wages had always been high in east Kent and they rose higher and faster there than in many other parts of the country when labour became scarce after 1348.[7] On the Christ Church manor of Ickham, where the monks spent £27 on the purchase of cart-horses, stots and oxen in 1366, the number of *famuli* ploughmen was reduced from ten to four, that is, from five teams to two. Such a move produced considerable savings for the priory and may have been the primary motivating factor behind the greater utilization of horses.

The fairly widespread reduction in manpower, however, may have been too drastic and some fields may not have been adequately prepared before seeding. In the second half of the fourteenth century crops in east Kent were very susceptible to fungoid diseases and weeds. To give but one example: on the Christ Church manor of Copton (in the parish of Preston) the yields of barley, sown at 6 bushels per acre (540 lit/ha), which had frequently reached four times seed in the 1340s, dropped to two times seed in the 1360s and 1370s.

---

[4] Cathedral Archives and Library, Canterbury, Bedels rolls, Monkton.
[5] *Calendar of Inquisitions Miscellaneous*, IV, no. 441.
[6] Public Record Office (henceforward PRO) SC6/899/10; SC6/900/5. Since a large acreage was under the plough this meant a stocking density of 33.96 and 39.20 livestock units per 100 sown acres—calculated using Bruce Campbell's weighting for the various types of animals: Campbell, this volume.
[7] Mate 1985a, 55–68.

Likewise wheat yields dropped from five times seed to just over three times seed. The local serjeant explained these poor harvests by referring to the chronic weediness of the fields or to an "excessive fall of mildew in the summer". The serjeant, however, had cut back on the number of full-time employees. In the 1290s, with 133 acres under cultivation, two ploughmen, a carter, a shepherd, and a pigman had been hired for the year. In the 1360s, with roughly the same area under the plough (116 acres) the serjeant hired two ploughmen for the year and a harrower part-time. The full-time carter, shepherd and pigman were no longer employed. Yet in the past these men, in addition to their regular duties, could have weeded and carted manure. Without their help, these tasks, as on some Norfolk estates studied by Bruce Campbell, may not have been done, or were done less frequently.[8]

Yet the drop in yields in the third quarter of the fourteenth century was so widespread, not just in Kent, but in other parts of England as well, that it suggests a significant shift in climate.[9] Fields which had been less intensively cultivated, however, may have been more vulnerable to outbreaks of fungoid disease. A reduction in the labour force did not inevitably lead to lower yields, but in years when the climate was such that weeds and fungoid diseases flourished, then the lack of staff made itself felt. On the other hand, no reduction in staff did not necessarily prevent a reduction in yields.

Grain and stock prices began to fall in the 1390s, although they could still rise to great heights during years of bad harvests such as 1438 and 1439. In the 1440s, however, prices began a slide that reached a low point in the early 1460s. According to the work of George Grantham, the response of agricultural productivity to price was likely to be "strongly positive". A high farmgate price encouraged farmers to expand output. Conversely low prices discouraged agricultural innovation, since farmers saw no profit in expanding labour and capital to raise food that either they could not sell, or would bring them little income.[10] In the 1450s and 1460s, therefore, demesne farmers seem unsure what policies to pursue since they could

---

[8] Cathedral Archives and Library, Canterbury, Bedels rolls. The mildew quote is from Copton for the year 1273. Campbell has suggested that lower yields of wheat and barley on Norfolk estates were the result of a reduction in the use of labour: Campbell 1983a, 38–9 and 1991.

[9] Campbell 1991, 163; Hallam 1984, 124–32; Thornton 1991, 194.

[10] Grantham 1989 and 1995.

not know how long the deflation would last. In northeast Kent they continued to rely on horse power and to sow a high percentage of legumes. On the other hand some seeding rates were cut back. On the Christ Church home farm of Northgate at Canterbury (the Barton) the seeding rate for wheat was reduced from 4 bushels to 3 bushels an acre (360 to 270 lit/ha) in 1445 and it remained at the lower level thereafter. The seeding rate for barley was reduced from 5 bushels an acre to 4 bushels an acre (540 to 360 lit/ha) in the 1440s, but in the 1450s it became variable, as barley was sown some years at 4 bushels and in other years at 5 bushels. Finally oats—the least valuable grain—which had been sown at 6 bushels an acre in the early fifteenth century, were sown at 5 bushels an acre in 1444 and 4 bushels an acre in 1447 before going back to 5 bushels an acre in the 1450s and finally returning to 6 bushels an acre in 1470–1.[11] Likewise on the Battle Abbey manor of Wye, where in 1410–11 oats and barley had been sown at 6 bushels an acre and wheat at 4 bushels, in 1446–7 wheat was sown at 3 bushels an acre and oats and barley at 5 bushels an acre.[12] This meant that although the yield per seed might remain high, the yield per acre was likely to be less.

The overall area under the plough also contracted. This allowed lords to leave land fallow for longer periods and cropping patterns became extremely flexible. Large fields might not be fully utilized. At the Northgate (Canterbury) the serjeant, in the early years of the fifteenth century, had frequently sown 80–7 acres of grain in Barton-field. In the four years 1444–8 it was sown with 15 acres of peas (1444), 44 acres of wheat and oats (1445), 72 acres of barley (1446) and 29 acres of beans and barley (1447). Small fields were also left fallow, or just sown with legumes for several years in succession. Furthermore, in places where convertible husbandry had been practised, the pasture, after being turned over, was cultivated for just a few years before being left fallow or with just an inhoking of legumes. As a result the fields at Northgate produced exceptionally good harvests. In 1445–6 they yielded 36 bushels an acre of oats (7.3 times seed; about 3,230 lit/ha), 26 bushels of barley an acre (6.6 times seed; 2,330 lit/ha) and 11 bushels an acre of wheat (3.75 times seed; 990 lit/ha). The following year the oats and barley both yielded 5.6

---

[11] Cathedral Archives and Library, Canterbury, *Caruca Bertona* rolls 18–43.
[12] PRO SC6/902/27; SC6/905/1.

times seed and the wheat yielded 4.38 times seed.[13] Yet, however good individual harvests might be, the land as a whole was clearly less productive, and being farmed less intensively, than it had been in the fourteenth century.

In addition the wool trade collapsed and farmers frequently could not find buyers for their wool at the accustomed price. They either had to let it accumulate unsold in their warehouses or sell it at a very low price.[14] Some demesne flocks were scaled down. At Wye in 1446–7 there were just thirty-two ewes and forty-six lambs.[15] Yet since the area of the demesne under the plough had also contracted, the stocking density remained virtually the same as in the late fourteenth century—34.06 units per 100 sown acres.[16] Moreover as the sheep of some of the tenants went with the sheep of the lord, their manure would help enrich the soil. Thus, as at Northgate (Canterbury) the yields of individual harvests might have been high, but without enough consecutive accounts one cannot plot any definitive pattern. The Canterbury monks likewise reduced the size of their flocks when they switched from keeping sheep for their wool to keeping them for their meat. In the early fifteenth century the home farm contained forty cows, sixteen horses, eight sows and 504 sheep— a stocking density of 45.3 units. By the 1450s and 1460s, however, the manor was specializing in fattening stock for the market. Each year the serjeant bought a number of yearling lambs. These were then fattened up and sold as wethers. Thus at the beginning and end of the year there were between 200–300 sheep on the manor, roughly half the earlier level.

This evidence supports George Grantham's thesis that lords were very aware of price levels for grain and stock and were willing to adjust their agricultural policies accordingly. In the late fourteenth century the rise in the price of wool and cattle encouraged the expansion of sheep flocks and cow herds. Conversely the deflation of 1440–65 and the general contraction of trade was accompanied by a cutback in seeding rates and a contraction in the amount of stock. Ecclesiastical lords such as the monks of Battle Abbey, who earlier had actively produced for the market, withdrew almost totally in the

---

[13] For further information on yields and cropping patterns, see Mate 1991b, 274–80.

[14] Mate 1987.

[15] PRO SC6/905/1.

[16] Using Campbell's method of calculation: see note 6 above.

mid-fifteenth century and concentrated on supplying their household.[17] With no surviving accounts from the late fifteenth century it
is impossible to tell when a revival occurred, but the city of Canterbury was very depressed at this time, with falling rents and unten-
anted houses, so it may have been the mid-sixteenth century before
agriculture in this area was carried out as intensively as earlier.

Mixed Farming with Cattle

In parts of northwest Kent and Surrey demesnes practised a form of
husbandry that was far less intensive than in northeast Kent. Oxen
remained the primary plough animal; stocking densities were generally low and legumes were not a prominent part of the crop rotation.
In some places the amount of legumes expanded after 1348. At
Brockham in Surrey in the late thirteenth century just a few acres,
5–7 per cent of the total acreage, was sown with vetches. In 1364–5,
18 per cent (23 acres) was cultivated with peas and vetches.[18] Likewise at Meopham, in northwest Kent, the acreage under legumes
doubled—from 20 to 40 acres—in the early 1380s and in 1384 went
up even higher to 51 acres. Some of the beans were used in the
grain livery to the *famuli* and the peas and vetch, as elsewhere, were
used for forage—for example, to feed the horses of the prior when he
visited the manor. How common this increase was, and whether it
continued into the fifteenth century, is not known, but on the archiepiscopal manor of Otford the percentage of legumes remained low;
in the four years for which information is available it fluctuated
between 5 and 11 per cent.[19]

Seeding rates were also lower than in northeast Kent and they
generally remained that way. In the late fourteenth century wheat
was sown at 2–3 bushels an acre and oats at 3–4 bushels an acre.
High seeding rates were so obviously beneficial that it is not clear
why they were not adopted at a time of high grain prices even by
lords such as Canterbury Cathedral Priory who were willing to in-

---

[17] On the Battle Abbey manor of Wye the number of cows was reduced from
forty-one to twenty-five. They were no longer leased to a cheeseman. No cheese
was sold on the market and the herd just supplied the needs of the abbey household. On the Battle Abbey manor of Apuldram the reduced cowherd was used to
breed oxen needed to run the manor. No cheese was sold. Mate 1987, 528.

[18] Arundel Castle Library A 488, A 491.

[19] Lambeth Palace Library ED 836, ED 853, ED 857, ED 863.

crease the rate of seeding for oats in the Romney marshes. The most likely explanation is that these estates were already producing enough to satisfy their own needs or for sale in local markets. The rise in the cost of carting with the general rise in wages after the Black Death may have made the cost of inland transportation of grain high enough to discourage production for more distant markets.

Arable Husbandry with Swine

In the Sussex Weald the heavy clay soils were unsuited to barley and often too wet for sheep folding. Farmers there tended to specialize in cow-keeping, pig-rearing, and coppice-cutting. In the late fourteenth and early fifteenth centuries some demesnes planted a significant portion of their land with beans. On the manors of Claverham and Chalvington, held by the Sackville family, 106 acres were under the plough in 1413–14, of which roughly one quarter (25 acres) were sown with beans.[20] The *famuli* who were responsible for most of the agricultural work on the manor, received a money wage and a grain livery of beans and oats. The rest of the beans, apart from seed, were used for fattening the pigs or sold. By the 1420s, however, the arable acreage had fallen to around 80 acres. The acreage under beans was reduced—first to 16 acres and then to 8 acres—as the bailiff started hiring ploughmen by the day for money and paid the remaining *famulus* his grain livery in oats. The beans that were still grown were used to fatten the pigs and in the 1420s the manor supplied the household at Buckhurst with twenty or more pigs and piglets each year. During the course of the 1430s, however, the bailiff gave up keeping large numbers of pigs, so the immediate need for beans disappeared. In the last surviving account, in 1444–5, he sowed 60 acres of oats, and 2 acres of peas, with no beans at all.[21] If he had sown the some amount of beans as earlier and then sold them, the soil would have benefited. That he did not do so, points not only to the sluggishness of the market, but underlines the suggestion made earlier that legumes were not *primarily* sown for their fertilizing properties.

---

[20] East Sussex Record Office (henceforward ESRO) SAS/CH 263.

[21] ESRO SAS/CH 251–285. In the mid-1440s the few remaining pigs were fattened with oats. On the manor of Rimpton, where beans were also used for grain liveries, the area under legumes likewise declined after the Black Death: Thornton 1991, 195.

Sheep-corn Husbandry

In parts of Sussex lords practised a form of sheep-corn husbandry on their demesnes. Little use was made of horses, even for carting, and oxen remained the primary plough animals throughout the whole period. Wages, however, were generally lower than in Kent, so there was less incentive to reduce the number of plough teams and with it the number of permanent staff. Moreover these Sussex demesnes had plentiful supplies of pasture so that oxen were economical to keep. Furthermore the oxen could be eaten. At some point in their working lives they were fattened up and then sold to local butchers, or eaten in seigneurial households. The manors also contained large numbers of sheep which were usually folded on the arable at night, so that the soil was enriched with their dung.

Before the Black Death these demesnes had sown at a high rate of seed and planted a significant amount of legumes.[22] In the late fourteenth century some lords started to cut back. At Bury, for example, in the 1390s, when the price of barley began to fall, the seeding rate was reduced from 6 to 5 bushels. At the same time the acreage under legumes was reduced from 21 to 18 per cent.[23] Such reductions became more severe over the course of the fifteenth century. On the Battle Abbey manor of Lullington, where in the late fourteenth century around 40 acres of legumes were sown (16 per cent of the total acreage), the percentage fell to 10 per cent in the early years of the fifteenth century and reached a low point in the late 1450s and early 1460s with just a few acres under peas—2 per cent. Although the acreage climbed slightly in the late 1460s to 7 per cent, that was still way below the fourteenth century level. Likewise on the Battle Abbey manor of Alciston a low point was reached in the mid-1470s when just 12 acres of legumes were sown—6.8 per cent of the area under cultivation.[24] The reasons behind this contraction remain unclear. The peas and vetch were used primarily as forage crops—to fatten pigs and to sustain sheep and lambs—and the monks may not have been fully aware of their fertilizing properties. In the fifteenth century, when the monks were eating more beef than pork, fewer pigs were being kept, so the need for legumes was less. Moreover with the

---

[22] Brandon 1969; 1972; Mate 1991a.
[23] Arundel Castle Library A 368–385.
[24] For further details, see Mate 1991b, 269–71.

general contraction in trade, there was no incentive to produce extra goods for the market. Thus, even if the monks were aware that grain sown on fields previously planted with legumes produced more, that may not have been sufficient inducement to plant the same percentage of legumes as earlier.

In the mid-fifteenth century both the area of the demesne under cultivation and the number of sheep on the manors had contracted significantly. At Alciston, for example, in 1399, with 406 acres under the plough, there were thirty cows, thirty-six oxen, 1053 wethers, 478 ewes and assorted young animals pasturing on the manor—a stocking density of 85.59 units per 100 acres.[25] In 1482-3, on the same manor, there were just 800 sheep, but the area under cultivation had also shrunk—to 225 acres.[26] Since the number of stots, oxen and cows had remained virtually the same, the stocking density had actually increased—to 93.1 units. As on the Kent manor of Northgate the unused land at Alciston was left fallow for longer periods between sowings. This extra fallowing must have helped offset the reduction in the area under legumes since there was no significant reduction in the yields per seed.[27] The overall productivity of the land, however, had clearly fallen. In the mid-1450s the serjeant at Alciston was sowing around 78 acres of barley a year—half the acreage of the early fifteenth century. Most of the yield was used on the manor, in payments to *famuli*, and making ale for harvest workers, leaving just a small quantity for dispatch to Battle. Even if the monks were receiving all the barley they needed, the serjeant could still have sown a larger acreage and sold the additional grain in local markets. He did not. One can only guess at the reasoning behind the policy decision, but the low price of barley—3s. to 3s. 4d. per quarter—meant that in years of average harvests the difference between the cost of production and sale revenues would be slender. Even in years of poor harvests grain prices did not rise very high, so that with a reduced yield the profit margin could disappear altogether. The monks may have decided that the potential reward was not worth the additional effort.

---

[25] ESRO SAS/G44/54.
[26] ESRO SAS/G44/129.

Extensive Mixed Farming

Demesnes practising "extensive mixed farming" existed primarily on
the coastal districts of Kent and Sussex. On the northern marshes of
Kent just a small area on each manor was sown with wheat, barley,
or legumes. The remaining land was left as pasture, which meant
that the pasture and arable were not well integrated. Thus, although
many manors had high stocking densities, if the sheep remained on
the marshes and were not folded on the arable fields, then their
manure would not benefit grain yields. Boxley Abbey, for example,
in the 1370s, had over 2,000 sheep and lambs on the marshes at
Sheerness, but ploughed just 38 acres of land. The value of the manor
lay in the revenues from the sale of wool and cheese and goods
supplied to the monastic household.[28]

Likewise the Sussex coastal manors were teeming with animals in
the late fourteenth and early fifteenth centuries. The accounts of the
manor of Pebsham, held by the Pelham family, clearly show the
flourishing trade that existed between the Cinque Ports and their
hinterland. Calves and bullocks from Pebsham were sold to Hastings
butchers, and reeds, cartloads of hay, wheat, and oats were sold to
merchants from Winchelsea and Hastings. Large numbers of oxen,
cows, and sheep pastured at different times on the manor. Pebsham,
however, was just a small cog in a complex system of transhumance,
so that meaningful stocking densities are hard to determine as the
animal population constantly fluctuated. At Michaelmas 1416, for
example, there were thirty-nine cows on the manor. During the year
another thirty-four cows came in from Herstlyngen and Wartling,
but by Michaelmas 1417 some of these cows had returned, leaving
thirty-six animals over.[29] The size of the sheep flock also expanded
and contracted as animals were bought and sold. In 1429–30 at one
point there were 805 wethers and ewes at Pebsham, but 322 animals
were sold and the year ended with 406 sheep on the manor. There
was only a small acreage under the plough—79 acres at Pebsham in

---

[27] For further details, see Mate 1991b, 278–80.

[28] PRO SC6/897/3–10.

[29] ESRO RAF Pebsham. At Herstlyngen in 1418 there were sixty-four cows at
the beginning of the year; another eighty-six came in from different places, and the
year ended with 122 cows on the manor. There were also three affers, nine mares,
eight foals, forty-three oxen, seventy-six bullocks and fifty calves. British Library
Additional Roll 31824.

1429–30—which was primarily used to supply the needs of the manorial household. Thus the manure of many of these animals had no immediate impact on grain yields, but would ultimately enrich the soil if it was ever ploughed under.

In the mid-fifteenth century, as in other parts of the region, agricultural operations clearly contracted. Demesne sheep flocks at Pebsham disappeared and the system of cattle transhumance was given up. In the 1460s a dairy herd of forty animals was kept at Pebsham and leased to a cheeseman, and the herd of forty cows at Herstlyngen was used to breed cattle for the Pelham table. For Heighton St. Clair there are no mid-fifteenth-century records, but in 1405 both the area under the plough and the amount of stock had been drastically reduced (see table 11.1). How long this contraction lasted is not known, but there are signs that the economy in the area had begun to revive by the middle of the sixteenth century.

At Heighton St. Clair just a single account, for 1548, has survived, but it sheds light on aspects of agricultural production, such as the percentage of legumes and seeding rates, that remain virtually undocumented for early- and mid-sixteenth-century demesnes. The account (table 11.1) shows that, although the area under cultivation

Table 11.1. Agrarian practices followed on the manor of Heighton St. Clair.

	1319	1405	1548
Total area	139 acres	97.5 acres	179 acres
% legumes	10%	17%	7.2%
Seeding rate			
wheat	3 b.*	3 b.	2.5 b.
barley	7 b.	6 b.	4.5 b.
oats	7 b.	4 b.	6 b.
Total number			
cows	30**	0	64 in, 34 left
oxen	20**	8	66 in, 94 left
horses	7	0	8
sheep/lambs	195	124	620
Stocking density	99 units	26.8 units	170.5 units

* b. = bushels   ** before the outbreak of cattle disease

Sources: East Sussex Record Office SAS/G1/44; SAS/G1/45; SAS/G1/48.

had expanded, the area under legumes had contracted. Seeding rates were also lower than in the early fourteenth century. How typical these changes were is not known, but they do show that agricultural policies were by no means fixed and could be altered in light of changing circumstances. The most significant change, however, was the dramatic increase in the amount of stock on the manor. The size of the sheep flock had tripled from that of the early fifteenth century. In addition cattle were clearly being raised for sale to local butchers as well as for consumption in the seigneurial household. In 1548, forty-eight oxen were either eaten or sold and ninety-four remained, of which forty-six were said to be available for husbandry and forty-eight were to be "fattened". Some of this stock was bred on the manor, but a good portion was bought in new each year (sixty-one oxen, eleven cows, twenty-three calves in 1547). If other estates in the area were also taking advantage of the growing market for meat, then, as was the case in Norfolk, livestock farming may have been the most dynamic sector of agriculture.[30]

Oats and Cattle

On the Romney marshes in the late fourteenth century horses were the primary plough animals. The area under legumes was frequently higher than on the east Kent manors—as much as 39 per cent at Dengemarsh in 1366—but the crop was usually beans, which were either sold or given as food to estate workers and not fed to animals. The fields themselves would be enriched by the presence of the legumes, but the available manure would not be enhanced. Oats was the major crop and after the Black Death the seeding rate for oats went up—to 7 bushels an acre on the Battle Abbey manor of Dengemarsh and 8 bushels an acre on the Christ Church manors of Ruckinge, Appledore and Ebony.[31] During this period oat prices were unusually high, partly as a result of the general inflation, and partly because of an increased demand with the greater use of horse power. Thus it looks as if the monks were seeking through the manipulation of their seeding rates to increase their yields in response to the market. They succeeded. In the 1360s and 1370s the oats sown on the Christ Church Romney marsh manors yielded 16–22 bushels an acre after seed.

---

[30] For similar developments in Norfolk, see Overton and Campbell 1992.
[31] PRO SC6/889/19; Cathedral Archives and Library, Canterbury, Bedels rolls.

Yet this grain, in addition to the high rate of seed, had been sown after a crop of beans, and on land that had been well manured with cow dung or was newly reclaimed marshland. All these factors undoubtedly contributed in some way to these high yields, but it is impossible to separate out which factors, if any, contributed the most. Whatever the reason, the monks were able to sell some of their surplus grain and send some to the monastic household at Canterbury or Battle, thus saving the high cost of purchased grain.

Over the course of the fifteenth century the arable component in these manors shrank and the cattle component became more important. When, for example, the archbishop of Canterbury's marsh of Bekard was leased for seven years in the 1480s the lessee agreed to plough and sow 80 of the 100 acres just twice within the term. So too in the case of the Cheyne marsh of 829 acres, the lessee agreed that, in the last year of the five year lease, he would not plough or sow any part beyond 20 acres.[32] It is not clear whether such lords were primarily concerned to maintain the fertility of the soil or the quality of the pasture. By the 1520s this intermittent cultivation may have been curtailed further or even ceased. In 1528 Sir Edward Guildford complained that farmers on the Romney marsh no longer dwelt on their farms breeding cattle and tilling the land, but simply used the land for grazing Welsh cattle.[33] What made this increasing specialization possible was the expanding market for meat, and especially for beef.

The impact that an expanding market could have on agricultural production comes out very clearly in the changes taking place on the Sussex manor of Mote. When Sir John Scott acquired his new manor on the outskirts of Rye in 1464, it had been leased out. Rye, however, was growing in size and importance.[34] Furthermore, when prices started to rise after Edward IV's recoinage, the profitability of agriculture increased. Sir John Scott determined to take advantage of the new market opportunities. The manor was taken back in hand. At first the bailiff sowed just a few acres of oats—seeding at 4 bushels an acre (360 lit/ha)—and the yields were merely average, around three times the seed. In 1473 he started on a programme of improvement—ditching, enclosing, and removing trees and underbrush. Over the next six years £27 was spent on this work of land clearance.

---

[32] Lambeth Palace Library ED 1200.
[33] PRO SP1/48, fo. 225.
[34] For full details of this expansion, see Mayhew 1987.

Once the trees and underbrush had been removed, he marled 32 acres, and started sowing wheat (at 2.5 bushels per acre). Finally he increased the seeding rates to 3 bushels an acre for wheat and 5 bushels an acre for oats. In addition the manor housed a large amount of stock, since it was used to breed oxen for Scott's household at Calais, as well as for sale. In 1477, for example, with 36 acres under the plough, there were sixty-one oxen, sixty-five *bovettus* and other young animals. The following year the bailiff received £9 19s. 4d. from the sale of grain—including ten quarters of wheat at 10s. a quarter—and £18 lls. from the sale of stock, not counting the value of the twenty-eight oxen that were shipped to Calais.[35] As Bruce Campbell suggested, there seems to be a clear correlation between a high economic rent and the introduction of intensive and productive husbandry. The reasons for the high yields at Mote (table 11.2) are harder to pinpoint. The bailiff made little use of legumes. No peas or vetch were sown and in 1476–7 just 2 acres of beans (5.5 per cent of the total acreage). Plenty of manure was available, but it is not clear whether it was regularly carried on to the fields or whether, as in the case of the "extensive mixed farming" regions, crop and animal production were mutually exclusive. It is likely, therefore, that the marling and the bringing of new land into cultivation were primarily responsible for the exceedingly high yields of 1477.

Table 11.2. Grain Yields at Mote, 1474–80.

	Oats		Wheat	
	Net yield per acre	Yield per seed	Net yield per acre	Yield per seed
	bushels		bushels	
1475	10.9	3.73 @ 4 b.	–	
1476	8.6	3	–	
1477	28.5	8.2	25	10.93 @ 2.5 b.
1478	17.5	4.5 @ 5 b.	17.23	7.89
1479	8.4	2.6	9.75	4.25 @ 3 b.
1480	15	4.32	20	7.6

Sources: East Sussex Record Office SAS/HC 178; SAS/HC 179/3; SAS/HC 179/5; SAS/HC 179/6; SAS/HC 179/8.

---

[35] ESRO SAS/HC 179/5.

Peasant Agriculture

There is not enough evidence about peasant agriculture to divide their farms into distinctive farming types. What evidence there is suggests that at least some peasants followed similar practices to those on demesnes in their area. Thus horses were more likely to be found in east Kent and oxen elsewhere. The Sussex peasant, Nicholas atte Nash, when he died in 1365 in an area of sheep-corn husbandry possessed two oxen, thirteen wethers, thirteen ewes and no horses.[36] In contrast the Kentish peasant, William Baker of Bridge, farming in an area of "intensive mixed farming" demesnes, possessed five horses, sixty sheep, three young oxen, and a number of piglets in 1362.[37] The oxen were probably being bred for their meat, with the horses being used for hauling and agricultural work. Baker held 11 acres from the abbot of St. Augustine's, Canterbury, and may have held other land from other lords. His example does show that not all peasant families were livestock deficient. Other peasants were also well-stocked. When an inventory was taken of the goods of John atte Wood, who was hanged at Canterbury in 1450, he had four oxen, six cows, three horses, and six pigs on his holding of 20 acres. If he regularly sowed around fifteen acres of his land, then he would have a stocking density of 104 units per 100 sown acres.[38] How typical this stocking density was is unclear. Campbell and Overton have found from their study of probate inventories that very small farms can produce very high livestock densities that are not representative of the majority of farms.[39] That seems to be the case in Kent. When three participants in Cade's rebellion of 1450 were hanged at Tenterden, a yeoman with 100 acres had just six oxen, ten cows, one horse, two sheep, eight pigs, four bullocks, six calves—a stocking density of 41.7 units per 100 sown acres if he normally cultivated, say, 70 acres in any one year. His fellows with holdings of 30 acres and sixteen acres had stocking densities of 73.8 units and 53.8 units respectively.[40] Since arable and pastoral husbandry were almost certainly

---

[36] ESRO SAS/G18/27.
[37] *Calendar of Inquisitions Miscellaneous*, III, no. 831.
[38] PRO E357/42.
[39] Overton and Campbell 1992, 386–8.
[40] PRO E357/42. In a Wealden tenement, with heavy clay soils, he was still using oxen rather than horses.

well integrated on these peasant holdings, the manure from these animals would help enrich the soil.

Peasant fields in east Kent might also benefit from a planting of peas and vetch. When Henry Farthing was outlawed in 1371 he held 3 acres from the prior of Christ Church, Canterbury, which was sown with corn mixed with vetches.[41] An inventory of the goods of Thomas Pynham in 1382, who held a messuage and 20 acres of land, showed that he had 200 copps of peas and vetch, in addition to stacks of oats and barley in sheaves and 18 acres sown with wheat.[42] Thomas Castelyn, on the day that he was killed in 1385, had 1 acre sown with rye and 2 acres sown with peas and vetch. This high percentage of legumes was probably needed for forage in support of small-scale cattle and sheep breeding. He possessed one cow, one heifer, two bullocks, two piglets, one ewe and two lambs.[43] In contrast William Baker of Bridge (mentioned earlier) on his 11-acre holding had sown 2 acres with wheat, 2 acres with spring barley, 2 acres with oats, 1 acre with vetch, leaving 4 acres fallow. Without more evidence one cannot say how representative any one of these peasants was of farming practices in the area, but it does seem likely that peasant land would have its stores of nitrogen replenished on a fairly regular basis. Indeed some peasant farms with a high stocking density, a high percentage of legumes, and a high labour input, may have been more intensively cultivated than the neighbouring demesnes. Such practices seem to have continued unchanged even during the fifteenth-century depression. In 1446–7 the prior of Christ Church, Canterbury, purchased 884 quarters of peas, beans, and vetch in the neighborhood (in patria).[44] Some, if not all, of these crops must surely have been grown by peasants.

Peasant plantings of legumes were fairly common elsewhere. In sheep-corn areas there are a number of examples of peasants growing vetch. John Bridgeman leased 1.5 acres from a widow. In the year that she died (1382) he had 1 acre sown with vetch and 0.5 an acre left fallow.[45] Philip Topyn leased 11 acres from his mother in

---

[41] *Calendar of Inquisitions Miscellaneous*, III, no. 849.
[42] *Calendar of Inquisitions Miscellaneous*, IV, no. 178. He may have leased additional land.
[43] PRO E153/1000, no. 5.
[44] Lambeth Palace Library ED 74.
[45] ESRO SAS/G18/29.

1399 and sowed them with wheat, barley and vetch.[46] When in 1391 William Page took over a bond tenement belonging to Battle abbey, he was given two mares, two oxen, and some land already sown— 7 acres with wheat, 3 acres with barley, 2.5 acres with vetch, and 0.5 acre with oats.[47] Indeed in areas of mixed farming with cattle peasants may have cultivated a larger percentage of legumes than some neighbouring demesnes. As noted earlier, on the archiepiscopal manor of Otford few legumes were sown. Yet in 1445 an Otford labourer had 3 rods of land sown with wheat and 2 acres sown with peas and vetch.[48] At Northfleet the archbishop, in 1408–9, leased small parcels of former demesne land to peasants to sow with peas, or peas and vetch.[49] The peasants received the benefit of the crop, but the lord reaped the benefit of the nitrogen. Their willingness to take these leases underlies their recognition of the value of legumin-ous crops, albeit perhaps primarily as a source of fodder.

On the other hand, in the oats and cattle area of the Sussex Weald tenant cultivation of legumes may have contracted as it did on the demesnes. In the late fourteenth century legumes were clearly grown. When 2 acres that had been leased without a license were seized in 1376 they had been planted with wheat, vetch, peas and oats.[50] In the fifteenth century, however, when leases specify what crops should be sown on the land, they specify oats. Complaints about the tres-pass of animals mention the loss of oats, with no reference to legumes.[51] Tithe data also reinforce the picture of oats as the major crop. In 1412–13 the rectory at Mayfield received in tithes 25 quarters and 6 bushels of wheat, 4 quarters and 2 bushels beans, 2 bushels of peas, and 182 quarters and 3.5 bushels of oats.[52] Pasture was plentiful within the Weald and pigs were allowed to feed in the woods during the season of pannage for a fairly modest fee. Tenants had less need of legumes for forage and they do not seem to have grown beans for their own sustenance. Fertility on the land was maintained by prac-tising a form of convertible or "up and down" husbandry, leaving the ground under grass for a year or two and then turning it over.

---

[46] ESRO SAS/G18/33.
[47] ESRO SAS/G18/31.
[48] PRO KB9/254, m. 62.
[49] Lambeth Palace Library ED 794
[50] British Library Additional Roll 31893.
[51] British Library Additional Rolls 31990, 31980, 32030.
[52] Lambeth Palace Library ED 696.

In 1426, for example, when John Bridge leased 12 acres from Thomas Chambre for four years it was on the understanding that John would sow the land for two of the four years.[53]

In contrast on the Sussex coastal marshes—an area of extensive mixed farming—beans seem to have been an important component of the peasant economy. When cattle and pigs at Brede in the early fifteenth century trespassed on the land of their neighbours, they consumed beans and peas as well as wheat and oats. So too in 1446 at least one tenant, John Fairman of West Dean, possessed beans, peas, oats, and barley, but there is no means of knowing how typical he was.[54] These beans and peas were probably used to fatten pigs and young cattle where tenants did not have ready access to pannage. By the end of the fifteenth century, if not before, peasants were clearly breeding stock to sell. On the Sussex manor of Mote demesne officials served as middlemen. They bought stock from small local farmers—eight different people in 1477—fattened it and then sold it to a local butcher. So, too, some of the stock that was bought each year at Heighton St. Clair could well have been bred by local peasant farmers, who could provide intensive care to pregnant animals and their issue.

Conclusions

The localization of farming practices did not change in the post-Black Death period and thus the distinctive farming regions were not modified significantly. Horses continued to predominate in northeast Kent and Romney marsh and not elsewhere. Yet this decision was almost certainly a response to local conditions. Wages were particularly high in these two areas. The switch to the more efficient horse was often accompanied by a reduction in the number of plough teams and thus a saving on the wage-bill. Such savings may have been at least part of the reason behind the change. Common pasture was also in short supply in east Kent, so that peasants found it difficult to feed the number of oxen needed for a full team and naturally turned to horses. The continued or expanded use of horses in east Kent in its turn undoubtedly encouraged the continued use

---

[53] British Library Additional Roll 31976.
[54] PRO KB27/748, m. 38v.

of legumes. Conversely in areas of mixed farming with cattle, where oxen remained the primary plough animal, and the overall animal density was low, there was less need of legumes for additional forage. Moreover, since wages in these regions tended to be lower than in east Kent and the Romney marshes, lords had less incentive to reduce the number of plough teams by switching to horse power.

Peas and vetches were primarily used as fodder crops, but were lords and peasants also aware of their fertilizing properties? Since lords always planted wheat—the most valuable crop—immediately after a field had been sown with legumes, this suggests that they were at least partly aware that legumes benefited the land. Peasants likewise seem fully aware of the usefulness of vetches and peas, but whether as fodder, or fertilizer, or both, is not clear. Yet one of the major changes that took place over the course of the fifteenth century—in four of the six farming types—was the reduction in the area under legumes, especially beans. The reasons behind the reduction, however, vary among the different farming types. Where beans and other legumes had been used to fatten pigs or to sustain sheep and lambs, their use was curtailed when sheep-flocks were reduced and demesne lords ate more beef than pork. Where beans had been given to *famuli* as part of their wages, they were no longer grown when workers acquired a higher standard of living and were able to demand payment of their grain livery in barley or oats. In either case the fertilizing properties of legumes were ignored when the decision was made to discontinue their sowing.

The determinants behind grain yields are also hard to gauge because of factors such as climate and fungoid disease. Nonetheless there is considerable support for a point made by Campbell that an increase in the stocking density did not inevitably lead to an increase in yields.[55] In areas of extensive mixed farming, where pasture and arable were not well integrated, any increase in stock did not necessarily affect the arable operations. In areas of intensive mixed farming the value of the extra manure might be offset by a reduction in labour inputs, so that the ground was less well prepared. Likewise a contraction in the area under legumes, such as occurred in the mid-fifteenth century, might not produce lower yields if it was counterbalanced by a widespread adoption of convertible husbandry or an

---

[55] Campbell 1991, 163.

extended use of fallow, even in regularly cropped demesne fields. Thus while the overall productivity of the land was undoubtedly lower in the mid-fifteenth century than in the late fourteenth, the yields of individual harvests were frequently higher.

This study clearly shows the strong correlation between intensive agricultural techniques and *periods* of high economic rent. Lords were aware of market opportunities. When the economy was buoyant and prices were high in the late fourteenth century, producers who were already engaged in intensive mixed farming were prepared to farm even more intensively than earlier. Similarly, on "extensive mixed farming" demesnes sheep flocks and cow herds were expanded. Likewise, the depression of 1440–60 had a major impact on agricultural practices. In the area of sheep-corn husbandry there was a widespread contraction in the area under demesne cultivation, in the size of the sheep flocks, and in the amount of legumes sown. Cow herds on extensive mixed farming demesnes also shrank. Even in the area of intensive mixed farming in northeast Kent seeding rates were reduced. The correlation between *places* of high economic rent and farming techniques is less clear-cut. Demesne farmers in northwest Kent and Surrey practised the least intensive husbandry of the whole region—mixed farming with cattle—even during periods of economic growth. Yet much of the area was within relatively easy distance of London. In addition, expanding towns such as Maidstone in Kent and Guildford and Croydon in Surrey should have been able to stimulate a demand that would have encouraged more intensive practices (although, given the scarcity of evidence for the late fifteenth century, there is little evidence that they did). On the other hand, the soils here were generally poor so that demesne lords who were already receiving what they regarded as an *adequate* return from their land may not have believed it was possible to achieve more, especially since neighbouring lords were following similar conservative techniques.

Peasant practices are much harder to document, but there is little evidence for the suggestion that peasant farmers, with few resources, were a source of inefficiency.[56] Evidence from inventories suggests that some peasants were keeping significant numbers of animals. This idea is further reinforced by the widespread use of vetch. In the mid-fifteenth century, even though the profitability of animal husbandry

---

[56] Outhwaite 1986, 15.

diminished, the stocking density on peasant lands may not have been cut back, because much of the cheese and meat could be used up by family consumption. Certainly by the early sixteenth century, when the economy began to revive, there is evidence that some peasants were specializing in the breeding of young stock. If they also continued to cultivate the land intensively, carting manure on to the fields, weeding frequently, and harvesting carefully, peasant agriculture with its greater labour inputs may have become more productive than demesne agriculture.

Bibliography

Biddick, K. 1989, "The link that separates: consumption of pastoral resources on a feudal estate". In *The social economy of consumption*, eds H. J. Rutz and B. S. Orlove, Lanham, Maryland, 121–48.

—— 1991, "Agrarian productivity on the estates of the bishopric of Winchester in the early thirteenth century: a managerial perspective". In *Land, labour and livestock*, eds B. M. S. Campbell and M. Overton, Manchester, 95–123.

Brandon, P. F. 1969, "Demesne arable farming in coastal Sussex during the later middle ages", *Agricultural History Review* 19, 113–34.

—— 1972, "Cereal yields on the Sussex estates of Battle Abbey during the late middle ages", *Economic History Review* 25, 403–20.

Calendar of Inquisitions Miscellaneous, 7 vols., London, 1916–68.

Campbell, B. M. S. 1983a, "Agricultural progress in medieval England: some evidence from eastern Norfolk", *Economic History Review* 36, 26–46.

—— 1983b, "Arable productivity in medieval England: some evidence from Norfolk", *Journal of Economic History* 43, 379–404.

—— 1991, "Land, labour, livestock and productivity: trends in English seigneurial agriculture". In *Land, labour and livestock*, eds B. M. S. Campbell and M. Overton, Manchester, 144–82.

Grantham, G. 1989, "Agricultural supply during the industrial revolution: French evidence and European implications", *Journal of Economic History* 49, 43–72.

—— 1995, "Time's arrow and time's cycle in the medieval economy: the significance of recent developments in economic theory for the history of medieval economic growth", unpublished paper given at the Fifth Anglo-American Seminar on Medieval Economy and Society, Cardiff, 1995 (cited by permission of the author).

Hallam, H. E. 1984, "The climate of eastern England, 1250–1350", *Agricultural History Review* 32, 124–32.

Langdon, J. 1986, *Horses, oxen, and technological innovation*, Cambridge.

Mate, M. 1985a, "Labour and labour services on the estates of Canterbury Cathedral Priory in the fourteenth century", *Southern History* 7, 55–68.

—— 1985b, "Medieval agrarian practices: the determining factors?", *Agricultural History Review* 33, 22–31.

—— 1987, "Pastoral farming in southeast England in the fifteenth century", *Economic History Review* 40, 523–36.

—— 1991a, "The agrarian economy of southeast England before the Black Death: depressed or buoyant?". In *Before the Black Death*, ed. B. M. S. Campbell, Manchester, 79–109.

—— 1991b, "Farming practices and techniques: Kent and Sussex". In *The agrarian history of England and Wales*, Vol. III, *1348–1500*, ed. E. Miller, Cambridge, 268–85.

Mayhew, G. 1987, *Tudor Rye*, Falmer.

Outhwaite, R. B. 1986, "Progress and backwardness in English agriculture, 1500–1650", *Economic History Review* 39, 1–18.

Overton, M. and Campbell, B. M. S. 1992, "Norfolk livestock farming, 1250–1740: a comparative study of manorial accounts and probate inventories", *Journal of Historical Geography* 18, 377–96.

Thornton, C. 1991, "The determinants of land productivity on the bishop of Winchester's demesne of Rimpton, 1208–1403", in *Land, labour and livestock*, eds B. M. S. Campbell and M. Overton, Manchester, 183–210.

12. WAS ENGLAND A TECHNOLOGICAL BACKWATER IN THE MIDDLE AGES?

John Langdon

This investigation stems from a casual conversation I had with a well-known historian of medieval technology at a molinological conference a few years ago. We were discussing the contribution of milling to the medieval economy, and he was vigorously extolling the revolutionary benefits of milling to the middle ages, particularly in relation to industry. His reaction to my view that the English experience did not seem to bear out the revolutionary pattern was to claim that—well, after all—England was a well-known technological backwater in the middle ages.

Although these remarks were more in the nature of a good-humoured and not overly serious interchange, they nonetheless had an important sub-text. All too often, we make rather bland assumptions and over-confident statements concerning technological development during the middle ages, which often blur the differences observed between regions. But these differences clearly existed and need to be taken much more seriously than perhaps has been the rule in the past. In the case of England, it is not only important to know how the country performed technologically in relation to other nations in Europe at the time, but it also has important implications of a methodological and historiographical nature. England has been blessed with an abundance of surviving records from which to judge questions regarding medieval technology, especially those relating to agriculture. It has enabled us to make important steps towards understanding the process of technological change in the medieval English economy.[1] These findings have to some extent been exportable to the degree that wider comments about European technological development have often been made on the basis of English evidence.[2] But

---

[1] For example, Langdon 1984, 1986, 1991; Mate 1985; Holt 1988; Campbell 1988a.

[2] This includes such important articles and books as Bloch 1935, White 1962 and Gimpel 1976, all of which used much English evidence. Regional studies elsewhere also often draw upon English experience when evidence for a particular

if England was out of step with mainstream continental Europe in relation to its technological development, then use of English evidence in this way might be seriously compromised.

But, first, it is useful to consider whether there are any *a priori* reasons for thinking that England would be technologically backward in the middle ages. On the surface, such an idea would seem to have little validity. Although that area we know as England today had been very much on the fringes of European social and economic activity in the Roman period and before, by AD 1000, with the marked shift in economic focus towards the northwest, England was very much an integral part of the European economy. By the twelfth and thirteenth centuries it had developed a political and ethnic cohesiveness that matched any other area in Europe, and its material resources, agricultural and non-agricultural, were already very impressive.[3] They were certainly of a degree for England to be a major political player throughout the middle ages, holding its own against other medieval states, such as France, suggesting that its technological state could not have been far below that of the continent. Also, England was very much patched into an international economy, by no means a new thing, that had its focus in the lower North Sea and drew southern England, northern France and the Lowland countries into a powerful economic nexus that inevitably bound these areas together not only in regard to economic activity but also technological development.[4] An excellent example of the strong disseminating force of this system is provided by the post windmill, as it suddenly appeared all around the lower North Sea at the end of the twelfth century, the original site of invention—France, Flanders or England—being very much disputed by historians.[5]

On the other hand, there were certainly elements of England's geographical position and institutional development that might have hindered its technological development. One was that England often tended to be on the high water mark in terms of the flow of technological ideas, which tended to radiate from the continental core

technical issue is scarce (for example, in relation to draught animal usage for Flanders: Verhulst 1990, 64–7).

[3] For example, Britnell 1993, ch. 5.

[4] For a good survey of the power of the northern medieval economy, see Postan 1973. This long-ranging economic network in northwestern Europe can be traced back at least to the age of *emporia* in the eighth and ninth centuries: for example, Hodges 1982.

[5] Bautier 1960; Philippe 1982; Bauters 1984; Kealey 1987.

out to areas like the British Isles. One only has to compare the erudition in, say, the Italian Pietro de Crescenzi's early fourteenth-century agricultural treatise, *Opus Ruralium Commodorum*, with the much more rudimentary tract of his near contemporary, Walter of Henley, to get some sense of how England might have lagged behind in terms of agricultural knowledge and sophistication.[6] In turning theory to practice, too, a more conservative turn of mind might well have characterized the medieval English farmer compared to his continental counterpart, as suggested perhaps by the slavish following of agricultural treatises shown by English lords and demesne managers at the end of the thirteenth century.[7]

England's institutional and managerial development was also ambivalent towards technological development. Although the rise of the market economy had clearly contributed to significant urban growth by 1300, the number of large towns in particular was still low by the standards of many other areas, particularly when compared to Flanders and northern France directly across the Channel.[8] On the other hand, the strong central government that developed at a precociously early stage in England did provide a political stability that aided economic and technological development (as long as these governments did not overreach themselves and embark upon overambitious and costly political schemes, a particular weakness of English kings). The strengthening of royal government over the twelfth and thirteenth centuries was also accompanied by managerial changes at the landlord level. Of particular interest here is the return to direct demesne farming that was so notable a feature of English agriculture during the thirteenth and fourteenth centuries and which was not paralleled to anything like such a high degree in any other area in Europe. The significance of this movement had several facets, but one of the most important was the monopolizing of such things as capital investment

---

[6] de Crescenzi 1304; Oschinsky 1971. See also Fussell 1972, 64ff.; Olson 1944. To be fair, most other areas were also probably behind the Italians to a considerable degree, for example, Comet, this volume.

[7] Mate 1985. Again, however, this attitude was also amply displayed elsewhere, for example, as indicated in de Crescenzi's close adherance to classical models: Fussell 1972, 64–5.

[8] For example, see the list of towns over 20,000 in Europe around 1300 provided by Bairoch 1988, 159, which shows sizeable English towns as being relatively rare compared to most other areas (although admittedly the revision currently going on about the size of English towns may alter this picture somewhat); see also Britnell 1991 for the low level of English urbanization compared to that of northern Italy.

and a much more immediate seigneurial presence in the development of agriculture and, by extension, agricultural techniques, than anywhere else in Europe. This also opened up a more distinct division between the demesne and peasant sectors than probably existed in the rest of continental Europe (where the lessees of demesnes were likely to have had closer connections with the peasantry than their lordly lessors), which may in turn have reinforced a very different technological climate in England than elsewhere.

Within the context of these general remarks, then, what sorts of observations can we make about the development of agrarian technique in England compared with that on the continent? It is the purpose of this paper to investigate the issues involved by comparing some aspects of agrarian technique in medieval England with practices on the continent, and in particular to focus on three areas: crop production, draught animal usage, and milling, where substantial evidence and a reasonable body of published work exists to make some meaningful comparisons. Also, the analysis for the most part will be limited to a comparison between southeast England—the most progressive area in terms of agricultural technique in the country during the thirteenth and fourteenth centuries—and that territory immediately across the Channel stretching from Normandy through the Paris basin to Flanders, and even possibly to that area we know today as the Netherlands,[9] all of which showed considerable technological progress in agriculture. Thus we are seeing both England on one side and the continent on the other at their best. It is recognized that other "progressive" areas could have been drawn into the analysis as well, particularly northern Italy (referred to below in relation to milling), as well as other areas of agricultural activity, such as livestock production and the development of agricultural tools, but again a more narrow framework of analysis has been chosen for the sake of simplicity and a relatively clear view of the problem.

Crop Productivity

The low level of crop production per unit area of land in medieval England is well known. Demesne farmers in most of England were lucky to clear as much as ten bushels an acre (about 900 litres per

---

[9] See Hoppenbrouwers this volume.

hectare) after seed and tithes.[10] Yet clearly the technical and agricultural know-how to improve yields and agricultural productivity generally existed, and some areas of medieval England did achieve crop yields well above the norm.[11] At their best the productivity of these lands rivalled anything on the continent,[12] and the promotion of intensive farming, manuring, marling, use of legumes, and other aids to agricultural productivity was remarkably vigorous here. The level of technological knowledge displayed by *some* medieval English farmers at least was thus impressive and likely to compare favourably with any displayed on the continent.

Yet how pervasive was this high level of technical expertise in the context of the totality of English agriculture? Recent work carried out by Bruce Campbell, James Galloway, Derek Keene and Margaret Murphy on a ten-county area around London has given us the most in-depth analysis of the issues of crop productivity and marketing for medieval English agriculture.[13] The results of this study are in many ways very surprising. Only one area in the ten-county region—eastern Kent—displayed that highly progressive, very productive agriculture that has been associated with various places on the continent, especially areas such as Flanders and Artois.[14] But geographically it was extremely limited. To this reader at least, one of the most surprising things about the Campbell *et al.* study was the relatively weak impact that London had upon the productivity of its hinterland.[15] Admittedly there was some establishment of productivity zones around the city, following the theoretical pattern established by J. H. von Thünen in the nineteenth century,[16] but by and large the effects were far less noticeable than in, say, eastern Kent, where agriculture was being affected by *both* the impact of London and the pull of cross-Channel markets.[17] A von Thünen-like model would probably fit better here if it was acknowledged more strongly that not only the metropolitan demand of London and other major urban centres in England

---

[10] For example, Titow 1969, 80–1; Campbell *et al.* 1993, 125–8.
[11] Brandon 1971; Mate 1980; Campbell, 1983.
[12] Campbell *et al.* 1993, 141.
[13] Campbell *et al.* 1993. The counties included in the study were Kent, Surrey, Berkshire, Oxfordshire, Northamptonshire, Buckinghamshire, Bedfordshire, Hertfordshire, Middlesex and Essex.
[14] Slicher van Bath 1963, 175–80.
[15] Langdon 1995.
[16] Von Thünen's theories are usefully summarized in Campbell *et al.* 1993, 4–8.
[17] Campbell *et al.* 1993, 179–80.

was influencing agricultural production, but also the metropolitan demand overseas from (principally) the great Flemish and Lowland cities.[18] Where these influences overlapped, the incentive for market production was correspondingly greater. In any case, from wherever the von Thünen ripples emanated, their impact seems clearly to have been dampened in the English case. There are any number of reasons for this,[19] but it is particularly notable that the heavily demesne-oriented data used by the Campbell *et al.* team were shot through with complications arising out of the practice of direct demesne farming—of particular importance was the inclination of many (particularly ecclesiastical) lords to bypass the market altogether and to produce almost solely for the needs of the household.[20] While production largely for subsistence did not necessarily mean low productivity,[21] it contributed little to the overall market nexus that is generally seen as the primary means for mobilizing high production in various regions.[22]

Looking at England as a whole, then, the areas of high grain productivity were very scattered and certainly not very cohesive in nature, for the most part being strung along England's eastern and southern shores—Holderness, eastern Norfolk, eastern Kent, south Sussex.[23] The reliance of these areas upon the overseas dimension to achieve their special status agriculturally in England is apparent from their location and scattered distribution. In contrast, the areas of high crop production across the Channel were remarkably cohesive. Flanders in particular was a highly productive area, with grain production consistently around the 1500 lit/ha level (equal to about 17 bushels per acre) or above for wheat and rye, a very high performance

---

[18] See Thoen and Hoppenbrouwers, this volume.

[19] Which would include such things as soils, the nature of communal restrictions, access to transportation networks, and so on: for example, see the discussion on the productivity of eastern Kent in Campbell *et al.* 1993, 141; also Campbell this volume.

[20] Campbell *et al.* 1993, 148–53.

[21] The lands of Peterborough Abbey, for example, an estate very strongly oriented towards household supply, nonetheless had high productivity rates for grain: Biddick 1989, 65–77; see also Campbell *et al.* 1993, 149–50.

[22] For example, Persson 1988, ch. 3; Campbell *et al.* 1993, 8–12; see also Campbell, this volume.

[23] See sources in note 11 above and Campbell *et al.* 1993, 126. Other areas such as the Nene valley in northern Northamptonshire (which benefited from access to the various Wash ports) may also have experienced this pull from overseas markets: Campbell *et al.* 1993, 126.

by medieval standards, and recent work has shown how this probably extended through much of northern France.[24] The high level of
urbanization and industrialization in the Lowlands-northern France
area created its own agricultural dynamism that, around *c.* 1300,
was clearly due more to internal rather than to external influences.
The strongest English agricultural performances, on the other hand,
were in effect upon the fringes and strongly dependent upon foreign
demand to lift their agricultural performance to a new plane. Here
it should be pointed out that the high grain production figures for
the Lowlands and northern France, cited above, generally apply to
both demesne and peasant farming, and the question then arises to
what extent the English demesne figures are masking a more progressive performance by English peasants. The answer to this question is still unresolved,[25] but the few cases of English peasant yields
that we have do not as yet strongly support the view that they performed better than the demesne.[26] There are also some other indications that peasant agriculture may have suffered under the shadow
of direct demesne farming. First of all, although direct demesne farming did not change the institutional framework that regulated lord-
peasant relations, it brought many of the effects of that framework
very strongly to bear, in particular in its concern to direct resources,
such as labour services and pasture requirements, more firmly towards demesnes. The disruptive effects that this might have had for
the peasant sector can only be guessed at, but the arbitrary way in
which lords exacted labour services or forced tenants to commute
these services to money payments could only have been unsettling to
peasant farmers.[27] Perhaps even more destructive was the tendency
for lords to set the tone for the nature of economic activity in medieval England. Not only did they often tend to short-circuit the market
network by sending grain directly to estate headquarters, but they
also tended to block a more fruitful interchange between agriculture
and industry. Of particular interest here is the failure of demesnes to

---

[24] Slicher van Bath 1963, 175–6; Verhulst 1990, 99–101; Derville 1987; Campbell
et al. 1993, 141; see also Persson 1988, 78–82, for a survey of highly productive
areas across Europe, including parts of Italy and the area of the lower Rhine.

[25] See, for example, Campbell, this volume.

[26] For example, in 1377, Walter Shayl's holding of 35 sown acres at Hampton
Lucy in Warwickshire yielded as little as half that for the demesne: Hilton 1975,
41–2.

[27] For example, Miller and Hatcher 1978, 125–8.

adopt industrial crops, such as flax and hemp, except occasionally in very small plots.[28] There are signs that the English peasantry might have moved more quickly in this direction,[29] but the degree was probably small compared to the continent.[30] Also, as discussed below, lords had a tendency to block investment in industrial hardware, particularly mills for fulling, forging and other manufacturing uses, through the jealous guarding of their demesne privileges. In other words, English lords' closer control of their demesnes during much of the thirteenth and fourteenth centuries, may have acted— both directly and indirectly—as a drag on the integration of agriculture both with the market and industry.

Draught Animal Usage

The use of draught animals in England shows a similar pattern to the continent. For Europe as a whole the introduction of horses to agriculture was an important development in improving the efficiency of ploughing, harrowing and, especially, hauling. For England, the impact was no less although possibly occurring a little later than the continent, beginning in the twelfth century rather than the eleventh century or earlier as in other parts of Europe.[31] By the thirteenth century the use of horses in agriculture was firmly established for the demesne sector and even more so among the peasantry, who played a strikingly innovative role in the adoption of the animal.[32] Indeed, the greater use of the horse was again a feature that affected those areas around the lower North Sea, where southeast England, the Lowlands and northern France all shifted to a much more horse-oriented agriculture during the twelfth and thirteenth centuries.[33]

---

[28] Hallam 1988, 380, 473.

[29] Hallam 1988, 311, 366; Miller 1991, 232-3.

[30] Where, for example, flax often comprised up to 10 per cent of the crops in Artois and Flanders: Derville 1987, 1414; Verhulst 1990, 122; see also Comet, Thoen and Hoppenbrouwers, this volume.

[31] Horses used in ploughing occurred in other parts of Europe as early as the ninth century (Norway) and judging from a comment by Urban II, plough-horses at the end of the eleventh century may have been fairly common in some places by then. In England, however, horses for ploughing seem only to have been a twelfth-century event: Langdon 1986, 19-20, 50-1.

[32] Langdon 1986, ch. 4.

[33] Langdon 1986, 264-5; Verhulst 1990, 66-7; Thoen 1993, 265-6; Duby 1968,

But there are some differences to be noted. The evidence is fragmentary but it does appear that the area from Normandy east to at least Flanders, including the area around Paris, went over much more whole-heartedly to the use of horses than did southeast England.[34] Horses in this part of the continent performed all facets of draught work, and appear to have been doing so from at least the thirteenth century if not before.[35] Although the last word on this has certainly to be said, there seems to have been much less in the way of the mixed ploughing teams of horses and oxen that featured so markedly on English demesnes and to a certain degree among the English peasantry.[36] It seems that northern France and the Lowlands cut more directly to the separation into predominantly horse-oriented and ox-oriented areas that would characterize England later in the fifteenth and sixteenth centuries.[37] In England, only in Norfolk and probably the Chiltern Hills was a wholesale shift to horses evident in all sectors of farming and this in many cases only from the middle of the fourteenth century.[38] Altogether the ambivalent attitude of English farmers to the use of draught animals, especially among demesnes and wealthier peasants, is perhaps symptomatic of the influence that direct demesne farming held over medieval English farming, in developing intermediary forms of draught animal use that suited the need for new farming arrangements but also looked back to the past. This is most forcibly expressed in Walter of Henley's famous advice given in the late thirteenth century, where, mainly on financial grounds, he advised English farmers to stick to the use of oxen as much as possible for ploughing, even if in mixed teams.[39]

109–11; Duby and Wallon 1975, 412; Bois 1984, 180–1, 190; Fourquin 1964, 77–8; Fossier 1968, I, 374–5, 377–81, 415–16; Small 1990, 57, 59.

[34] Sources as in note 33 above.

[35] For Picardy Fossier places the more or less complete transition from oxen to horses in the late twelfth century: Fossier 1968, I, 377–81. Similarly, Guy Bois states that "[the] choice of horse as draught animal [in Normandy] must have been ancient": Bois 1984, 190.

[36] Langdon 1986, 50–3, 105–12, 217.

[37] Langdon 1986, 205–12.

[38] Langdon 1986, 100–5; Campbell 1988b, 91–3. Again, as mentioned above, the peasantry as a whole in England used more horses than the demesne and in East Anglia particularly had shifted mostly to horses by c. 1300 (Langdon 1986, 205), although even here oxen maintained a presence among the wealthier peasants in particular (for example, see the experience for Blackbourne Hundred in Suffolk in 1283: Langdon 1986, 192–4).

[39] Oschinsky 1971, 318–19.

Milling

The impact of direct demesne farming was perhaps most evident in milling. The adoption of watermills and windmills was certainly as active in England as it was on the continent. The dissemination of mills into England again may have been a little later. None are known in Anglo-Saxon England before the late seventh century,[40] whereas functioning mills on the continent seem to have continued right through from the end of the Roman era to the early medieval period.[41] But this lag in English milling, if it ever existed,[42] was gone by the eleventh century. The mills at Domesday numbered at least 6,000, indicating how well established water-powered milling was in England by 1086. Nor was England slow to adopt the windmill a century later. By 1300, the proportion of mills to population in England was probably comparable to anywhere else in Europe.[43]

Yet, the pattern of investment in English milling showed some striking peculiarities. In particular, until the fifteenth century, there seems to have been a noticable reluctance to invest in industrial mills in England. Fulling mills generally consisted of far less than 10 per cent of all water- and wind-powered mills, while mills for other non-agricultural purposes were virtually non-existent.[44] While grain mills dominated in other areas in Europe as well, the presence of industrial mills was much more prominent. For example, around Florence in the early fifteenth century, there is evidence for 711 grain mills, 60 fulling mills, 7 "iron" (that is, forging) mills, 8 saw mills, 3 tanning mills, and 5 honing (or tool sharpening) mills.[45] In the area across the Channel we are mostly using for comparison in this study, while harder to quantify, Picardy in particular seems to have adopted a wide range of industrial mills, for fulling, tanning, crushing woad for dyeing, and crushing oleaginous plants (probably the plant called

---

[40] Holt 1988, 3.

[41] Philippe 1984, 108–9.

[42] We do know of several watermills existing in Roman Britain: Wikander 1985; Spain 1984.

[43] In England, at the beginning of the fourteenth century, there were probably between 10,000 and 15,000 watermills and windmills: Langdon 1994, 5. Assuming the population of England was around 6 million, this gives a figure of one mill for every 400–600 people. In comparison, France allegedly had 40,000 mills by 1300 (Philippe 1994, 107); at a population of around 17.6 million (Pounds 1974, 159–60) this yields a figure of 440 persons per mill, roughly equivalent to the English case.

[44] Langdon 1991, 434–6; Langdon 1994, 13–14.

[45] Muendel 1981, 98, 104–5.

rabette, an early form of rape, which was seemingly common in northern France).[46] Altogether, both the number and range of industrial mills is well beyond that evident in England.[47]

While the existence of industrial mills does not bear directly on the question of agrarian technique, again it suggests that Europe generally had a more fruitful mix of agricultural and industrial activities, and that investment was much more forthcoming for things like industrial crops and industrial mills. In the case of milling, I have recently argued that in England lords in the thirteenth and early fourteenth centuries maintained a strong and very conservative hold over milling investment.[48] Thus, because of the reduced revenues that came from industrial mills (as compared to mills for grinding corn or malt) lords were generally unwilling to invest in them. That they did so to some extent for fulling speaks more for the strength of local cloth industries than the entrepreneurial inclinations of lords. Similarly, mills for other industrial purposes such as the forging of iron or the crushing of bark or oak-galls for tannin in the leather industry were almost totally ignored by lords. Most crucially, it seems clear that lords not only invested conservatively but also blocked investment by the lower orders of society, which was often much more geared towards these industrial mills.[49] Again, the establishment of direct demesne farming, along with the reclamation of many mills into the demesne sector,[50] was instrumental in the creation of this very conservative investment policy.

---

[46] Fossier 1968, I, 383, n. 84; Bois 1984, 197. For the use of industrial mills more generally in northern France, see Comet, this volume.

[47] See, for example, Blaine 1966, chs. 3–5.

[48] Except where otherwise indicated, the material for this paragraph comes from Langdon 1994, especially 9–17.

[49] A good example of this is indicated when occasionally lords did allow a tenant to build a tool-sharpening mill, which they do not appear to have considered a great infringement upon their seigneurial rights, but under condition that the tenant did not violate the lord's right of monopoly on other types of mills (*sub tali conditione quod non faciat aliquod aliud molendinum nec aquaticum [here meaning a corn-mill] nec fullereticum*: from a 1449–50 account for the Soke of Winchester, Hampshire Record Office, 11M59 B1/187, m. 16v). As the date of this reference indicates, it was an attitude lords continued to reflect right through the middle ages.

[50] Holt 1987, 7–11, 13–22.

Conclusion

It seems likely, then, that in terms of the three technical areas examined in this paper—crop productivity, draught animal use and
milling investment—England was significantly out of step with
much of western Europe. This was not on the level of technological
knowledge, however, since few, if any, of the techniques familiar in
Europe were unknown in England. The expression of this technical
know-how (by such as Walter of Henley) may have seemed rudimentary when compared to de Crescenzi or earlier classical models such as
Columella, Cato, Varro and Palladius, but the proliferation of these
sorts of agricultural manuals in thirteenth- and early fourteenth-century
England, in number far outstripping any other area in Europe at the
time,[51] shows clearly how concerned lords and demesne managers
were with the practical issues of farming. Rather, the factors behind
this difference in agrarian technology were probably more philosophical
in nature, particularly in relation to the readoption of direct demesne
farming. The renewed interest of English lords in returning to a more
hands-on approach to agriculture tended to guide many aspects of
farming in a specific direction. Generally speaking, this direction was
far from being totally market-oriented, as many lords indeed turned
to direct demesne farming as a means of reducing costs for foodstuffs for the household.[52] The conflicting aims of direct demesne
farming resulted in a considerable ambivalence towards technological
change,[53] the rather constant nature of household needs, for example,
emphasizing the need for a stable output. This was an attitude strongly
reflected in the various agricultural treatises of the time, which emphasized the need to keep outputs more or less at the same level, as in
the case of Walter of Henley's remarks about strictly regulating production according to extents or surveys made on the land.[54] Similarly, although investment in things like mills certainly occurred at
an impressive level, it was also accompanied by a tendency to choke
off investment by other sectors of society towards a greater degree of

---

[51] Fussell 1972, 73.

[52] For estates that looked to direct demesne farming for subsistence purposes, see
Biddick 1989, 72–7; Campbell *et al.* 1993, 148–53.

[53] For example, Britnell 1993, 118.

[54] Perhaps most obvious in Walter of Henley's remarks about strictly regulating
production in relation to extents or surveys made on the land: Oschinsky 1971,
312–15 (cc. 16–24); see also Mate 1985, 31.

industrialization. In other words, direct demesne farming acted as a technological filter that tended to maintain the traditional agricultural basis of society. The more control lords had over the processes of production and investment the more society remained in its traditional agricultural pattern. A more fruitful combination of agricultural, industrial and urban development was thus to some extent held back.

Was England a technological backwater in the middle ages then? The answer is yes and no. It certainly was not slow in taking on innovations, such as the windmill, that suited the strongly agrarian orientation imposed upon it by the ruling mentality of the time. But that ruling mentality also tended to block innovations that might have led to a more vital mobilization of the economy along the lines of, say, Flanders. It should be said, however, that the effects of direct demesne farming were limited essentially to two centuries—the thirteenth and the fourteenth. When the wholesale leasing of demesnes took place in the late fourteenth and into the fifteenth centuries, the characteristics of many technological developments began to change. Thus, in terms of draught animal use, there was a noticeable polarization in the use of horses and oxen, at both the demesne and peasant level, as some areas went over solely to horses while others intensified their use of oxen.[55] Similarly, as the seigneurial monopoly of milling investment was relaxed over the fifteenth century, there are signs of a much more flexible attitude to the uses of mills, with people much lower down in the order beginning to invest in such things as fulling mills, forging mills, tool sharpening mills and bellows mills than ever before.[56] At this point, it would appear that England began to move more in line with the technological mainstream in Europe.

Does this particularly English technological development pose problems for using English evidence about medieval technological development in a wider context? Again, the answer is yes and no. On one level, because the magnitude of the English return to direct demesne farming in the thirteenth and fourteenth centuries was fairly unique in Europe, it helped to establish a distinctive technological pattern. As a result, one should be cautious in claiming that technological patterns seen in thirteenth- and fourteenth-century England would necessarily have had wide acceptance elsewhere. Secondly, it

---

[55] Langdon 1986, 97–100, 205–12.
[56] Particularly in places like the Weald and Cornwall: Holt 1988, 150; Miller 1991, 740–3.

can be dangerous to assume technological generalities based on the experience of only one sector. In this case, much of what we know about medieval English agricultural technology comes from demesne-oriented material. But as is becoming increasingly clear, the peasant sector in particular was also very active technologically.[57] Altogether, it underlines the importance of looking at technological development in a multi-sectoral way.

Perhaps more important, though, it is not the patterns of technological development in any particular area that provide models for other areas but the mechanisms by which they occurred. Thus the technological activity of peasant sectors, which is gradually becoming clearer in the English evidence, provides a model applicable across Europe, as do the restrictive effects of lords or other groups—for example, government, royal or borough—that attempted to direct economic and technological development to their own ends. Also, it is important to keep in mind the regional differences in technological development, particularly as studies in medieval technology have suffered from a tendency to apply blanket descriptions to all of Europe.[58] Although the existence of these differences is widely recognized, the detail across Europe is only just emerging. In this regard, the relative lack of research into technological development in medieval agriculture is all too obvious. The less cohesive and informative documentary sources for the rest of Europe compared to England has meant that detailed study in agrarian techniques has progressed rather fitfully. Similarly, although some excellent work on medieval agrarian technique has been done from archaeological sources,[59] this area of research is still by and large in its infancy. Most obviously lacking are systematic studies from the huge amount of iconographic material available, still a largely untapped source and one that has tended to be poorly integrated into this subject.[60] Finally, to echo the sentiments of other contributors to this volume,[61] only when more

---

[57] See especially Langdon 1986, 291–2; Campbell 1988a, 204–5.

[58] Most notably in the works of those in the "technology as revolution" camp: for example, White 1962; Gimpel 1976.

[59] Most notably Myrdal 1985.

[60] The most famous of these studies from iconographic sources for the medieval period is still Lefebvre des Noëttes 1931, but see also McNeill 1979. Scholars of the Roman and Greek periods have always been better served in this regard, for instance, the works of Georges Raepsaet based on iconographic evidence: Raepsaet 1985 and 1987.

[61] Especially Comet, this volume.

work is completed across the entire spectrum of sources will adequate comparisons among the various experiences relating to medieval agrarian technology be possible.

Bibliography

Bairoch, P. 1988, *Cities and economic development*, Chicago.

Bauters, P. 1984, "The oldest references to windmills in Europe". In *Transactions of the fifth symposium of the International Molinological Society*, Saint-Maurice, 111–24.

Bautier, A.-M. 1960, "Les plus anciennes mentions de moulins hydrauliques industriels et de moulins à vent", *Bulletin Philologique et Historique*, Brussels, 567–626.

Biddick, K. 1989, *The other economy: pastoral husbandry on a medieval estate*, Berkeley.

Blaine, B. B. 1966, "The application of water-power to industry during the middle ages", unpublished PhD thesis, University of California, Los Angeles.

Bloch, M. 1935, "Avènements et conquêtes du moulins à eau", *Annales E. S. C.* 7, 538–63.

Bois, G. 1984, *The crisis of feudalism: economy and society in eastern Normandy c. 1300–1550*, Cambridge.

Brandon, P. F. 1971, "Demesne arable farming in coastal Sussex during the later middle ages", *Agricultural History Review* 19, 113–34.

Britnell, R. H. 1991, "The towns of England and northern Italy in the early fourteenth century", *Economic History Review* 44, 21–35.

—— 1993, *The commercialisation of English society 1000–1500*, Cambridge.

Campbell, B. M. S. 1983, "Agricultural progress in medieval England: some evidence from eastern Norfolk", *Economic History Review* 36, 26–46.

—— 1988a, "The diffusion of vetches in medieval England", *Economic History Review* 41, 193–208.

—— 1988b, "Towards an agricultural geography of medieval England", *Agricultural History Review* 36, 87–98.

——, Galloway, J.A., Keene, D. and Murphy, M. 1993, *A medieval capital and its grain supply: agrarian production and distribution in the London region c. 1300*, Historical Geography Research Series, 30.

de Crescenzi, P. 1304, *Opus ruralium commodorum* (available on microfiche in a 1471 printed edition by J. Schüssler of Augsburg).

Derville, A. 1987, "Dîmes, rendements du blé et 'revolution agricole' dans le nord du France au Moyen Âge", *Annales E. S. C.* 42, 1411–32.

Duby, G. 1968, *Rural economy and country life in the medieval west*, London.

—— and Wallon, A. (eds) 1975, *Histoire de la France rurale*, vol. I, Paris.

Fossier, R. 1968, *La terre et les hommes en Picardie jusqu'a la fin du XIIIe siècle*, 2 vols, Paris.

Fourquin, G. 1964, *Les campagnes de la région Parisienne à la fin du Moyen Âge*, Paris.

Fussell, G. E. 1972, *The classical tradition in west European farming*, Newton Abbot.

Gimpel, J. 1976, *The medieval machine: the industrial revolution in the middle ages*, New York.

Hallam, H. E. (ed.) 1988, *The agrarian history of England and Wales*, vol. II, *1042–1350*, Cambridge.

Hilton, R. H. 1975, *The English peasantry in the later middle ages*, Oxford.

Hodges, R. 1982, *Dark age economics*, London.

Holt, R. 1987, "Whose were the profits of corn milling? An aspect of the changing relationship between the abbots of Glastonbury and their tenants 1086–1350", *Past and Present* 116, 3–23.

—— 1988, *The mills of medieval England*, Oxford.

Kealey, E. J. 1987, *Harvesting the air: windmill pioneers in twelfth-century England*, Berkeley.

Langdon, J. 1984, "Horse hauling: a revolution in vehicle transport in twelfth- and thirteenth-century England?", *Past and Present* 103, 37–66.

—— 1986, *Horses, oxen and technological innovation*, Cambridge.

—— 1991, "Watermills and windmills in the west Midlands, 1086–1500", *Economic History Review* 44, 424–44.

—— 1994, "Lordship and peasant consumerism in the milling industry of early fourteenth-century England", *Past and Present* 145, 3–46.

—— 1995, "City and countryside in medieval England", *Agricultural History Review* 43, 67–72.

Lefebvre des Noëttes, R. 1931, *L'attelage et le cheval de selle à travers les âges*, Paris.

Mate, M. 1980, "Profit and productivity on the estates of Isabella de Forz (1260–92)", *Economic History Review* 33, 22–31.

—— 1985, "Medieval agrarian practices: the determining factors", *Agricultural History Review* 33, 22–31.

McNeill, C. A. 1979, "Technological developments in wheeled vehicles in Europe, from prehistory to the sixteenth century", unpublished PhD thesis, University of Edinburgh.

Miller, E. (ed.) 1991, *The agrarian history of England and Wales*, vol. III, *1348–1500*, Cambridge.

—— and Hatcher, J. 1978, *Medieval England: rural society and economic change 1086–1348*, London.

Muendel, J. 1981, "The distribution of mills in the Florentine countryside during the late middle ages". In *Pathways to medieval peasants*, ed. J. A. Raftis, Toronto, 83–115.

Myrdal, J. 1985, *Medeltidens Åkerbruk: agrarteknik i sverige ca. 1000 till 1520*, Stockholm.

Olson, L. 1944, "Pietro de Crescenzi: the founder of modern agronomy", *Agricultural History* 18, 35–40.

Oschinsky, D. 1971, *Walter of Henley and other treatises on estate management and accounting*, Oxford.

Persson, K. G. 1988, *Pre-industrial economic growth: social organization and technological progress in Europe*, Oxford.

Philippe, R. 1982, "Les premiers moulins à vent", *Annales de Normandie* 32, 99–120.

—— 1984, "L'église et l'énergie pendant le XIᵉ siècle dans les pays d'entre Seine et Loire", *Cahiers de Civilisation Médiévale* 27, 107–17.

Postan, M. M. 1973, "The trade of medieval Europe: the north". In M. M. Postan, *Medieval Trade and Finance*, Cambridge, 92–231.

Pounds, N. J. G. 1974, *An economic history of medieval Europe*, Harlow.

Raepsaet, G. 1985, "L'attelage ruraux de nos régions dans l'antiquité", *Revue de l'Agriculture* 38, 1423–44.

—— 1987, "Archéologie et iconographie des attelages dans le monde Grèco-romaine: la problématique économique". In *Histoire économique de l'antiquité*, eds T. Hackens and P. Marchetti, Louvain-la-Neuve, 29–48.

Slicher van Bath, B. H. 1963, *The agrarian history of western Europe AD 500–1850*, London.

Small, C. M. 1990, "Grain for the countess: the "hidden" costs of cereal production in fourteenth-century Artois", *Proceedings of the Annual Meeting of the Western Society for French History* 17, 56–63 .

Spain, R. J. 1984, "The second-century Romano-British watermill at Ickham, Kent", *History of Technology* 9, 143–80.

Thoen, E. 1993, "The count, the countryside and the economic development of the towns in Flanders from the eleventh to the thirteenth century. Some provisional remarks and hypotheses". In *Studia historica oeconomica: liber amicorum Herman van der Wee*, eds E. Aerts, B. Henau, P. Janssens, and R. van Uytven, Leuven, 259–78.

Titow, J. Z. 1969, *English rural society*, London.

Verhulst, A. 1990, *Précis d'histoire rurale de la Belgique*, Brussels.

White, L. 1962, *Medieval technology and social change*, Oxford.

Wikander, O. 1985, "Archaeological evidence for early watermills—an interim report", *History of Technology* 10, 151–79.

13. MEDIEVAL FARMING AND TECHNOLOGY: CONCLUSION

Christopher Dyer

This conclusion will pose a series of general questions about medieval agricultural technology, and will provide answers based on the state of knowledge reflected in the essays in this book. The problems to be examined include: considering whether it is still possible to talk of a technological revolution in medieval Europe; defining the chronology of change; exploring the economic and social contexts of technical development; and explaining the unevenness of technological change.

Was there a Technological Revolution in Medieval Europe?

We all carry in our collective memory a model of technical innovation which centres on the heroic inventor, arriving at a new idea by accident, or doggedly persisting in experiments until a moment of revelation or achievement—Watt, Stephenson, Volta, Edison, the Wright brothers are some obvious examples. Or we remember from our school text books that the agricultural and the industrial revolutions derived from a sequence of discoveries—horse-hoeing husbandry, turnips, spinning jennies, water frames, and so on, most of them linked with the name of a great inventor. Historians of modern technology now regard this view as something of a myth, and there is no prospect that we could ever write the history of medieval science and technology in such terms—the inventors are usually unknown, the moment of discovery obscure.[1] And in any case change came not through a series of "breakthroughs", to use a journalistic term, but mainly by small modifications and adaptations which added up to a significant technological shift. Often there was no "discovery", but rather a reintroduction of a long-known technique, or the diffusion of some method which had previously been confined to a particular

---

[1] On the gradual nature of technical change, Mokyr 1990, 273–99.

region. The innovation frequently lay not in the adoption of some complete novelty, but rather through a new combination of existing practices, which together gave specific advantages to the cultivators.

The word "technology" suggests mechanical devices of some kind, and in the past historians of the subject have focused their attention on constructions of wood and iron such as ploughs and mills, but the fundamental changes often concerned ways of organizing agricultural production. The medieval technological revolution could well have been "imagined", as Raepsaet suggests in his essay, and we may be justified in speaking only of a revolution in method.

Various contributors to this book have emphasized that innovations in medieval agriculture should be viewed in combination, as a "technological package" or "technological complex". If we review the state of farming methods at the end of the thirteenth century, we find an impressive array of techniques which either had been introduced within the previous four centuries, or had in the same period gone through some important modification. The agrarian landscape had been moulded by the almost universal trend towards the extension of grain cultivation, for which the German term *Vergetreidung* is used here by Hoppenbrouwers and Widgren. By *c.* 1300 this resulted in some regions having more than 70 per cent of the land surface devoted to cereals and legumes. Forms of crop rotation or field system were in place which seem to have been suited to local soils and resources. Various authors here discuss the wide range of types: some followed conventional patterns, such as the two-course or three-course rotation, by which each piece of land was fallowed every two or three years. Or the land could have been left for a longer period without cultivation, when a continuously cropped infield was combined with outfields that were planted occasionally. Thoen shows that in parts of Flanders in the thirteenth century an "up-and-down" husbandry was being practised, in which a piece of land was continuously cropped for some years, and then put down to grass for a period, before being cultivated again. At the other end of the range of practice the area under crops was increased by sowing part of the fallow field, or by adopting intensive methods which virtually eliminated the need to leave any land uncultivated.

The mix of crops had been adapted to local needs, and no doubt in some places the same types of grain were sown for many centuries. Innovations include new varieties of wheat, as shown by Astill and Comet; the cultivation of rye had expanded in some regions (a point emphasized here by Poulsen and Hoppenbrouwers); and in many parts of Europe the amount of land used to grow legumes—peas, beans, lentils, and vetches—had been increasing (receiving much attention from Campbell and Thoen). A good deal of effort was devoted to the control of weeds, not just the laborious task of hoeing and removing thistles in the early summer, but also the more systematic ploughing of the fallow, sometimes two or three times in the summer months.

The preparation of the seed bed began with ploughing. In the late thirteenth century two main types of ploughing implement were to be found in Europe, the ard or scratch plough especially characteristic of the Mediterranean world, and the heavier mouldboard plough of the north. In both England and Scandinavia in the early middle ages both types were in use at the same time, and the heavy plough extended its territory, to the total exclusion of the ard in the case of England (Astill, Myrdal, Poulsen, and Widgren). The heavy plough had evolved into many local variants: some had wheels, and others were swing ploughs in which the direction and depth of the furrow depended on the strength and skill of the operator. Ploughs were fitted with iron shares and coulters, and these could involve the use of substantial quantities of metal (Myrdal). Breaking the clods and covering the seed was usually done with harrows drawn by horses or oxen, but sometimes hand tools were brought to this task. Similarly hand methods might be used in sowing, like the dibbling of beans by which seeds were put into individual holes. When the normal broadcast method was used, cultivators sowed in varying densities, calculating sometimes that a thick growth of grain would reduce the numbers of weeds, and, if the soil was fertile enough, produce better yields for each unit of land.

An alternative technology, mentioned here by a number of writers, but emphasized by Comet, involved the use of hand tools in cultivation. In the case of horticulture, vines, and industrial crops such as hemp, flax, teazels, and dye plants, the seedbed of small plots was often prepared using spades and mattocks—iron-shod implements. But grain and legumes were also grown on a horticultural scale by smallholders and cottagers, who turned the soil by hand and foot because they had no beasts or ploughs, and their parcels of land were the wrong size and shape for a plough to manoeuvre. Hand cultivation was also practised in mountainous regions, a point made by Widgren. Horticulture involved many other small-scale and intensive husbandry practices of course, such as pruning, grafting, replanting, and pest control.

In many parts of northern Europe drainage techniques were developed. In the remarkable case of the marshes, fens, and other wetlands, reclamation and the subsequent protection of hundreds of square kilometres of dyked land had become a fine art, perhaps one of the greatest achievements of medieval technology. The more widespread task of removing excess rainwater from heavy soils was tackled by less spectacular but still effective methods: digging ditches on the edges of fields (Myrdal), and ploughing in ridges. The latter practice deliberately built up accumulations of soil along the strips of ploughland, from which the water would be carried off along the adjacent furrows. The authors in this volume are most concerned with the skill of medieval hydraulic engineers in drainage and reclamation works, but it is worth adding that they were successful managers of water by bringing it into ponds and moats, and penning it behind dams and weirs, as a power source for mills, and in order to breed fish.[2]

Successful cultivation of grain depended on difficult decisions about the integration of arable and animal husbandry. Fallowing, cultivation of fodder crops, and the management of permanent pastures and meadows all contributed to the feeding of stock. Some authors in this book refer to the exploitation of woodlands, and we must not underestimate the importance of woods as a source of food for animals, both the grass in the clearings, and the foliage and fruits of trees (such as acorns and beechmast). Woodlands required careful control, to give stock access to pasture among the trees, while coppices were fenced to allow wood to be grown for fuel, fencing, and other products on a regular cycle.[3] The pulling power of animals, and their manure, were essential for grain cultivation. Animals became more efficient contributors to the agricultural economy because of developments in harnessing methods, design and construction of vehicles, and the adoption of horse- and oxshoes (Raepsaet has most to say here). Where horses were adopted as beasts of burden they gave faster pulling power than did oxen. Campbell shows that grain and legumes were used either to give working animals more energy, or to keep stock through the winter, or to fatten beasts for the table.

---

[2] Both aspects are treated in Crouzet-Pavan and Maire-Vigueur 1994; see also Aston 1988.
[3] Rackham 1976; Stamper 1988.

We could add that special diets, notably bread baked for the purpose, were given to horses taking loads long distances by road, and a variety of food supplements were given to young lambs. In general the technology of animal husbandry deserves more attention, as the authors here follow historiographical tradition in their preoccupation with the cultivation of plants. Veterinary skills could be cited, such as employing mercury and tar to combat disease among sheep. The ingenuity of medieval pastoral farmers is shown by selective breeding, which resulted in changes in the size and quality of animals. In the provision of shelter such as cow sheds, stables, and sheepcotes agricultural managers applied superior techniques of construction for the welfare of their stock, and one only needs to observe the still-surviving medieval dovecots to appreciate the efforts made for one relatively minor species. In England a new semi-domesticated animal, the rabbit, was introduced in the twelfth and thirteenth centuries, and was kept in carefully protected warrens. Deer, especially those in parks, were managed systematically, and were fed on browse, hay, and even oats.[4]

Animal manure provided one of the most important means of maintaining or increasing soil fertility, and there were different ways of conveying the manure to the land—from folding the stock in the field to carting from animal houses. Many techniques were employed for treading, digging, ploughing, and harrowing the muck into the soil. We are also aware of various methods, some of them ancient and some innovative, by which the fertility or texture of the soil could be improved—digging or ploughing in turf as green manure, or beat burning (paring the turf, burning it, and spreading the ashes on the field), or the cultivation of legumes, or carting of urban night soil, or digging and spreading marl, or burning and spreading lime (most of these are mentioned by Comet, Hoppenbrouwers, and Poulsen).

In the grain harvest, as a number of our authors demonstrate, a great variety of sickles, "hooks", and scythes were used, and the "Flemish hook" was introduced. As shelter for animals developed, so did crop storage, with the construction of barns and granaries. In northern Europe protection from damp and rodents was improved with the adoption of stone walls or at least stone foundations (Astill),

---

[4] Trow-Smith 1957; Ryder 1983; Armitage 1982, 50–4; Dyer 1995; Bailey 1988; Birrell 1992.

to which might be added the raising of the floor of the granary or barn above the ground on staddle stones.[5] Grinding grain involved the use of mills, driven by hand, animals, water, and wind. Grapes, apples, and other fruits had their juice extracted in mechanical presses. On a smaller scale the plunge churn was introduced into butter making (Myrdal) and hand tools of iron were made for preparing flax, hemp, and other crops. Once harvested, gathered, processed, or fattened, much agricultural produce went to market, and its progress was aided by developments in farm traction. The technical contribution of the state and local government does not receive much attention in these essays, but the provision of a transport infrastructure deserves to be mentioned, by the construction and repair of bridges, roads, wharfs, and canalized rivers.[6]

This discussion has necessarily consisted of a litany of methods and processes, and in spite of its length will inevitably have omitted some of the techniques, crops, and products of the hey-day of European agriculture in the late thirteenth century. Even an incomplete résumé of agricultural technology serves a purpose in indicating the many stranded character of the "technological package" or "complex". The techniques were often interrelated, and connected with general trends in the economy and society. While we must be suspicious of earlier attempts to identify simple and direct connections between some key developments, such as ploughs, crop rotations, and horse traction, or to make extravagant claims for medieval industrial revolutions or the advent of a mechanical age, the list of developments is impressive, and demands interpretation and explanation. This survey of technology is of course based on a period—the late thirteenth century—when administrative documents, and other sources, such as representations of daily life in painting and sculpture, are relatively plentiful, and we remain unclear about how long these methods had been practised before they were first recorded.

The distinction between innovation and diffusion deserves emphasis. For example, the regular open-field system, operating on a two- or three-course rotation, can be identified with some confidence from documentary evidence for specific places in the tenth century, but some communities did not adopt such a system, or convert recently cleared land into a regular field system, until the twelfth or thir-

---

[5] Dyer 1984, 42–5.
[6] For example Harrison 1992; Bautier 1989.

teenth century. We must regard the spread of such field systems as an important ingredient in overall technological development, just as much as their first use. Changes accumulated by increments, as with the provision of barns of substantial construction in northern Europe. Such buildings were not usually given stone foundations until the thirteenth century, at a time when timber framing was advancing in sophistication. Roofs of ceramic tiles or stone slates were not provided, if at all, until a century or two later. No-one "invented" a barn, but building workers and their employers by gradual improvements added to the productivity of farming by preventing wastage in the months between the harvest and the consumption or sale of the crops.

This emphasis on incremental change and long term diffusion should not eliminate from our thinking the occasional intervention of an "heroic inventor" and the rapid dissemination of a brilliant idea. We do not know his name, but it seems very likely that a millwright or group of such craftsmen, working near the shores of the North Sea in the late twelfth century, realized that the machinery of a watermill, reduced in scale and housed in a wooden box, could be mounted upside down, and turned by wind driven sails.[7] This was not the result of imitating practices outside Europe, as the windmill of the middle east was based on different principles. Nor did the windmill evolve by stages, but appeared suddenly and fully formed. Wind-powered milling spread rather hesitantly at first, but within a century thousands of mills had been built over northern Europe. Like all successful innovations, we can see that it arrived at the right moment. Watermills had been built in every feasible place, but had encountered a technical blockage because some streams were simply too small or sluggish to give reliable power. We find small mills being installed with excessive optimism in the twelfth century on inadequate streams, and eventually being abandoned because they were not working efficiently. Now windmills harnessed a new source of energy, and could be built on uplands and in the fens, where swiftly running water was lacking. Also around 1200 lords were anxious to increase their revenues, to counteract inflation, and to take advantage of commercial growth. New lordships were being formed, and the minor lords (mainly smaller religious houses and knights) were especially anxious to expand their limited assets. Those whose corn was ground

---

[7] Holt 1988, 20–35.

in the new mills, though they no doubt grumbled that the tolls were excessive, were relieved of the time wasted in carrying corn to distant watermills, or even of the drudgery of hand milling. The labour saved could be put to more productive use at a time when peasants and artisans were profiting from the expanding market. But although these long-term social and economic factors help to explain the enthusiasm with which the windmill was eventually adopted, we can still recognize the daring and ingenuity of the anonymous inventor or inventors.

We must conclude, with all the necessary qualifications about the pace of change, and the absence of a single transforming innovation, that the "package" or "complex" which had formed by the thirteenth century amounted to a significant shift in agricultural methods. Perhaps "revolution" is too strong a term to describe a development with so many incremental changes, and spread over a number of centuries, but embedded in the gradual and evolutionary movement we can glimpse, as in the case of the windmill, an occasional great leap forward.

Chronology of Change

The research summarized in this book strengthens the characterization of *c.* 900–1300 as a major period of technical change. In fact most of the features in the "technological complex" described above were in place by 1200, including the windmill, but the thirteenth century still saw much dissemination of techniques. More debate must surround the definition of the beginning of the process. While we can be confident that a number of elements of the "complex" were functioning in some localities in the tenth century—such as regular crop rotations, harrows, extensive drainage schemes—some of the authors in this book (notably Poulsen and Widgren) point to the ninth century as giving the first dated examples, and we begin to wonder if we are observing techniques long after their inception. We are excessively dependent on chance documentary references or single archaeological finds in the period before 1000 which lacks plentiful written and material evidence. Indeed there are strong grounds in England, as outlined here by Astill, for regarding the period 650–850, with its new crops and advanced mills, as an important earlier phase of innovation. And as Comet and Raepsaet argue most vociferously in these essays, technologies that can be observed in the tenth century or later were present either in the Roman period, or the late Iron Age, including such vital elements in the "complex" as heavy ploughs and watermills.

Some important discontinuities, however, prevent us from dismissing early medieval technology merely as an extension of the Roman past. The Gallic *vallus* or harvesting cart, which disappeared in the post-Roman period, shows that some of the Roman techniques were simply not appropriate for later application. Some medieval practices were undoubtedly new, such as the permanent shoeing of horses; and in other cases, such as watermills, the technology, though known in classical times, became much more widely disseminated in the early middle ages.

Four or five centuries (*c.* 800 or *c.* 900 to *c.* 1300) seems a protracted span of time within which to locate our innovations, and is rather too long, as Myrdal argues, to be designated a revolution. Again there are problems with evidence that does not allow us to date with any precision either the introduction or dissemination of methods or techniques. It is possible that concealed within the relatively long period that has been identified, there could have been two shorter episodes of more rapid change, one around the tenth century (which saw the introduction of regular field systems and the application of water control systems) and the other in the decades around 1200 (when the windmill was invented, and the horse-drawn cart proliferated).

Finally, in the light of the impressive evidence presented here, we should not underestimate the pace of innovation in the last century of the middle ages, when very different circumstances prevailed after the famines, the Black Death, and other factors had put an end to the growth in population and the generally expansive tendencies in the cultivated area and the commercial economy. In some places the trends of earlier periods continued, with still very intensive and productive husbandry systems in the Netherlands and Flanders, reported here by Hoppenbrouwers and Thoen. Turnips were introduced as food for humans and animals in Flanders, and some parts of Holland turned to dairying. Poulsen shows that reclamation of land continued on parts of the North Sea coast. In much of Europe, the drop in population led to new specialisms, notably the adoption of large-scale pastoral husbandry which was appropriate in circumstances of new demands for meat, and the shortage of labour (Mate and Poulsen). In Jutland and England horses were replacing oxen on many farms (Langdon and Poulsen). A number of authors dealing with northern Europe mention the growing use of scythes rather than sickles in harvesting, and the manufacture of plough shares and horse shoes with greater quantities of iron. The functions of mills shifted—in England to more industrial uses, and in the Netherlands windmills were employed for the first time as water pumps to protect farm land drained in earlier centuries (Hoppenbrouwers). We recognize from these observations that technical change might have been stimulated by economic expansion in the tenth to thirteenth centuries, but after the Black Death, with less population pressure, and reduced prices for primary products, peasants might search for some advantage, such as scaling down their investment of labour, and also feel free when they were no longer threatened by starvation to take risks and experiment with new methods. For example, the reduction in the intensity of cultivation, and the creation of a new balance between arable and pasture might favour the introduction of some form of "up-and-down" husbandry.

Economic and Social Contexts of Technological Change

The "technological package" of *c*. 900–1300 coincides with a remarkable period of general growth in the economy, and the formation of institutions. The increase in population by a factor or two or three times was associated with both an extension of the cultivated area, and the emergence in certain regions of the nucleated village as a distinctive settlement form. The modern division in settlement patterns between regions dominated by villages and those where the majority of the inhabitants live in dispersed hamlets and farmsteads was already established by the thirteenth century. These differences had implications for land use and management, as the nucleated village was more likely to be associated with extensive grain production and regular field systems. A closer organization of rural society went with these changes, with more intensive seigneurial control in the tenth and eleventh centuries as the size of estates diminished, and new powers of lordship were imposed by a numerous aristocracy. The French version of this, involving the imposition of private jurisdiction on the peasant population, has been called the "feudal transformation".[8] Elsewhere there was more emphasis on the levying of rents, and demands for labour services for the cultivation of lords' demesnes. But as Poulsen reminds us, in some parts of Europe such as Scandinavia the developments in settlements were not necessarily coincident with the establishment of strong lordship. The parish system reflected these changes in secular administration, and new churches were built across the countryside. The peasant communities developed powers of internal discipline and self-government. The new rural landscape, of churches, castles, compact territories, and planned villages and hamlets, was closely linked with extensive, subdivided, and regulated field systems, and other signs of land management, such as coppiced woods, pastures with stints of specified numbers of animals, and meadows in which shares were rigorously measured.

In the same period new towns proliferated, and the proportion of the population living in towns more than doubled. Commerce linked the towns with the countryside, and much industrial production, though not all, was located in the towns. The power and function of governments grew at the same time, claiming legal power, levying taxes, and waging war. From the former core of western Europe, that is the old Carolingian Empire (and its small-scale imitation in late ninth and tenth century England), the institutions of lordship, castles, towns, and villages spread in the twelfth century into areas previously organized on different principles, such as eastern Germany, southern Spain, and western Britain.[9]

We can easily see in a general way how the "technological package" fitted into this general picture of a larger economy and more regulated society, but the precise connections are more difficult to define. Perhaps the link between technology and commercial growth is most clearly identifiable. A number of innovations, like the horseshoe, harnessing, and iron-tyred horse-drawn cart, were designed to give agricultural producers more convenient access by road to market. Having sold goods, they were stimulated to buy, notably the products of both urban and rural artisans, some of which could contribute to the efficiency of cultivation. One example is the increased use of iron in agricultural implements. Craft specialization gave the

---

[8] Bois 1992.
[9] Bartlett 1993.

rural population access to skilled woodworkers able to make better wheels, ploughs, cart bodies, and timber-framed buildings.

The growth of the market must help to explain the degree of specialization evident in the local combinations of crops and animals observable by the thirteenth century, and the occasional reliance of cultivators on a single product, such as livestock in the uplands of northern Europe and wine in the south. Wine, wool, and dye plants were directed specifically at urban consumers and urban industries. And the general increase in productivity was partly aimed to provide a surplus for sale, as well as to satisfy the food needs of the rural cultivators. We can observe examples where a particular agrarian regime was geared to meet demand from nearby towns, like the horticulture and viticulture practised in the outskirts (Comet), or the intensive production of relatively cheap crops such as oats within a short carting journey, as predicted in von Thünen's model of zones of landuse around towns (Campbell and Thoen). The combinations of crops analysed by Mate in a later period were similarly sensitive to price movements. Changes in consumer habits and innovations in the preparation of food and drink had their impact on rural producers, of which the best documented example must be the introduction of beer in the towns of north western Europe, which led to the cultivation of hops, and no doubt created local demand for brewing grains near the centres of manufacture.

Technological change and population growth used to be seen as involved in a negative relationship, because techniques were said to have been static, and production could not therefore cope with the increased demand. That view receives some justification from the dearths and famines in the decades around 1300, which must indicate some failure of the agricultural system. But there is no doubt that in some circumstances demographic growth stimulated new methods, as in the more densely peopled regions, like the Low Countries and East Anglia in the thirteenth century, where the abundance of labour favoured the adoption of intensive farming. The careful preparation of the seed bed for cereals, with repeated ploughings, harrowings, and the use of hand tools, repaid the work with high yields per hectare, and in addition specialist horticultural and industrial crops were cultivated in greater quantities. After the drop in population in the fourteenth century in many regions the opposite tendency can be observed—a reduced intensity of cultivation, and the choice of products requiring less work, such as beef cattle. Some of the new practices of this period, such as the harvesting of corn with scythes, can be seen as labour-saving devices.

Not all of the developments in technology are easily linked with the economic environment. In the Netherlands and Flanders, for example, intense cultivation continued after the Black Death, which in theory would have reduced demand for produce and the supply of labour. The kaleidoscope of local farming practices like the eight agricultural systems identified by Campbell in thirteenth- and fourteenth-century England ought to be explicable in terms of responses to the market for produce and labour, proximity to towns, soil types, and other factors, but this is not always possible. An important ingredient in the formative centuries after *c.* 900 was the pursuit of orderly and planned models, most clearly evident in the regular and geometric plans found in villages and towns all over Europe which were either founded or reorganized at this time. This adoption of a standard pattern is found also among the constitutions of the new towns, which followed the example of a precursor—such as Magdeburg or Breteuil. A regular field system could well have been perceived in the same way, as a model of organization, imitated from one village to another. A field system was an idea, often part of the regional culture, as well as a practical way of growing crops, and when we see the lord of a manor like Cuxham in Oxfordshire observing without deviation the laws of three-field cultivation every year for decades we can appreciate the force of models and customs in agricultural practice.[10]

Finally, the historical trends of the period 900–1300, and particularly the first two centuries of that period, included the formative stages of the seigneurial regime. The lords enjoyed great power and influence, and everyone agrees that they must have had a role in technical change. This represents one of the most controversial problems for historians of medieval technology, though it is not often the subject of any sustained debate. Those who presume that lords were strong innovators can point to their active role as agricultural managers during the period when the "technological package" was developing. Monastic landlords, and in particular the Cistercians, are often presumed to have been at the cutting edge of technological "progress". They had the larger farms, the capital to invest, and the resources to take risks. Above all, their superior intellectual training

---

[10] Harvey 1965, 164–5.

gave them the capacity to devise and plan new technologies, and as lords of great estates they had the power to implement them. Many of the techniques and developments discussed in this book are known from the records of the great lords, but that in part simply reflects their role as producers of documents. They are likely to have made their main contribution as investors in major buildings, of which mills and barns are the best examples. Not only did they build to high standards on their own farms, but may well have encouraged small-scale imitation among the peasants.

We can speculate that lords stimulated peasants to change techniques, both directly, by insisting that labour services be performed with heavy ploughs and well constructed carts, and indirectly by demanding rents which forced peasants to produce a saleable surplus of crops. On the other hand sufficient stimuli for change could have been generated within peasant society—by independent contacts with the market for example. The impression given by most custumals is that lords were seeking to exploit the existing resources of their tenants in terms of animals and equipment, and not to impose changes on the peasant economy. If tenants began to use horse-drawn rather than ox-drawn vehicles, the custumals changed—but the initiative came from the peasants. In some circumstances lords could have acted as a drag on change—their conservative attitude to the application of mill power to industry has been cited here by Langdon. And as is well known, at certain periods high rents and dues would have prevented peasants from accumulating capital. One indicator of the direct interest of lords in agricultural methods, and their willingness to apply their minds to technological problems might be their interest in agricultural literature. In this respect the relatively prolific English writers—Walter of Henley and his like—seem to have no parallel in Europe. Indeed, even the English books help to prove the point, because they are symptomatic not of a "technological revolution" but of the "managerial revolution", in that they are concerned primarily with auditing accounts and the supervision of administrative staff, and information about agricultural method is often incidental to that main purpose.

These doubts about the leading role played by lords in technology cannot be a preface to an unequivocal statement that the peasants were the real source of new methods. Change was risky for people with slender resources, and their priority was often satisfying subsistence requirements rather than maximizing market returns. Their contribution to technology came not so much from the class struggle (which according to one view forced lords, with the decline of slavery, to introduce new methods of exploitation, such as the mill) but from their management of their own holdings and tenure of their lords' assets. Mills, for example, though often claimed as a seigneurial monopoly, are found in many parts of Europe to have been built by consortia of peasants and, in those areas where they were founded by lords, they often came into the hands of peasant tenants. Peasant attitudes contributed to the success of the mill. The profits of milling might derive not so much from the enforcement of the suit of mill, but from the decisions of peasant consumers, who could be wooed away from rivals by convenient transport and a competitive rate of toll.[11]

Perhaps a major inhibition on peasant initiative came from the conservative village community, which undoubtedly enforced routines of cultivation and punished individuals who attempted to better themselves? But some of the most daring enterprises of the middle ages, the large scale drainage of wetlands, were carried out by village communities, and some of the wholesale changes in rotations and field systems—planting on the fallow field, for example—were only possible when the community opted for a bold innovation.

On those occasions when we probe into the internal workings of the peasant holding, as in the examples from southeast England quoted from court records here by Mate, we find some of them carrying larger densities of stock and planting more legumes than their lords, clearly pursuing their own strategies. Peasants were more likely than lords to plant industrial crops such as flax, hemp, and dye plants (Campbell). As we are reminded by Thoen, quoting the Flemish evidence, large farms (such as lords' demesnes) were different from small holdings not just in size but also in their technologies. The reconstruction of peasant budgets, inevitably based on the yields recorded on well-documented demesnes, may be underestimating the productivity gains of intensive application of labour by peasant families, and this may explain the remarkable ability, against all predictions, of large populations to support themselves on limited quantities of land.

One dimension of the peasant holding which helps to explain the survival of families on apparently inadequate resources was the "cottage economy" presided over by the women of the household, by

[11] Muendel 1981, 97–9; Holt 1987; Langdon 1994.

which every possible benefit was squeezed out of horticulture, poultry keeping, dairying, brewing, and petty trade, and much free food, fuel, and raw materials was gathered from commons, woods, and hedgerows. A great deal of collective experience must have accumulated among village women about such skilled processes as fermenting ale and making cheese, but this lore was transmitted orally and therefore has left little trace. These "low technology" activities saw some changes, such as those affecting butter churns, and the spread of spinning wheels must have had a major impact on the pattern of rural domestic work. Mechanical milling relieved women of the drudgery at the quern or hand mill, and freed them for other tasks.

The conclusion of a comparison between lords' and peasants' contributions to technical change must be that peasants were more likely than the lords or their officials to be in intimate contact with practical problems and their solutions. Trial and error, the method by which most new methods developed, can sometimes be observed in the records of demesnes, but must also have occurred on peasant holdings. Towards the end of the middle ages the proportion of land managed by peasants, in England but also in parts of the continent, increased with the leasing out of demesnes, and then the opportunities increased for peasants to take technological initiatives.

Finally, in reviewing those responsible for technical change, the state deserves some consideration. In the conduct of war kings, nobles, and their advisers showed a lively awareness of new developments in fortification, armour, and weapons such as siege engines and guns. There were peacetime spinoffs from military technology, visible for example in Edward I's Welsh campaigns, which involved in the course of conquering and pacifying a difficult terrain, road and bridgebuilding, and the construction of harbours and quays. It was presumably partly for military reasons, at least in origin, that states supervised the maintenance of bridges and roads, but these had obvious economic benefits. Likewise states anxious to defend law and order, and to collect the profits of justice, if successful created a peaceful climate in which property was protected, thereby encouraging investment and economic growth.

Uneven Technological Change

Historians of medieval technology are often disappointed by the failure of innovations to be rapidly adopted, or by the reluctance of one region to follow the practices of another. The language used to describe these apparent obstacles to growth becomes quite moralistic. Technical change is judged to be "progressive", and those people and localities that resist innovations are regarded as "backward". Medieval attitudes to novelty are often dismissed as "conservative", which is attributed either to the prevailing culture (the otherworldliness of the intellectuals, or the mindlessness of the peasants), or the social structure which prevented investment. Yet "conservatism" did not prevent change in other spheres of life, such as government or education, and the blanket term cannot be applied to a society which accepted such a daring invention as the windmill with relative speed.

We should not slip into Whiggish or judgmental attitudes towards medieval technology, leading us to disparage the people of the time for their resistance to change. There was no single "technological package", but a whole range of them depending on the region. Campbell identifies eight types of farming system in England in the thirteenth and fourteenth centuries, which if extended over the whole continent would no doubt grow to twenty-eight or even more. Each system was formed out of a particular physical or social environment, and employed the technologies which were thought to be appropriate. The unifying factors came from the influences of rising populations and urbanization in the centuries before 1300, and the demographic decline and economic recession of the fourteenth and fifteenth centuries. We cannot regard the Flemish system as "progressive", and lament the failure of other countries to adopt the methods used in Flanders, the Low Countries and parts of eastern England. These systems of intensive husbandry, appropriate for densely peopled and strongly urbanized regions, could not be exported. If they had been employed in Brandenburg or Aveyron they would have caused an ecological disaster. For similar reasons, but with a cultural as well as an economic dimension, the very productive agriculture of the Moslem inhabitants of Sicily and southern Spain was not transferred easily to their Christian neighbours, nor indeed was continued immediately after the Christian conquest of these lands.[12] We should set aside our prejudice in favour of "improvement", and appreciate the positive nature of the decision not to adopt some new technique. It was not through "ignorance" or "conservatism" that Scandinavian peasants continued to use the ard, or English villagers practised the two-field system. In each case they must have been aware of the alternatives,

---

[12] Watson 1995.

but decided that their soils did not need to be turned by a mouldboard, or that a three-course rotation would have reduced the grazing for animals and endangered their yield ratios.

A model often used to explain regional differences is that of the core and periphery. That is of some value in analysing the patterns of technological development of medieval Europe, as we can identify the countries around the North Sea as an innovatory core. It was in the Low Countries, Flanders, and East Anglia that we find the most intensive husbandry systems and the first windmills. These were also inventive regions in industrial and maritime technology, and there may be a connection between ship-building skills for example, and those used to build windmills. There is evidence also for the diffusion of some techniques, notably the penetration of heavy ploughs and regular field-systems into Scandinavia. But our earlier comments on regional differences should encourage the belief that similar techniques may have developed at the same time in various places, because of local stimuli rather than because of the spread of ideas from "advanced" to "backward" areas. As Langdon shows in his essay, England, which has traditionally been regarded as lagging behind in its medieval technology, was not in fact so "backward", and in any case the methods that its cultivators adopted were well suited to their circumstances.

The Whiggish view that gives strong approval to new techniques and thinks ill of those who stick to tradition, does not always take into account the welfare and opinions of those who operated the methods. For example, the intensive agriculture of Flanders and Norfolk involved much hard work, with small rewards for wage earners or smallholders. We can scarcely blame late medieval peasants for preferring low intensity and pastoral farming, if these methods gave them adequate returns, and a more leisured and enjoyable life. Another feature of Whiggish attitudes is that they accord importance to those methods which were to have the most impact on modern agriculture. Because "Dutch" and Flemish farming were much discussed and imitated in the seventeenth and eighteenth centuries, and because the Norfolk rotation was to have such an influence on the English agricultural revolution, they are selected for most study and admiration. But medieval people put a great deal of effort into developing such ventures as rabbit warrens and fishponds, which have had limited importance in modern times. They worked well in the sense that they gave much profit and employment, but are now regarded as mere historical curiosities. We should appreciate the whole range of medieval technologies, and not impose our modern priorities too strongly on the past.

Medieval agriculture saw no revolution comparable with that of eighteenth-century England, but over many centuries cumulatively created new "technological packages" appropriate to each region's society and environment. In most years the population could be fed, including the substantial minority who lived in towns and did not produce their own food. The technology was imperfect, hence the occasional famines and continuing poverty. We can still recognize the skills, adaptations, and inventions of which the unnamed precursors of Tull, Townshend, and Young were capable.

Bibliography

Armitage, P. 1982, "Developments in British cattle husbandry", *The Ark* 9, 50–4.

Aston, M. (ed.) 1988, *Medieval fish, fisheries and fishponds in England*, Oxford.

Bailey, M. 1988, "The rabbit and the medieval East Anglian economy", *Agricultural History Review* 36, 1–20.

Bartlett, R. 1993, *The making of Europe. Conquest, colonization and cultural change 950–1350*, Harmondsworth.

Bautier, R. 1989, "La circulation fluviale dans la France médiévale". In *Recherches sur l'économie de la France médiévale*, Paris, 7–36.

Birrell, J. "Deer and deer farming in medieval England", *Agricultural History Review* 40, 112–26.

Bois, G. 1992, *The transformation of the year one thousand*, Manchester.

Crouzet-Pavan, E. and Maire-Vigueur, J.-C. (eds) 1994, *Water control in western Europe, twelfth–sixteenth centuries*, Milan.

Dyer, C. 1984, "Evidence for helms in Gloucestershire in the fourteenth century", *Vernacular Architecture* 15, 42–5.

—— 1995, "Sheepcotes: evidence for medieval sheep farming", *Medieval Archaeology* 39, 136–64.

Harrison, D. 1992, "Bridges and economic development, 1300–1800", *Economic History Review* 46, 240–61.

Harvey, P. D. A. 1965, *A medieval Oxfordshire village. Cuxham 1240–1400*, Oxford.

Holt, R. 1987, "Whose were the profits of corn milling? An aspect of the changing relationship between the Abbots of Glastonbury and their tenants 1086–1350", *Past and Present* 116, 7–23.

—— 1988, *The mills of medieval England*, Oxford.

Langdon, J. "Lordship and peasant consumerism in the milling industry of early fourteenth-century England", *Past and Present* 145, 3–46.

Mokyr, J. 1990, *The lever of riches. Technological creativity and economic progress*, New York.

Muendel, J. 1981, "Mills in the Florentine countryside". In *Pathways to medieval peasants*, ed. J. A. Raftis, Toronto, 83–115.

Rackham, O. 1976, *Trees and woodland in the British landscape*, London.

Ryder, M. L. 1983, *Sheep and men*, London.

Stamper, P. 1988, "Woods and parks". In *The countryside of medieval England*, eds G. Astill and A. Grant, Oxford, 128–48.

Trow-Smith, R. 1957, *A history of British livestock husbandry to 1700*, London.

Watson, A. M. 1995, "Arab and European agriculture in the middle ages: a case of restricted diffusionism". In *Agriculture in the middle ages. Technology, practice, and representation*, ed. E. Miller, Philadelphia, 62–75.

INDEX

Compiled by Ann Hudson

Notes: The index covers the main text and illustrations only. Page references in italics refer to illustrations. A page reference such as "268–71 *passim*" indicates that the subject is referred to frequently but not continuously on these pages.